Symmetry, Orbitals, and Spectra

Symmetry, Orbitals,

and Spectra (S.O.S.)

Milton Orchin and H. H. Jaffé

Professors of Chemistry

University of Cincinnati

Wiley-Interscience, a Division of John Wiley & Sons, Inc.

New York · London · Sydney · Toronto

Library of Congress Catalog Card Number: 76–136720

ISBN   0   471   65550   3

Printed in the United States of America

10 9 8 7 6 5 4 3 2 1

Preface

The graduate programs in chemistry at many universities are characterized by extensive formal instruction in standard advanced courses in traditional fields of chemistry. There is now general recognition that, despite such formal divisions of chemical knowledge, the increasing unity of chemistry is blurring the distinctions between subdisciplines. The identification and integration of common core material in advanced courses would not only result in saving a considerable amount of instructional time but would also emphasize the essential unity of chemistry.

It is our opinion that the three pervasive and integrating concepts that provide an inextricable link through much of modern chemistry are symmetry, molecular orbital theory, and absorption spectroscopy. The present book is designed as a one-semester textbook to implement this opinion and hence the title

Symmetry, Orbitals, and Spectra, or S.O.S.

Its use is intended to guide the reader through introductory quantum mechanics and molecular orbital theory, the free-electron method and the calculation of ultraviolet spectra, symmetry, group theory and its applications, the structure, bonding, and ultraviolet spectra of inorganic and organometal complexes, selection rules governing the intensities of absorption bands, the fundamentals of infrared spectroscopy, noncomputer methods for Hückel molecular orbital calculations and their applications, and the basic concepts of photochemistry and excited-state chemistry, including the Woodward-Hoffman rules for the conversation of orbital symmetry in concerted reactions. Students with a strong background in chemistry will be able to cover the first three chapters rather quickly, but we hope that for others these chapters will fill some of the gaps and dispel the misunderstandings that frequently occur because of short or inadequate treatment of these important topics. The last chapter provides a condensed summary of the various treatments which extend beyond Hückel molecular orbital theory and which require computer assistance.

The background assumed for an understanding of our book is the equivalent of a B.A. or B.S. in chemistry with no advanced training. At many universities and colleges, students in their senior year may have sufficient background to handle the material. Our choice of level was also dictated by our desire to make the book suitable for self-study. We have made a conscious effort to interest those who completed their formal chemical education five or more years ago and who want a painless introduction to newer theory; accordingly nonmathematical treatments predominate.

The omission of references to the original literature is deliberate, but in no way reflects our independence of previously published material; on the contrary we rely very heavily on it and are grateful to the many unacknowledged authors who helped us write this book. However, the task of identifying, evaluating, and assessing credit for the original contributions in so many areas is a formidable undertaking and is oblique to our goals. We are more concerned with concepts and their generalization and application than with completeness or assignment of credit or priority. Hence despite our heavy indebtedness to so many authors we have not generally identified them and humbly ask their understanding of this omission.

The problems at the end of each chapter are designed to illustrate and expand the subject matter of the chapter. Detailed answers to these problems are contained in a separate supplement.

Appendix 1 provides the background required for understanding the method of descending symmetries, a procedure commonly employed but rarely explained in sufficient depth to give meaning to correlation tables. Appendix 2, in the pocket of the back cover, consists of the character tables for common point groups.

It is always a pleasure to acknowledge the aid of students and colleagues because discussions with them sharpen ideas and help eliminate error. Although a list of all such helpful persons would be very long indeed, we particularly want to thank our outstanding student collaborators, Nelson Phelan, Gary Kuehnlenz, Richard Ellis, Len Spaulding, and Janet Del Bene, and our faculty colleagues Darl McDaniel, David Morgan, John Worley, Marshall Wilson, and Roger Macomber. Especial thanks are due to Mrs. Sandra Bozeman, who typed our entire manuscript and then retyped it twice. We are, of course, also indebted to our patient and understanding families for the countless hours they have permitted us to be away from family responsibilities.

One of us (M.O.) is grateful to the Israel Institute of Technology in Haifa (and particularly to Professors David Ginsburg and Frank Herbstein) and to the University of California at Berkeley (and particularly to Professors

Henry Rapaport and Bruce Mahan) for visiting professorships which provided opportunities to discuss the material with a wider body of students and faculty.

Milton Orchin
H. H. Jaffé

Cincinnati, Ohio
June 1970

Contents

Symmetry, Orbitals, and Spectra

1 Light and light energy

1.1 The nature of light

The electromagnetic nature of light was proposed by Maxwell in the 1860's and confirmed by the experiments of Heinrich Hertz in 1887 and 1888. As the word electromagnetic implies, both an electric field and a magnetic field are associated with all light waves. These fields move in the same direction as the light beam and are at right angles to each other. In spectroscopic work we are concerned almost exclusively with the electric field part of the electromagnetic radiation.

In 1900 Max Planck, in considering the energy distribution curves of a radiating black body, assumed that the radiation occurs in discrete packets of energy which he called *quanta* and that the magnitude of the quantum, that is, its energy, E, was directly proportional to the characteristic frequency, v, of the vibration:

$$E = hv \qquad (1.1)$$

The proportionality constant, h, in equation 1.1 is a universal constant, now called *Planck's constant*.

Planck's hypothesis was concerned with the discontinuous energy content of vibrating molecules. It was Einstein who, in his work in 1905 on the photoelectric effect, extended the quantum theory to the nature of light. He assumed that light energy is absorbed in concentrated packets, each of energy equal to hv. These light quanta are called *photons*. Einstein assumed that the energy of the light quantum or photon is so concentrated that when it enters a substance it can transfer all of its energy to a single electron. The concept of the photon is difficult to explain. It is not a material particle carrying energy, as Newton advocated in his corpuscular theory of light. It is a packet of energy which, when it transfers its energy to a molecule, ceases to exist. It appears to have a dual nature since, in the interaction of light with matter, the photon appears to be a particle, but its energy is determined by the frequency of a wave. As a matter of fact, it is the Planck-Einstein relation,

1

equation 1.1, which allows one to convert quantitatively from the concept that radiation is a wave to the concept that it is a stream of quanta or photons.

Does light consist of waves or of particles? Classical theory says waves, and quantum theory is unable to resolve the impasse as to whether light should be described as waves or particles. Starting in the 1920's, scientists began to reconcile these two theories, which became unified in the new theory of wave mechanics. In 1924, de Broglie showed that waves could be associated with very small particles, and the equations describing the behavior of such waves were provided in 1926 by the wave mechanics of Erwin Schrödinger. Alternative treatments were almost simultaneously advanced by Werner Heisenberg and Paul Dirac.

1.2   The Electromagnetic Spectrum; Classification and Units

All electromagnetic radiation, be it gamma rays, x-rays, ultraviolet, infrared, or radio waves, travels at the same velocity, namely 2.96×10^{10} cm sec, or 186,000 miles/sec (in vacuum).

Electromagnetic radiation is characterized by two properties, amplitude and periodicity; the latter may be described in terms of three different quantities: wavelength, wavenumber, or frequency.

The wavelength is the distance between crests, or between any two points which are in phase, on adjacent waves, Fig. 1.1. Many units of length are used

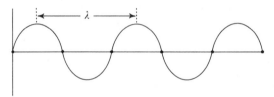

Fig. 1.1   A wave with a wavelength of about 1.5 in (approx. 3.7 cm)—the microwave region.

to describe the wavelength, λ, and since the complete electromagnetic spectrum, Table 1.1, covers a large range of wavelengths, various units are commonly employed for describing the different ranges. In absorption spectroscopy, λ is most often expressed in the infrared region by microns (μ), and in the ultraviolet by millimicrons (mμ), now officially called nanometers (nm) because 1 nm = 10^{-9} m, or by angstrom (A) units:

$$1 \mu = 1000 \text{ nm} = 10,000 \text{ A} = 0.0001 \text{ cm}.$$

The frequency, ν, is the number of waves that pass a particular point in unit time, most commonly the second, and ν therefore has units of reciprocal seconds (sec^{-1}); the official designation of this unit now is the *hertz* (Hz).

Frequency and wavelength of electromagnetic radiation are related through the equation $\nu = c/\lambda$, where c is the speed of light. Thus, light in the

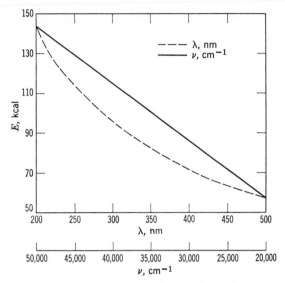

Fig. 1.2   Relation between energy, wavenumbers, and wavelength.

ultraviolet region at 3000 A corresponds to a frequency of:

$$\nu = \frac{c}{\lambda} = \frac{2.96 \times 10^{10} \text{ cm/sec}}{3 \times 10^{-5} \text{ cm}} \approx 10^{15} \text{ sec}^{-1} \text{ or } 10^{15} \text{ Hz}$$

The frequency and the wavelength are inversely related, as a simple analogy makes clear. If a tall runner and a short one run at the same speed, in order to keep pace the short one obviously has to take more of his shorter strides than the tall one takes of his long strides.

The wavenumber, $\tilde{\nu}$ (Greek nu tilde), is the number of waves per unit distance, usually 1 cm, and hence has units of reciprocal centimeters (cm^{-1}). Like the frequency, $\tilde{\nu}$ is inversely related to λ. A wavelength of 3000 A thus corresponds to

$$\tilde{\nu} = \frac{\text{number of angstroms in 1 cm}}{\lambda} = \frac{10^8}{3 \times 10^3} = 33{,}333 \text{ cm}^{-1}$$

or 33,333 reciprocal centimeters (sometimes inaccurately called 33,333 wavenumbers). The wavenumber and the frequency are directly related by the speed of light, and hence frequently the two are used interchangeably. The relationship is:

$$\tilde{\nu} \text{ (cm}^{-1}) \times \text{speed of light (cm/sec)} = \nu \text{ (sec}^{-1})$$

The region of the electromagnetic spectrum of greatest interest to most spectroscopists extends from 10^2 to 10^{10} A. The official designations of the various regions are given in Table 1.1.

TABLE 1.1
The Regions of the Electromagnetic Spectrum

Region	Wavelength, μ	Wavelength, nm	Wavenumber, cm^{-1}	Frequency,* Hz $\times$ 10^{-12}
Far ultraviolet	0.01–0.2	10–200	1,000,000–50,000	30,000–1500
Near ultraviolet	0.20–0.38	200–380	50,000–26,300	1500–787
Visible	0.38–0.78	380–780	26,300–12,800	787–385
Near Infrared	0.78–3	780–3000	12,800–3333	385–100
Middle infrared	3–30	3000–30,000	3333–333	100–10
Far infrared	30–300	30,000–300,000	333–33.3	10–1
Microwave†	300–1,000,000	300,000–1,000,000,000	33.3–0.01	1–0.0003

* A unit called the fresnel, f, is equal to 10^{12} cycles/sec, or 10^{12} Hz.

† Microwave spectroscopists use the unit of megacycles or megahertz (MHz), equal to 10^6 cycles/sec.

1.3   Energies Associated with Radiation

Planck's equation, $E = h\nu$, tells us that electromagnetic radiation of frequency ν occurs in small bundles, called *quanta*, and that the amount of energy any one of these quanta contains, $h\nu$, is proportional to the frequency. The proportionality constant, h, called *Planck's constant*, has a value of 6.625×10^{-27} erg sec. If light is absorbed or emitted by a molecule, the energy of the quantum (absorbed or emitted) is the difference in energy between two states of the molecule (Bohr's condition). We have already shown that a wavelength of 3000 A (300 mμ or nm) corresponds to a frequency of 10^{15} sec^{-1}. Substitution of this value into the above equation gives:

$$E = 6.625 \times 10^{-27} \ (h \text{ in erg sec}) \times 10^{15} \text{ sec}^{-1} = 6.625 \times 10^{-12} \text{ erg}$$

Since chemists are accustomed to dealing with moles rather than with molecules and with kilocalories rather than ergs, the above energy is converted to the desired units of kilocalories per mole by the appropriate arithmetic:

$$6.625 \times 10^{-12} \text{ erg} \times \frac{6.023 \times 10^{23} \text{ molecules/mole}}{4.184 \times 10^{10} \text{ ergs/kcal}} = 95.4 \text{ kcal/mole}$$

Thus absorption of light at 300 nm corresponds to an energy of 95.4 kcal/mole.

Another way of expressing the energy is in terms of the einstein. Since the einstein is Avogadro's number of photons, the energy corresponding to light absorption at 300 nm is 95.4 kcal/einstein. It is convenient to have a simple formula whereby absorption in the ultraviolet, frequently given in nanometers, and absorption in the infrared, frequently given in microns, can be converted to energy in kilocalories per mole:

$$E \text{ (kcal/mole)} = \frac{28.635 \times 10^3}{\lambda \text{ (nm)}} = \frac{28.635}{\lambda \, (\mu)} = 28.635 \times \tilde{v} \text{ (cm}^{-1}) \times 10^{-4}$$

$$(1.2)$$

It is important to remember that frequency and wavenumbers are directly proportional to energy, and so the same number of reciprocal centimeters represents the same energy anywhere in the spectrum. Thus, a spectral shift of 700 cm^{-1} anywhere in the spectrum corresponds to 1.95 kcal/mole. On the other hand, wavelength is inversely proportional to energy and hence the relationship is not linear, Fig. 1.2. At 200 nm, for example, an energy change of 1.95 kcal/mole corresponds to a shift of 2.7 nm, but at 800 nm the same energy change corresponds to a shift of about 4.4 nm. For those more familiar with units of electron volts (eV), an approximate formula that is easy to remember is:

$$E \text{ (eV)} = \frac{12,345}{\lambda \text{ (A)}}$$

$$(1.3)$$

PROBLEMS

1.1   Calculate, in kilocalories per mole, the energy corresponding to (a) 240 nm, (b) 480 nm, (c) 6.1 μ, (d) 12.2 μ, (e) 1700 cm^{-1}, (f) 3400 cm^{-1}, and (g) 34,000 cm^{-1}.

1.2   Calculate, in electron volts (eV), the energy corresponding to (a) 220 nm, (b) 7.2 μ, (c) 2100 cm^{-1}, and (d) 34,000 cm^{-1}.

1.3   A solution of acetone in ethanol absorbs at 260 nm, but in hexane absorption occurs at 272 nm. What is the energy difference, in kilocalories per mole, of the two absorptions?

1.4   Calculate the energy, in kilocalories per mole, which corresponds to the absorption of light at 589.3 nm by sodium vapor.

GENERAL REFERENCES
1. "Suggested Nomenclature in Applied Spectroscopy," *Anal. Chem.*, **24**, 1349 (1952).
2. G. Socrates, "SI (Système Internationale) Units," *J. Chem. Educ.*, **46**, 710 (1969).
3. "Report on Notation for the Spectra of Polyatomic Molecules," *J. Chem. Phys.*, **23**, 1997 (1955).

2 Electronic configuration and orbitals of atoms

2.1 The Periodic Classification

One common procedure for illustrating the electronic structure of any element is to construct a series of boxes or circles, one for the 1s orbital, one of higher energy for the 2s orbital, a series of three for the 2p orbitals, etc. Then a number of electrons corresponding to the atomic number of the element are distributed among the circles or boxes in accordance with certain rules.

In generating the correct box or "orbital" arrangement, three principles are utilized. The first of these is the *Aufbau* or building-up *principle*, which states that electrons are put into orbitals in the order of increasing orbital energy; the lowest-energy orbitals are filled before electrons are placed in higher-energy orbitals. The order of orbital energy in neutral atoms has been established as 1s < 2s < 2p < 3s < 3p < 4s < 3d .... The second principle is the *Pauli exclusion principle*, which states that a maximum of two electrons can occupy an orbital and then only providing that the spins of the electrons are paired or opposed. The third principle is *Hund's rule*, which states that *one* electron is placed in *each* of the orbitals of equal energy (degenerate orbitals) before *two* electrons are placed in *any one* of the the degenerate set, and the electrons in the singly-occupied orbitals have parallel spins. The three rules can be illustrated by considering the atomic structure of the nitrogen atom, $1s^2 2s^2 2p^3$, Fig. 2.1.

The rules that are employed for building up the electronic configuration of atoms are also used for building up molecular configurations. Before considering molecules and molecular orbitals, let us examine in greater detail the meaning of an atomic orbital, which the box or circle in Fig. 2.1 is supposed to represent. The obvious place to start such a discussion is with the hydrogen atom because this atom has only one electron and in the ground state the single electron occupies a 1s orbital.

6

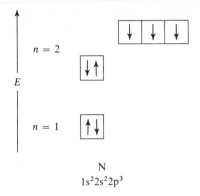

$n = 2$

E

$n = 1$

N
$1s^2 2s^2 2p^3$

Fig. 2.1   The electronic configuration of the nitrogen atom.

2.2   The Probability Density of the Electron in the Hydrogen 1s Orbital

The naïve notion that the electron of the hydrogen atom travels in a fixed orbit around the nucleus under the action of classical laws of motion gave way in the 1920's to the quantum-mechanical description of electron motion. The electron has properties of a wave, and the distribution of the electron in space can be described by an atomic orbital.

We can visualize the electron of the hydrogen atom as moving in all directions but being more frequently in some places than in others. If we could take instantaneous consecutive photographs of a cross section of the atom and the electron showed up as a dot, the relative density of the dots in any particular region would represent the probability of finding the electron in that region of space. This relative density, called the electron density, is shown in Fig. 2.2.

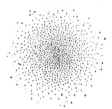

Fig. 2.2   Probability density cross section.

From this picture we could correctly conclude that the probability (the probability density or charge density) of finding the electron in one particular, infinitely small volume of space, $dx\,dy\,dz$, or $d\tau$, is greatest at the nucleus, Fig. 2.3. However, there are fewer of these small volumes of space at the nucleus than in any thin layer around the nucleus. If the space surrounding the nucleus was divided into uniform thin shells of thickness dr analogous to the layers of skin in an onion, we can see that the volume of such shells

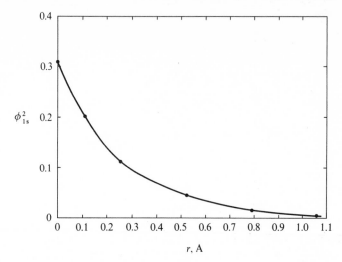

Fig. 2.3 Plot of the probability distribution function, ϕ_{1s}^2, for the ground state of the hydrogen atom vs. r.

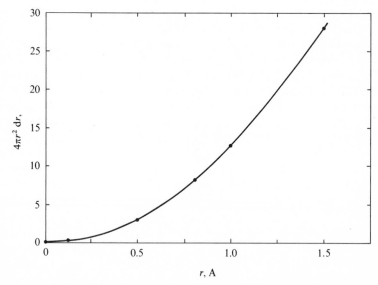

Fig. 2.4 Volume contained in a thin shell of unit thickness, dr, vs. distance of shell from nucleus, r.

would increase rapidly, Fig. 2.4, as the distance, r, from the nucleus increased, since the volume of such a shell would be equal to $4\pi r^2\, dr$, the area of the surface of the sphere times the thickness of the shell.

The probability of finding the electron at any distance r from the nucleus is a function of two opposing trends. The first of these, the electron density,

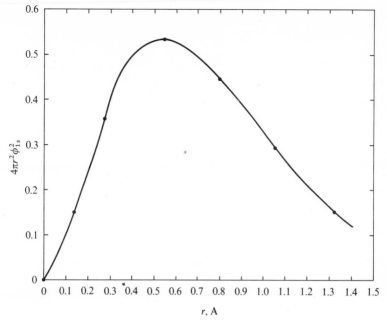

Fig. 2.5   Radial distribution function, $4\pi r^2 \phi_{1s}^2$, vs. r.

is greatest near the nucleus, where r is small, and decreases as r increases, Fig. 2.3. On the other hand, the second factor, the volume of the shell, increases as r increases, Fig. 2.4. The result of multiplying these two functions together is shown in Fig. 2.5. This curve is thus a plot of the radial probability distribution, giving the probability that the electron is found between r and $r + dr$ as a function of r; it shows that the electron is not likely to be found at the nucleus or at large distances from the nucleus, but at some intermediate position. However, the probability approaches zero asymptotically as r approaches infinity and never becomes zero at any finite r. In the hydrogen atom, the maximum in the curve occurs at 0.53 A and corresponds to the radius of the classical Bohr model of this atom and is consequently called the Bohr radius.

Since we are interested in the numerical value of the probability of finding the electron within a certain sphere with radius r, an even better way for depicting the probability of finding the electron at a certain distance from the nucleus is by the use of probability contours. We know that, as we increase r, the probability of finding the electron inside the sphere increases. In Fig. 2.6 are shown seven concentric circles, which are the cuts of spheres, with increasing values of r corresponding to various distances from the nucleus. The value listed on the circumference of each individual circle is

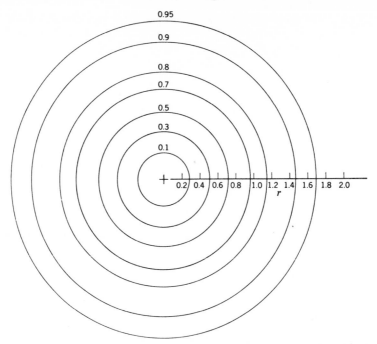

Fig. 2.6   The charge distribution in the normal hydrogen atom; Contours of equal Probability in the 1s orbital. From G. W. Wheland, *Resonance in Organic Chemistry*, John Wiley and Sons, Inc., 1955, by permission.

the probability that the electron will be found within the corresponding sphere; the circles are contours of equal probability. It is customary to represent the 1s orbital by one of the single-contour circles. Thus, for example, if the outer contour circle of probability 0.95 is chosen, the electron is nineteen times as likely to be inside the corresponding sphere with a radius of 1.7 A as it is to be outside that sphere. When a circle, Fig. 2.7, is arbitrarily drawn to represent the 1s orbital, it is implied that the circle represents a contour sphere such that the electron spends a high proportion of its time inside that sphere. The circle can be drawn to any size as a matter of convenience of display; in all cases it is implied that the circle is a cut of a sphere inside of which there is a very high (unspecified) probability of finding the electron, and hence the actual size of the circle is simply a matter of the scale used for the radius.

   The statistical character of the knowledge of small particles like electrons is a basic postulate of wave mechanics. This postulate defines the probability that the electron is in the infinitesimally small volume $dx\,dy\,dz$ ($d\tau$) surrounding the point x, y, z (i.e., in the element of volume between x and $x + dx$, y and $y + dy$, and z and $z + dz$) as $\rho\,d\tau$. There is a greater proba-

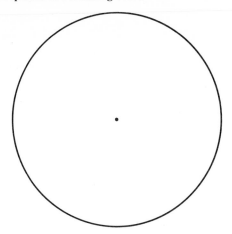

Fig. 2.7   The 95% contour line in cross section.

bility of finding the electron in a region where ρ is large than in one where ρ is small. Since the electron must be found somewhere, integration of this probability function over all space must be unity, that is,

$$\int_{-\infty}^{+\infty} \int_{-\infty}^{+\infty} \int_{-\infty}^{+\infty} \rho(x, y, z)\,dx\,dy\,dz = \int \rho\,d\tau = 1$$

Any integration over $d\tau$ is assumed to be over all space. The probability functions of the electron which we have just discussed have their origin in the quantum-mechanical description of the motion of electrons. It is necessary therefore to proceed to a discussion of this quantum-mechanical basis.

2.3   The Wave Properties of a Standing Wave

All types of wave motion—for example, the wave emanating from a stone thrown into a pool of water, or the waves generated by plucking a string fixed at both ends, or the difficult-to-imagine electromagnetic waves of light radiation—are characterized by the transmission of energy from one point to another without permanent displacement of the medium. The properties of all waves can be described by similar wave equations. The wave equation deals with the amplitude of the wave as a function of the distance from the generating source. Electron waves have characteristics similar to those of a so-called standing or stationary wave. A standing wave can be generated, for example, by stretching a string between two fixed points and plucking it. At one moment the individual segments of the string are displaced upward relative to the fixed positions at the ends, as shown by the solid line in Fig. 2.8a. At the next instant the string moves downward and reaches the position shown by the broken line.

A wave such as that shown in Fig. 2.8*a* is a fundamental wave; that is, the initial displacement or amplitude, which is upward everywhere, continuously increases from either end, reaches a maximum, and decreases again. Then the wave starts downward and undergoes the same behavior in the opposite direction. In such a fundamental wave there are no nodes between the fixed ends, that is, there are no points at which the string has zero displacement. All segments of the wave are always moving in the same direction. The fixed ends of the string are technically nodes, and they must be if the waves are well behaved, but we are primarily concerned with nodes between the ends.

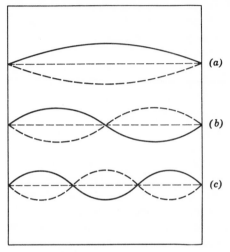

Fig. 2.8   The fundamental wave (*a*), and the first (*b*) and second (*c*) overtones with one and two nodal points respectively. From E. Cartmell and G. W. A. Fowles, *Valency and Molecular Structure*, Butterworth's, 1950, by permission.

The first overtone of a fundamental is a wave with one node, Fig. 2.8*b*. The first overtone (harmonic) has a node, a point where the displacement is zero, exactly at the midpoint of the fundamental. It is important to recognize that the two segments of the first overtone together form a single wave. The two parts are not two waves; they are not independent, since the displacement of any one point in the wave is related to the displacement at any other point in a fixed manner. Thus, the two maximum displacement points, one up and one down, come at exactly equal distances from the ends and are reached at precisely the same moment; this is so because the two portions of the wave act together and are part of a single wave. In this wave, when some of the segments are moving up, some are also moving down; it is customary to call upward motion "plus" and downward motion "minus." There is a change in sign as the wave passes through the nodal point, where the amplitude is always zero.

In Fig. 2.8c is shown the second overtone of the fundamental; it is derived from the fundamental by inserting two nodal points spaced so that the total wave is divided equally into three parts. Again Fig. 2.8c represents a single wave and not three independent waves. The three segments of the wave are dependent and must change their displacements in precise correlation with one another. Again at each nodal point there is a change in sign, or a change in direction of displacement as the wave goes from plus to minus or minus to plus, through the zero or nodal point.

2.4   The Wave Properties of an Electron Wave

The waves we have shown thus far are one-dimensional, and one feature of them is that the wave motion or amplitude is a function of the distance along the wave path. However, the electron, instead of moving in any one particular direction like the standing wave, can move in any direction. In three dimensions, the simplest wave is a spherically symmetrical motion. A three-dimensional wave is difficult to visualize. In a limited analogy it has been compared to the behavior of the surface of a balloon as the balloon swells and deflates, but of course there is no progression in this example, as there is in a real wave such as the circular wave generated by dropping a stone into a pool of water. The motion of an electron can be described by a spherical motion and must be analyzed in terms of three dimensions rather than along one direction. Of course, the spherical orbital, for example, the 1s orbital, does not describe the path of an electron. We can only consider where the electron appears to spend most of its time and not how it gets from one part of the orbital to another part. In the case of the 1s orbital we have already indicated that we can draw a circle to represent a cut of a sphere in which there is a high probability of finding the electron.

Our problem, then, is one of attempting to locate the position of the electron with respect to the nucleus. Since we are dealing with a spherical wave, it is advantageous to use polar coordinates r, θ, and φ. The relationship between Cartesian coordinates and polar coordinates is shown in Fig. 2.9, and the algebra required for mutual interconversion of points in the two systems is detailed below the figure. An orbital is a description of the location of the electron; since this description is a wave description, we call the description a *wave function* and indicate it, in the case of an atomic orbital, by ϕ. The location of the electron can then be expressed by the wave function, $\phi(r, \theta, \varphi)$.

2.5   The 2s Orbital

Just as the first overtone of the one-dimensional standing wave has a nodal point, we would expect the first overtone of the spherical standing wave to have one node. In three dimensions, there are several ways in which the

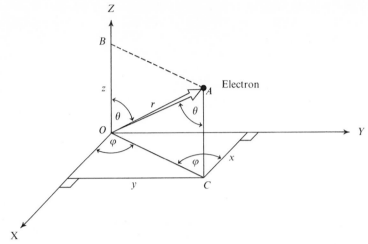

Fig. 2.9 The relationship between Cartesian and spherical coordinates. The projection of r on the z-axis is $z = OB$. OBA is a right angle, and hence $\cos \theta = z/r$, and $z = r \cos \theta$.

$\cos \varphi = x/OC$, but $OC = AB = r \sin \theta$; hence $x = r \sin \theta \cos \varphi$.
$\qquad \sin \varphi = y/AB, \quad y = AB \sin \varphi = r \sin \theta \sin \varphi$.

Accordingly, a point x, y, z in Cartesian coordinates is transformed to the spherical coordinate system by the following relationships:

$$z = r \cos \theta$$
$$y = r \sin \theta \sin \varphi$$
$$x = r \sin \theta \cos \varphi$$

fundamental wave can be broken into portions by inserting nodes. The first and easiest (lowest-energy) way is to have a spherically symmetric node or a nodal surface concentric with the fundamental wave. The wave that is associated with this next-higher-energy orbital, the 2s orbital, is the wave which describes this first overtone of the fundamental. The 2s orbital is commonly represented in the same way as the 1s orbital, that is, as a circle which is the cut of a sphere representing a high probability of finding the electron within such a sphere. In fact, however, the 2s orbital consists of two portions: an inner sphere the surface of which is a node—a region where the electron density is zero—and an outer spherical shell. Figure 2.10a shows the probability density of the 2s orbital in cross section, and Fig. 2.10b is a graph of electron probability in these spherical shells as a function of the radius of the shells.

The question is frequently asked, How is it possible for the electron to get across the nodal boundary if the probability of finding the electron in the nodal boundary is zero? The unsatisfactory answer is that the electron has wave properties and should not be thought of in this connection as a particle

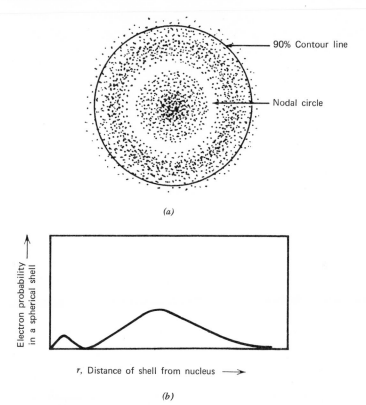

(a)

(b)

Fig. 2.10 (*a*) Probability density 2s orbital. (*b*) Charge or electron probability in spherical shell 2s orbital (distance scale greater than in Fig. 2.5). From F. A. Lambert, *Chemistry*, Vol. **41**, 1968.

in motion. Just as the standing wave can move past a nodal point, so the spherical wave moves past the nodal surface with a change in sign. In wave mechanics we do not consider that the electron is moving in a path, as it is considered to do in classical mechanics; instead we consider the probability of finding it in a particular region of space through which it moves in accordance with wave behavior.

The 2s orbital represents the first overtone of the fundamental wave. The two concentric spheres of the orbital are two portions of the same wave; they are not two independent waves, nor are they a combination of the 1s and 2s orbitals. The 1s and 2s orbitals do not overlap with each other, although they appear to occupy the same space. It is common practice to represent the 2s orbital by a single circle, but we see now that this is a shorthand way of representing the true orbital. Perhaps this practice can be justified by the fact that at bonding distances only the outer portion of the spherical 2s orbital is of interest.

2.6   The 2p Orbitals; Quantum Numbers

The necessity for the three p orbitals, all of equal energy (degenerate) in the absence of an external force, resides in quantum mechanics. We make no attempt here to discuss this theory in detail, but shall simply summarize some of its results. When we state that an electron is in a $2p_z$ orbital, we are really expressing the wave function, ϕ, for the electron. If we neglect the spin of the electron, the wave function is the product of three factors: the radial function, $R(r)$, and two angular functions, $\Theta(\theta)$ and $\Phi(\varphi)$. The radial function depends only on the radial distance r from the nucleus, and the two angular functions depend on the angles θ and φ, Fig. 2.10. The wave function is then:

$$\phi = R(r) \cdot \Theta(\theta) \cdot \Phi(\varphi)$$

Each of these factors depends on quantities called *quantum numbers*.

In the first approximation, the total energy of an electron in an orbital depends only on the *principal quantum number, n*. The radial function, $R(r)$, depends on n and the azimuthal quantum number, l. The quantum number l occurs in the $\Theta(\theta)$ factor of the wave function. Because it depends on the azimuthal angle θ, Fig. 2.10, it is called the *azimuthal quantum number*. The quantum number l determines the angular momentum of the electron due to its orbital motion around the nucleus; for this reason it is alternatively called the *angular momentum number*. It may have any integral value from 0 to $n - 1$, and its value determines the shape of the orbital. Thus, in the second quantum level, where $n = 2$, l may be 0 or 1, corresponding to 2s and 2p orbitals. As a matter of fact, the nomenclature of orbitals in the s, p, d, f, etc., notation corresponds to the l values of 0, 1, 2, 3, etc.; when $l = 3$, we are dealing with an f orbital, etc. The third quantum number, the *magnetic quantum number, m*, is associated with different orientations of the vector representing angular momentum, and hence of the orbital with reference to some defined direction; it occurs in both the $\Theta(\theta)$ and $\Phi(\varphi)$ parts of the wave function. It may have integral values from $+l$ to $-l$. Thus, when $n = 2$, $l = 0$, $m = 0$, we have the 2s orbital. When $n = 2$, $l = 1$, m may be $+1$, 0, or -1, and we have the three 2p orbitals. The magnetic quantum number of the electron, because it represents the projection of the angular momentum vector (quantized according to l) onto the z-axis, is frequently indicated as m_l.

The length of the angular momentum vector, $\mathbf{L}$, is

$$|\mathbf{L}| = \frac{h}{2\pi} \sqrt{l(l + 1)}$$

Since the projection of this vector on the z-axis, $m_l \leqslant |l|$, must be shorter than $\mathbf{L}$, the latter cannot ever be parallel to the z-axis. For $l = 1$, $m_l = +1$, 0, or -1, and $\mathbf{L}$ is at 45°, 90°, or 135° to the z-axis. Unfortunately, these angles of the angular momentum vector cannot be directly correlated with any planar rotational motion of the electron. The five orientations of the orbital angular momentum vector $\mathbf{L}$ for $l = 2$, of length $|\mathbf{L}| = \sqrt{6}$, are given in Fig. 2.11.

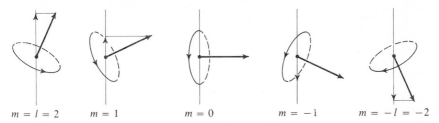

| $m = l = 2$ | $m = 1$ | $m = 0$ | $m = -1$ | $m = -l = -2$ |

Fig. 2.11   The m_l values for $l = 2$. (The heavy arrow is the angular momentum vector.)

There is also a fourth quantum number, namely, the *electron spin quantum number*, s, which is a measure of the angular momentum due to the spin of the electron. Although it is now known that the electron does not actually rotate on its axis, a classical treatment in terms of a rotating charged particle gives the same mathematical description of the spin angular momentum and provides a satisfactory classical analog. The value of the electron spin quantum number is $s = \frac{1}{2}$, and its projection on the z-axis, m_s, can have the values of $+\frac{1}{2}$ or $-\frac{1}{2}$ only. If two electrons occupy the same orbital, their spins must necessarily be opposed, because no two electrons can be described by four identical quantum numbers (the Pauli principle).

The orbital angular momentum of an electron can interact with the spin angular momentum of the electron to give spin-orbit coupling. When we are dealing with only one electron, the spin-orbit coupling energy is small and is neglected. However, as we shall see, when we have many electrons, this inter-action cannot always be neglected. In the many-electron situation, it will also be necessary to take account of the l and m_l values of each electron; as we shall see in Chapter 7, when we consider the addition of these values, a single electronic configuration can give rise to many so-called states.

A crude approximation of the shape of the p orbitals as a function of the product of $R(r)$ and the two angular functions can be arrived at as follows. We indicated earlier that the lowest-energy first overtone of the spherical wave had a spherical nodal surface giving rise to the 2s orbital. We may also have an overtone with a different kind of node, namely, a planar node which cuts through the fundamental wave, Fig. 2.12. The planar node for the p_z orbital is the xy-plane; for the p_y orbital is the xz-plane; and for the p_x orbital is the yz-plane. Passing a plane through the sphere bisects the sphere into two hemispheres, and the wave then consists of two interdependent hemispherical portions. The three resulting bisected spheres resemble the three p orbitals, all of equal energy. Although they are also first overtones of

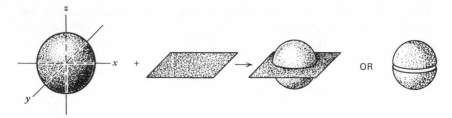

Fundamental spherical wave + Nodal plane *(xy)* ⟶ First overtone wave with one planar node

Fig. 2.12 First overtone; the 2p orbital. From F. A. Lambert, *Chemistry*, Vol. **41**, 1968.

the fundamental spherical wave, they are of higher energy than the first overtone obtained by insertion of a spherical node. We are familiar with the idea that in the second quantum level, where *l* can equal 0 or 1, we have two subshells, the 2s and the 2p shells, with the 2s being of lower energy. Of course the bisected sphere is not an accurate representation of the p orbital. When the contour lines are accurately drawn (Section 2.8 and Fig. 2.14), we find that we have an oval surface inscribed in each hemisphere.

2.7 The Wave Equation and the Schrödinger Equation

We have qualitatively examined the characteristics of classical wave motion. In order to have more quantitative information about wave motion and to make the transition to quantum mechanics (wave mechanics), we now must take a more detailed look at the physics of wave motion.

We examined earlier the characteristics of a standing wave, which we showed as being one-dimensional, say in the *x* direction. The amplitude of this wave is a function of the distance along the wave; that is, amplitude = $f(x)$. The equation which describes the motion of a standing wave is a second-order differential equation:

$$\frac{d^2 f(x)}{dx^2} + \frac{4\pi^2}{\lambda^2} f(x) = 0 \tag{2.1}$$

where λ is the wavelength. The electron wave which resembles the standing wave extends in all directions, and therefore we would be concerned with a function that depends on three coordinates. The function $f(x, y, z)$ in the wave equation in three dimensions is thus analogous to $f(x)$, which describes the amplitude behavior of the one-directional wave resulting from the vibration of a stretched string; hence $f(x, y, z)$ is properly referred to as an *amplitude function*. Now equation 2.1 can be transformed to describe the motion in three dimensions:

$$\frac{\partial^2 f}{\partial x^2} + \frac{\partial^2 f}{\partial y^2} + \frac{\partial^2 f}{\partial z^2} + \frac{4\pi^2}{\lambda^2} f = 0 \tag{2.2}$$

While $\partial^2 f/\partial x^2$ is the second partial derivative of f with respect to x, it is often convenient to talk about the expression $\partial^2/\partial x^2$. This is an operator which gives the instruction to partially differentiate twice, with respect to x, whatever follows behind. We further abbreviate the sum of the second differential operators with respect to three Cartesian coordinates as ∇^2 (del squared), the Laplacian operator:

$$\nabla^2 = \frac{\partial^2}{\partial x^2} + \frac{\partial^2}{\partial y^2} + \frac{\partial^2}{\partial z^2}$$

and then

$$\nabla^2 f + \frac{4\pi^2}{\lambda^2} f = 0 \qquad (2.3)$$

We can now replace λ by h/p, where h is the familiar Planck's constant and p is the momentum of the electron. The equality $\lambda = h/p$ is the so-called de-Broglie relationship, arising from the discovery that very small particles possess wave characteristics. Since momentum is equal to mass times velocity, we can convert equation 2.3 to the wave equation:

$$\nabla^2 f + \frac{4\pi^2 m^2 v^2}{h^2} f = 0 \qquad (2.4)$$

The wave equation may be used to calculate energies of electrons and, in the simplest case, the energy of the hydrogen atom with its one electron. If E represents the total energy, and V the potential energy, of the electron in the H atom, the kinetic energy, $\frac{1}{2}mv^2$, is given by $E - V$. Substituting $v^2 = 2(E - V)/m$ in equation 2.4, we get:

$$\nabla^2 f + \frac{8\pi^2 m}{h^2} (E - V)f = 0 \qquad (2.5)$$

We now replace the classical amplitude function, f, by the unspecified wave function, ϕ, which has the properties of the orbitals discussed earlier, and we obtain:[a]

$$\nabla^2 \phi + \frac{8\pi^2 m}{h^2} (E - V)\phi = 0 \qquad (2.6)$$

which is the familiar Schrödinger equation describing the behavior of the electron of the H atom.

---

[a] It might be recognized that this does *not* represent a derivation of the Schrödinger equation. The Schrödinger equation may be taken as a basic postulate of quantum mechanics and has validity only because results obtained from its use agree with experiment. In this respect it is similar to the three laws of thermodynamics.

We must impose certain restrictions on the wave function, ϕ. Since its square is to represent a probability, it must be continuous, single-valued, and normalizable. This means that it must be a smooth function not having a discontinuity at any point, that at any point in space it must have only one value, and that it must be possible to normalize ϕ. These three conditions actually are responsible for the quantization of energy.

The one-dimensional wave equation is readily solved for the function $f(x)$, which obeys the differential equation 2.1. The Schrödinger equation in three dimensions (equation 2.6) for the H atom may also be solved without too much difficulty, yielding the solution

$$\phi_{n,\,l,\,m} = NR_{n,\,l}(r)\Theta_{l,m_l}(\theta)\Phi_{m_l}(\varphi)$$

which we described earlier as the orbitals. For more complicated atoms and for molecules (with the exception of H_2^+ and H_2), it is not possible to find functions ϕ which are exact solutions of the Schrödinger equation. Consequently all of molecular quantum mechanics uses a long series of approximations, many of which we will encounter as we move through subsequent chapters. The one we must introduce at this point is the following: Whenever we consider the many-electron wave function, Φ, of an atom other than hydrogen, we assume that Φ can be expressed as a product of one-electron functions, ϕ, such as were discussed in the last few sections, and that these have the form of the exact solution of the H atom; hence they are often called hydrogen-like orbitals.

2.8 The Wave Function ϕ and the Probability Function ϕ^2:
Graphical Descriptions

In most wave phenomena it is the square of the amplitude of a wave that has physical significance; the amplitude by itself can take on positive and negative values and has only mathematical significance. We mentioned earlier that the probability of finding an electron in an element of volume $d\tau$ was $\rho\,d\tau$ and that, since the electron must be found somewhere, $\int\rho\,d\tau = 1$. A probability function has physical significance and here can be identified with the square of the amplitude; hence

$$\int\rho\,d\tau = \int\phi^2\,d\tau = 1$$

When the integral of the square of the wave function is equal to unity, the wave function is said to be *normalized*; when it is not, the wave function is multiplied by a number N such that the integral becomes equal to unity. For most purposes, the ϕ's of atomic orbitals are assumed to include the normalization factor, and hence $\int\phi^2\,d\tau = 1$. We will discuss the process of normalization again in Chapter 3. It is obvious that a probability has to have a value between 0 and $+1$; it cannot be negative or exceed 1. Since $\rho = \phi^2$, the probability function will be positive everywhere except at the nodes, where it is 0.

The fact that the wave equation for the "motion" of an electron—or, better, for the probability distribution of an electron—is a function of three variables:

$$\phi\,(r,\,\theta,\,\varphi) \;=\; NR(r)\Theta(\theta)\Phi(\varphi)$$

means that four dimensions are required to represent ϕ or ϕ^2 graphically. Since we usually attempt representations in two dimensions, we are presented with a difficult problem. Fortunately, each of the three factors in the wave function can readily be represented in two dimensions. The s orbitals have no angular dependence; ϕ is only a function of $R(r)$. In the case of the 1s orbital, there is no node, that is, there is no change in sign anywhere and hence no problem in representing the orbital in two dimensions and no problem in relating ϕ and ϕ^2. In the case of the 2s orbital, there is a node; and if we correctly draw a planar cut of the probability spheres, we will have a circle and around it a circular shell, both positive when representing ϕ^2, and one negative and the other positive when representing ϕ. The common planar representation of the 2s orbital as a single circle neglects the spherical node close to the nucleus and correctly focuses on the character of the orbital at bonding distance.

In the case of the p orbitals, the angular dependence is probably the most important single feature. Hence, it is common to represent a p orbital by a planar section through a three-dimensional representation of $\Theta(\theta)\Phi(\varphi)$. The resulting diagram for the $2p_z$ orbital is a pair of circles (or a cut through a pair of spheres) touching at the origin, Fig. 2.13a. An alternative representation is a plot of $\Theta^2(\theta)$ (or a cut through $[\Theta(\theta)\Phi(\varphi)]^2$). This plot produces the well-known dumbbell shapes, Fig. 2.13b. In both plots of Fig. 2.13 the expression for the wave function for the $2p_z$ orbital is $\phi = NR(r)\cos\theta$. The p_z orbital is cyclindrically symmetrical around the z-axis, and hence $\Phi(\varphi)$ is constant and is included in the constant N (the normalization constant). Of course $R(r)$ is held constant, and $NR(r)$ is taken as $\sqrt{6}/2$.

Most textbooks show the dumbbell surface ϕ^2, Fig. 2.13b, as the p orbital and give the lobes plus and minus signs, but such a representation is a hybrid one that is really wrong. In order to have plus and minus signs, the representation of the angular dependence function of the p_z orbital should be the two spheres, Fig. 2.13a; if the dumbbell shape ϕ^2 is shown, both lobes should actually be positive.

If a complete representation of $\phi = R(r)\Theta(\theta)\Phi(\varphi)$ is desired, we are reduced to the usual problem of trying to represent a function of three variables in three-dimensional space. A relatively simple expedient to achieve at least some graphic representation of this problem is to draw contour lines, that is

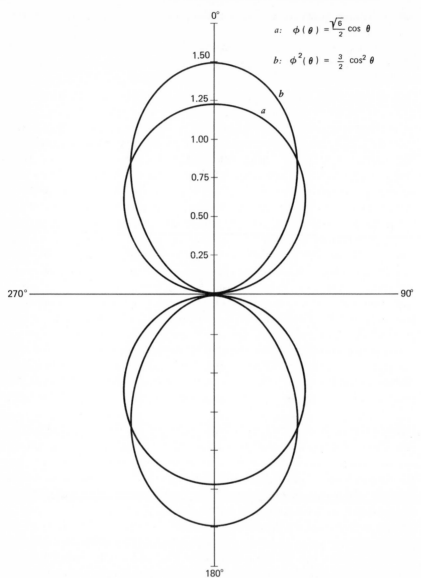

Fig. 2.13 Plots of the angular dependence of the $2p_z$ orbital of the hydrogen atom. (The numbers on the ordinate are units of Bohr radii.

lines connecting all points at which the function has the same value. When a series of such contours for a set of predetermined values of the function is drawn, we have a contour diagram. This is ideal for the representation of a function of two variables. Fortunately, the φ dependence is usually so simple that we can get by with two-dimensional cuts through the necessary three-dimensional contour surfaces.

A convenient way to represent contours is to draw lines for values of the function which are, on each contour, a given fraction of the maximum value. Such a contour diagram for a p_z orbital is shown in Fig. 2.14a. Similar contours for the total probability function $\rho = \phi^2$, again as a fraction of its

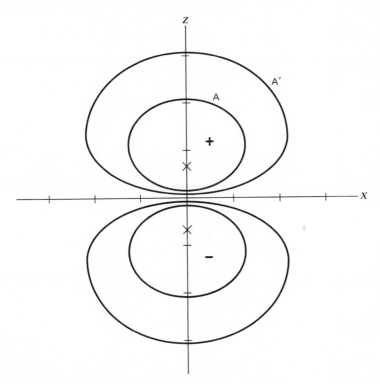

Fig. 2.14a Contour diagram for $\phi(p_z)$. Coordinate axes are in units of Bohr radii, 0.53 A. A: $\psi = 0.316\, \phi_{max}$; A': $\psi = 0.10\, \phi_{max}$.

maximum value, are given in Fig. 2.14b. In these two diagrams, the maximum (absolute) value (of ϕ or ϕ^2, respectively) is at some point inside the two oval-like shapes (marked $\times$ in Fig. 2.14), the values decreasing as one goes away from this point in any direction. The values of course become zero in the nodal surface (the xy-plane). Each of the contour surfaces of Fig. 2.14b generated by rotating the figure around the z-axis encloses a region in space having a given probability of finding the electron. This probability increases as our surface encloses larger and larger areas, that is, for contours of ever-decreasing value of ρ or ϕ^2.

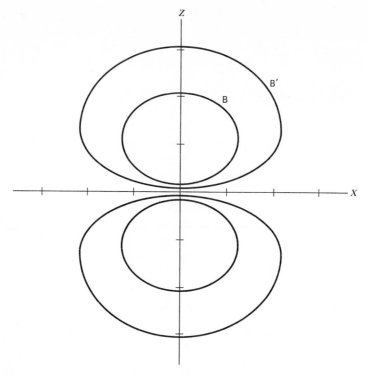

Fig. 2.14b   Contour diagram for $\phi^2(p_z)$. Abscissa in units of Bohr radii, 0.53 A. B: $\rho = 0.10$ ϕ^2_{max}; B′: $\rho = 0.01\ \phi^2_{max}$.

PROBLEMS

2.1   Given $x = r \sin \theta \cos \varphi$, $y = r \sin \theta \sin \varphi$, and $z = r \cos \theta$, express r, θ, and φ in terms of x, y, and z.

2.2   Plot ϕ, ϕ^2, and the radial distribution function, $4\pi r^2 \theta^2$, against r for the functions 3s, 3p, and 3d. (See Fig. 2.5 for a plot of the radial distribution of the 1s orbital.)

2.3   Plot ϕ^2 for the waves in Fig. 2.8.

2.4   What are the m_l values for $n = 4$ and $l = 0, 1, 2, 3$ (see Fig. 2.11)? Show schematically the angular momentum vector **L** for each of the l values and indicate the projection on the z-axis, that is, the m_l for each **L**.

2.5   What is the physical significance of the atomic orbital (one electron wave function) ϕ ?

GENERAL REFERENCES

1. J. C. Slater, *Quantum Theory of Atomic Electronic Structure*, Vol. I, McGraw-Hill Book Company, New York, 1960.

2. L. Pauling and E. B. Wilson, *Introduction to Quantum Mechanics*, McGraw-Hill Book Company, New York, 1935.

3. B. E. Douglas and D. H. McDaniel, *Concepts and Models of Inorganic Chemistry*, Blaisdell Publishing Company, Waltham, Mass., 1965.

4. L. F. Phillips, *Basic Quantum Chemistry*, John Wiley and Sons, New York, 1965.

3 Molecular orbital theory

3.1 The Hydrogen Molecule

As two individual hydrogen atoms at quite a large distance from each other are brought closer and closer together, the nucleus of each atom will start to attract the electron originally associated solely with the other atom. The change in energy of the system as a function of distance is called a *potential energy curve* and is frequently approximated by a Morse curve, Fig. 3.1 (cf. Section 9.2). When the distance separating the nuclei is at or near the bonding distance, the two electrons in the system are both associated with both nuclei; and instead of the original atomic orbital (AO) on each atom, we have a molecular orbital (MO) which results from the combination of AO's. When one electron is near one nucleus, the MO (or wave function)

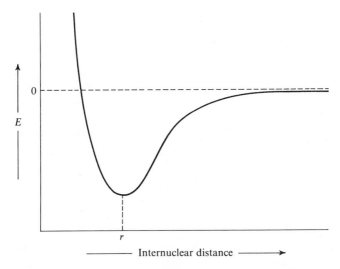

Fig. 3.1   The potential energy for the hydrogen molecule (the Morse curve).

25

may be assumed to resemble the AO (or wave function, ϕ_A) of that atom. Similarly, when the electron is in the neighborhood of the other nucleus, the MO resembles ϕ_B. Since the complete MO has characteristics separately possessed by ϕ_A and ϕ_B, it is approximated as a linear combination (linear combinations are combinations made by simple addition or subtraction of the functions to be combined) of the AO's. In this case we can indicate the molecular orbital (ψ) formed by the addition of the two atomic orbitals as $\psi_b = \phi_A + \phi_B$. This approach, called the *linear combination* of *atomic orbitals* or *LCAO* method, was first suggested by R. S. Mulliken.

The addition of the two AO's represented by the ψ_b molecular orbital implies that the two electrons in the hydrogen molecule are now shared by, or interact with, both nuclei, that is, the MO is bicentric. The addition further implies that both electrons spend most of their time between the two nuclei and hence will help to bond the two H atoms together; hence the subscript b in ψ_b indicates that the MO is *bonding*, and the orbital is called a *bonding* orbital.

The process of adding the two AO's together may be represented graphically in several ways. One simple way is to consider the boundary surface of the MO as made from the overlap of the two boundary surfaces of the individual AO's, Fig. 3.2. The plus sign in the MO indicates that the wave function is

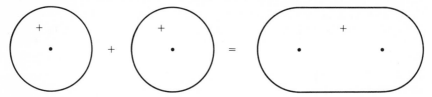

Fig. 3.2   The addition of two 1s atomic orbitals.

positive everywhere, and consequently there is no node. We will continue to use the physically significant probability boundaries (ϕ^2 or ψ^2) to represent the shapes of orbitals but will (incorrectly but conveniently) include signs to emphasize nodal properties.

In the linear combination explained above, we added the two separate AO's. It can be shown by quantum-mechanical arguments that the LCAO method applied to diatomics requires that a second MO be generated from the AO's by another linear combination of AO's. In LCAO theory, the combination of k atomic orbitals must generate k molecular orbitals. A rationalization of the necessity of two MO's from two AO's can be made as follows: Each AO can accommodate a maximum of two electrons; if two AO's combine, the combination must result in generating a sufficient number (two) of MO's to accommodate a maximum of four electrons. This is sometimes known as the law of conservation of orbitals.

The second linear combination of hydrogen AO's can be obtained by sub-tracting one AO from the other: $\psi_a = \phi_A - \phi_B$. We can represent this process by the boundary surface procedure as shown in Fig. 3.3. In this MO the probability of finding the electrons at exactly half the distance between the nuclei is zero, that is, there is a nodal plane in the MO, shown as a dashed line in Fig. 3.3. On the average, electrons in ψ_a are farther from either of the

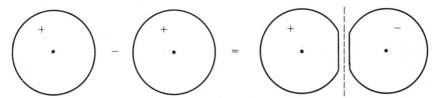

Fig. 3.3   The subtraction of two 1s atomic orbitals.

nuclei than they would be in the isolated atoms, and hence the atoms would be in an energetically more favorable condition if they were separated than if they were close together. Because electrons in such an orbital would tend to separate the atoms, the orbital (or wave function) is called an *antibonding* orbital: hence the subscript a in ψ_a.

We saw above that the LCAO method applied to two hydrogen atomic orbitals, ϕ_A and ϕ_B, led to two molecular orbitals, ψ_a and ψ_b. We can repre-sent this by an energy diagram, Fig. 3.4. The horizontal lines at the edges of the diagram represent the (equal) energy of each AO; the levels[a] in the middle represent the energies of the bonding and antibonding MO's. The resulting diagram is called a *molecular orbital energy diagram* (MOED).

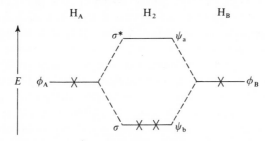

Fig. 3.4   The molecular orbital energy diagram for H_2.

[a] In a common approximation (neglect of the so-called overlap integral; *vide infra*), it is found that the energy of the antibonding orbital lies as much above the isolated AO as the energy of the bonding orbital lies below it.

3.2 Sigma and Sigma Starred Orbitals

The boundary surface of $\psi_b = \phi_A + \phi_B$ is shown in Fig. 3.2, and the plus sign inside this surface indicates that the orbital is plus everywhere, that is, there is no node. It is very desirable to describe a bond between two atoms in terms of the symmetry of the orbital occupied by the electrons bonding the atoms together. Specifically, a description of the symmetry of the orbital with respect to the bond axis is desired. If we take the egg-shaped orbital of Fig. 3.2 and rotate it around the internuclear axis, say by an angle of 180° (or any other angle for that matter), the resulting shape will be indistinguishable from the original. We have performed a symmetry operation. Furthermore, since the orbital was plus everywhere, the rotation has not affected the sign of the orbital anywhere—all the plus parts (which in this case were everywhere) have been transformed under the symmetry operation into plus parts. When a symmetry operation is performed and the resulting orbital has the same signs everywhere as the original, we say the orbital is *symmetric* with respect to that operation. In the present example, ψ_b is symmetric with respect to rotation around the internuclear axis. Because rotational symmetry in all directions characterizes the s atomic orbital, we call the MO which is symmetric with respect to rotation around the internuclear axis a σ (sigma, Greek equivalent of s) molecular orbital.

Now let us examine the antibonding orbital ψ_a pictured in Fig. 3.3. We see that, with respect to *rotation* around the internuclear axis, this orbital is also symmetric. Although the orbital has two equal halves of different signs, the signs do not interchange on rotation around the axis joining the centers of each half. Accordingly, this orbital is also a sigma orbital. However, it is desirable to distinguish this high-energy sigma orbital, ψ_a, from the lower-energy sigma orbital, ψ_b. Whereas the orbital ψ_b was positive near both atoms A and B, ψ_a is characterized by opposite signs near the two atoms, and consequently a node between them. To emphasize that both orbitals are sigma orbitals, they are commonly denoted by σ instead of ψ; to emphasize the antibonding nature of ψ_a, it is customary to add an asterisk: σ^*.

If the atoms A and B are like atoms, the distinction between σ and σ^* can be reduced to a *symmetry* distinction. If we assume a mirror halfway between A and B, we see that the *shapes* of both σ and σ^* are unchanged by *reflection* in the mirror. Also, the sign of σ is unchanged, that is, it is symmetric with respect to this reflection. On the other hand, the sign of σ^* is reversed (plus replaced by minus and vice versa), and hence σ^* is called *antisymmetric* with respect to reflection in the plane.

3.3 The Nonexistent Helium Molecule

Having developed the MO's from the combination of the 1s orbitals, ϕ_A and ϕ_B, of the two hydrogen atoms in the hydrogen molecule, we placed the two electrons of H_2 in ψ_b of Fig. 3.4. Now suppose that we wish to indicate the molecular orbital energy diagram of two helium atoms combined. Each He atom has two electrons in 1s orbitals, and so the combination of He_2 has four electrons. The combination of the 1s orbital of the two He atoms leads to the same MOED, Fig. 3.4, as in the hydrogen molecule; now, however, we place two of the four electrons into ψ_b, and, according to the Aufbau and Pauli principles, we must place the remaining two in the σ^* orbital, ψ_a. In the common approximation of neglect of overlap integrals, the bonding and antibonding levels are evenly spaced below and above the isolated atomic level so that equal electron occupation of the two levels results in exact cancellation of the bonding effect of σ by the antibonding effect of σ^*. The net result is that He_2 is not a stable species; as a matter of fact, in a somewhat less crude approximation, if the overlap integral, S, is not neglected, the antibonding orbital is more antibonding than the bonding orbital is bonding, and hence He_2 is unstable.[b]

3.4   Pi and Pi Starred Orbitals

In the preceding discussion we considered the combination of two s atomic orbitals to form σ and σ^* molecular orbitals. Similarly, two p orbitals on adjacent atoms pointing at each other combine colinearly to form σ and σ^* orbitals (cf. Section 3.7).

Let us now turn our attention to the combination of two *parallel* p orbitals. We again make our linear combination by the addition and subtraction of the orbitals, Fig. 3.5. Each of the original atomic p orbitals has one node; the resulting $\psi_b = \phi_A + \phi_B$ has one node, but $\psi_a = \phi_A - \phi_B$, where the ϕ's are the 2p atomic orbitals, has two nodes. The new node now appears between the two nuclei, and hence ψ_b is bonding and ψ_a antibonding, as before. Since rotation about the bond axis left the MO's formed from s orbitals unchanged, we called these MO's σ orbitals. However, in the present case, ψ_a and ψ_b each change sign on rotation by $180°$ about the internuclear axis. Because of this property these orbitals are designated as π (pi, Greek equivalent of p) orbitals; for ψ_b and ψ_a, we have π and π^* respectively.

Just as with the σ and σ^* orbitals, π does not have a node but π^* does have a node *between* the atoms. Again, if the atoms are alike, π is symmetric and π^* is antisymmetric, with respect to a mirror plane between the atoms. In order to classify orbitals further we introduce at this point a third type of

[b] The energies of the two orbitals including overlap are expressed by the equations:

$$E_{\psi_b} = E_\phi + \frac{\beta - SE_\phi}{1 + S}, \qquad E_{\psi_a} = E_\phi - \frac{\beta - SE_\phi}{1 - S}$$

Here, E_ϕ is the energy of the electron in an atomic orbital ϕ_A (or ϕ_B) of an isolated atom; and $(\beta - SE_\phi)/(1 \pm S)$ is the splitting of ψ_b and ψ_a from this level, as shown in Fig. 3.4. Because $1 > S > 0$, the demoninator is larger in the energy equation for ψ_a than in that for ψ_b, placing the antibonding orbital level higher above the isolated atom than the bonding orbital is below it.

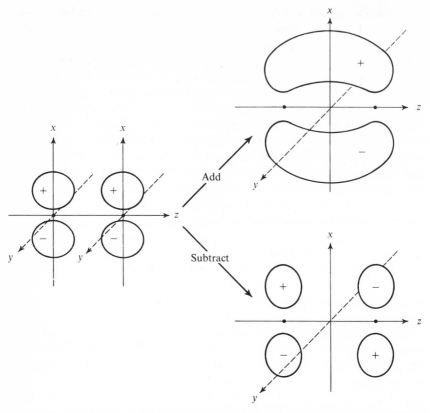

Fig. 3.5   The addition and subtraction of two 2p atomic orbitals.

symmetry operation[c] in addition to the operations of rotation around an axis and of reflection from a mirror, both discussed earlier.

The symmetry operation to be considered briefly here is called *inversion*. If we are dealing with a molecule, this operation consists of reflecting every atom through the center of the molecule. If on such reflection an equivalent atom is encountered, the molecule has a center of symmetry. Thus, ethylene, *trans*-dichloroethylene, and *s-trans*-butadiene all have centers of symmetry, and an inversion can be performed. In the case of orbitals, we are interested at this point in whether the orbital is symmetric or antisymmetric with respect to inversion. We can determine this by taking a point anywhere in the orbital, drawing a line from it through the center of the molecule, and

---

[c] We postponed discussion of this operation until now because, in considering σ orbitals, it could not have been distinguished from reflection at a mirror plane. A more detailed discussion of symmetry and symmetry properties will be found in Chapter 5.

continuing an equal distance in the same direction. If at this new point, which is a point at which the amplitude function has the same magnitude as at the old one, the sign of the orbital is identical with that of the original point, the orbital is symmetric with respect to inversion; if the new point has an opposite sign, the orbital is antisymmetric with respect to inversion. The German words *gerade* (even) and *ungerade* (uneven) are used to denote symmetric and antisymmetric behavior, respectively, with regard to inversion. Thus, we can call the π orbital u (ungerade) and the π^* orbital g (gerade). The g and u are frequently used as subscripts. Reference to Figs 3.2, 3.3, and 3.5 shows that, if A and B are identical atoms, we have σ_g, σ_u^*, π_u, and π_g^* orbitals.

The energy diagram for the combination of two p atomic orbitals can be written analogously to that for the combination of two s atomic orbitals, Fig. 3.6. Again, the splitting in the neglect-of-overlap-integral approximation

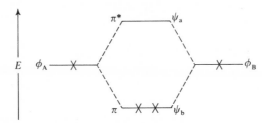

Fig. 3.6   The molecular orbital energy diagram for the interaction of two parallel 2p atomic orbitals.

is equal, that is, the antibonding orbital is as much antibonding as the bonding orbital is bonding. This level diagram would describe correctly the π system in ethylene, Fig. 3.7. In the ground or normal state, the two π electrons are

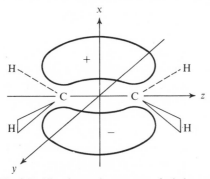

Fig. 3.7   The electronic structure of ethylene.

paired and are in the lowest-energy orbital, ψ_b. If the ethylene molecule is subjected to just the right amount of light energy, that is, if it is excited, an electron in ψ_b will be promoted to ψ_a in what is called a $\pi \to \pi^*$ (pi to pi starred) transition. The energy difference (or vertical distance in Fig. 3.6) between ψ_b and ψ_a in ethylene is considerably less than the analogous energy difference between ψ_b and ψ_a in the hydrogen molecule. Experimentally, we find that ethylene absorbs at about 185 nm, corresponding to an energy of approximately 155 kcal/mole. In the above excited electronic configuration, one electron remains in ψ_b and the other electron is now in ψ_a, or the electronic configuration of this excited state may be written $\pi\pi^*$ or $\pi_u\pi_g^*$. In the same system of notation, the ground state is expressed as π^2 or π_u^2. We are now in a position to consider the electronic structure of some simple diatomic molecules, and we will discover that, even in the ground (or normal) state, electrons can and do occupy antibonding orbitals.

However, before proceeding to examine the molecular structure of some simple diatomic molecules, we will find it useful to examine two concepts which are basic to a more complete MO treatment, namely, normalization and orthogonality.

3.5 Normalization

Although we can never know where to find a small particle like an electron at any particular moment, we can determine the probability of finding it in any given region in space. The position of the electron can be defined by a probability function. This function is usually called ρ, and is proportional to the square of the wave functions (ϕ^2 or ψ^2) discussed in Chapter 2. The electron is most likely to be found in the regions of space where ρ is largest. If we define a small volume of space in terms of infinitesimal distances along three axes, $dx\, dy\, dz = d\tau$, than $\rho\, d\tau$ is the probability of finding the electron in the small volume $d\tau$. Since, as we learned in Chapter 2, the electron must be found somewhere in space, if we integrate the probability from minus infinity to plus infinity, that is, over all space, we should get unity:[d]

$$\int_{-\infty}^{+\infty}\int_{-\infty}^{+\infty}\int_{-\infty}^{+\infty} dx\, dy\, dz = \int \rho\, d\tau = \int (N\phi)^2\, d\tau = 1$$

where N is the normalization constant. When the square of a wave function, ϕ, integrated over all space is equal to unity, $\int \phi^2\, d\tau = 1$, the wave function is called *normalized*. We will recall from Chapter 2 that all atomic orbitals, designated by ϕ, include the normalization constant and are therefore already normalized, so that $\int \phi^2\, d\tau = 1$.

In Section 3.4 we saw how the combination of two AO's, for example, the AO's or wave functions of two hydrogen atoms, H_A and H_B, combine to form two MO's:

$$\psi_b(_g) = \phi_A + \phi_B \quad \text{and} \quad \psi_a(\sigma_u^*) = \phi_A - \phi_B$$

---

[d] Whenever, for simplicity, we write the integral symbol, $\int$, with the differential d^τ, we mean integration over all space, that is, from $-\infty$ to $+\infty$, for x, y, and z.

These molecular wave functions are not normalized and must now be separately normalized, that is, multiplied by a factor N such that $\int (N\psi)^2 \, d\tau = 1$. In order to do so, we write for ψ_b:

$$
\begin{aligned}
\int [N(\phi_A + \phi_B)]^2 \, d\tau &= N^2 \int (\phi_A + \phi_B)^2 \, d\tau \\
&= N^2 \int (\phi_A{}^2 + 2\phi_A\phi_B + \phi_B{}^2) \, d\tau \\
&= N^2 \left(\int \phi_A{}^2 \, d\tau + 2 \int \phi_A\phi_B \, d\tau + \int \phi_B{}^2 \, d\tau \right) \\
&= 1
\end{aligned}
$$

Since the atomic orbitals ϕ_A and ϕ_B are already normalized, $\int \phi_A{}^2 \, d\tau = 1 = \int \phi_B{}^2 \, d\tau$. The third integral, $\int \phi_A\phi_B \, d\tau$, is called the *overlap integral* and is given the symbol S. The overlap integral is a measure of the extent of overlap of the atomic orbitals ϕ_A and ϕ_B, and the numerical value of S varies from zero to positive (or negative) unity. In ethylene, for example, the overlap integral for the $2p\pi$ orbitals is about 0.3. If in the above equation we replace the overlap integral by the symbol S, we have:

$$
N^2(2 + 2S) = 2N^2(1 + S) = 1
$$

and hence

$$
N = \pm \, \frac{1}{\sqrt{2(1 + S)}}
$$

Since we usually want our wave functions positive, we take the plus sign and drop the minus sign; however, this choice is arbitrary and immaterial. In much work in molecular quantum mechanics, the very crude approximation $S = 0$ (see footnote b, p. 29) is made. In that case, the last equation becomes $N = \pm \, 1/\sqrt{2}$, and the full wave function for the bonding MO of the hydrogen molecule is then:

$$
\psi_b = \frac{1}{\sqrt{2}} \phi_A + \frac{1}{\sqrt{2}} \phi_B \tag{3.1}
$$

At first glance it may seem strange to neglect the overlap integral when the overlap is actually responsible for π bonding between the carbon atoms. However, it has been shown that this neglect, which cannot be justified mathematically, leads to essentially the same results as are obtained when overlap integrals are included, at least as far as relative energies, charge densities, and bond orders are concerned. Actually, in the simple methods in which overlap integrals are neglected, such as the Hückel MO method, many other and more serious approximations are made.

The constants $1/\sqrt{2}$ in the molecular wave function above are called coefficients, c, and they tell us to what extent each AO is participating in the

MO. The preceding wave function is a specific example of the generalized linear combination wave function:

$$\psi = c_1\phi_1 + c_2\phi_2 + \ldots + c_n\phi_n$$

and in terms of these coefficients our requirement for normalization is (assuming $S = 0$):

$$\sum_i c_i^2 = c_1^2 + c_2^2 + \ldots + c_n^2 = 1$$

And, checking our bonding wave function for the hydrogen molecule, we find:

$$\left(\frac{1}{\sqrt{2}}\right)^2 + \left(\frac{1}{\sqrt{2}}\right)^2 = 1$$

If the same procedure is carried out for the antibonding wave function, we should obtain:

$$\psi_a = \frac{1}{\sqrt{2}}\phi_A - \frac{1}{\sqrt{2}}\phi_B \tag{3.2}$$

3.6 Orthogonality

According to quantum mechanics, any two molecular wave functions, in order to be proper wave functions of the same system, must be orthogonal to one another. The mathematical condition for orthogonality is as follows: the product of the two wave functions, integrated over all space, must be equal to zero, that is, $\int \psi_1\psi_2 \, d\tau = 0$. Let us take the two molecular wave functions for the hydrogen molecule and test their orthogonality:

$$\int \psi_b\psi_a \, d\tau = \int \left[\left(\frac{1}{\sqrt{2}}\phi_A + \frac{1}{\sqrt{2}}\phi_B\right)\left(\frac{1}{\sqrt{2}}\phi_A - \frac{1}{\sqrt{2}}\phi_B\right)\right] d\tau$$

$$= \int (\tfrac{1}{2}\phi_A^2 - \tfrac{1}{2}\phi_B^2) \, d\tau$$

$$= \tfrac{1}{2}\int \phi_A^2 \, d\tau - \tfrac{1}{2}\int \phi_B^2 \, d\tau$$

and, since $\int \phi_A^2 \, d\tau = \int \phi_B^2 \, d\tau = 1$, we have

$$\int \psi_b\psi_a \, d\tau = \tfrac{1}{2} - \tfrac{1}{2} = 0$$

Perhaps a graphical representation will make the concept clearer. If we represent the separate atomic orbitals of H_2 in Fig. 3.8a, the graphical representations of the molecular orbitals, ψ_b and ψ_a (Figs. 3.8b and c), follow from the form of their wave functions. The product of ψ_b and ψ_a is graphically represented in Fig. 3.8d, and we see that the area under the curve of the left half (which is positive) is exactly equal to the area under the right half of the curve (which is negative); hence the net area under the curve (the integral of the product) is exactly zero.

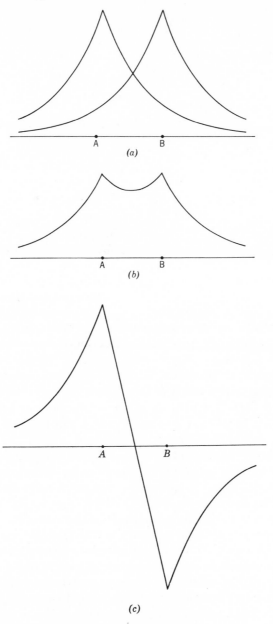

(a)

(b)

(c)

Fig 3.8   (*continued overleaf*)

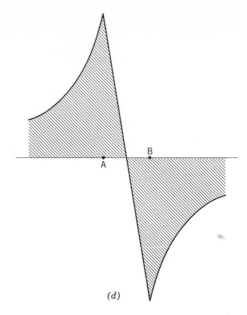

(d)

Fig. 3.8 Atomic and molecular orbital wave functions plotted against internuclear distance. (a) ϕ_A and ϕ_B; (b) $\psi_b = \phi_A + \phi_B$; (c) $\psi_a = \phi_A - \phi_B$; (d) $\psi_b\psi_a = (\phi_A + \phi_B)(\phi_A - \phi_B)$.

Finally, it should be mentioned that wave functions which are both orthogonal and normalized are called *orthonormal* functions.

3.7 The Structure of Nitrogen

The electronic configuration of the nitrogen atom, it will be recalled, is $1s^2 2s^2 2p^3$ (Fig. 2.1). When two N atoms combine, the same orbital types combine if they are of equal or approximately equal energy. Thus, the 1s, 2s, $2p_x$, $2p_y$, and $2p_z$ orbitals on one nitrogen combine with the similar orbitals on the other nitrogen to give, in each case, two MO's. The five AO's on each atom give rise to ten MO's in the molecule. Because each atom contributes seven electrons, the total of fourteen electrons in the molecule will distribute themselves in seven of the ten MO's.

Since the electrons will occupy the orbitals in the order of increasing energy, we must arrange our MO's in an energy sequence so that we can place our fourteen electrons properly. One of the most instructive ways to do this is by means of the molecular orbital energy diagram method, Fig. 3.9. We have already discussed the combination of two s orbitals to give σ and σ^* molecular orbitals. In the nitrogen molecule we have the combination of both 1s and 2s orbitals to give four σ type orbitals, two bonding and two antibonding, each of them occupied by two electrons. Because the 1s electrons are not valence

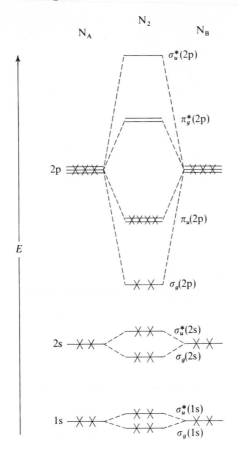

Fig. 3.9   The molecular orbital energy diagram for N_2, unhybridized.

electrons, we usually pay little heed to them. The combination of the 2s orbitals does not result in any net bonding, and these electrons can be crudely identified with the lone pair on each nitrogen in the familiar valence bond method of writing nitrogen as $:N{\equiv}N:$. Another way of representing the fact that the 2s orbitals do not interact appreciably is to show relatively little splitting in the MOED, that is, the distance of $\sigma_g(2s)$ and $\sigma_u^*(2s)$ from the isolated atomic 2s level is not very large.

The three atomic p levels in the isolated atom are of equal energy (degenerate); but when we bring one atom into the field of the other (cf. Fig. 3.10), the p_z orbitals pointing toward the other atom start to interact to form a σ bond between the two atoms, Fig. 3.11a. Of course, the corresponding σ^* orbital must also be generated, Fig. 3.11b. These orbitals $\sigma_g(2p_z)$ and $\sigma_u^*(2p_z)$,

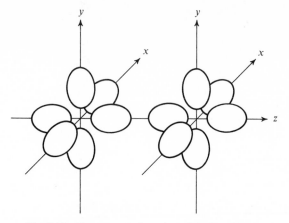

Fig. 3.10 Customary orientation of the 2p orbitals.

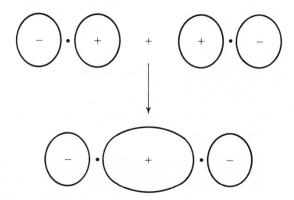

Fig. 3.11a Formation of the molecular orbitals from the 2p$_z$ atomic orbitals: σ_g(2p).

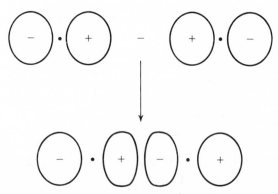

Fig. 3.11b Formation of the σ molecular orbitals from the 2p$_z$ atomic orbitals: σ_u^*(2p).

although generated from the combination of atomic $2p_z$ orbitals, are in fact σ molecular orbitals because they are symmetric with respect to rotation around the internuclear axis (the z-axis of Fig. 3.10). The bonding interaction is quite large, and hence the splitting is also relatively large.

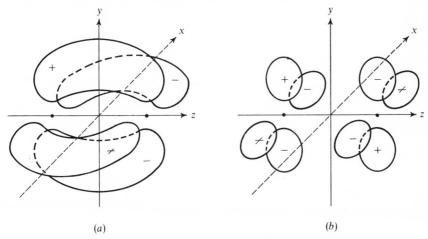

(a)   (b)

Fig. 3.12   The π molecular orbitals of the nitrogen molecule. (a) $\pi_u(2p)$. (b) $\pi_g^*(2p)$.

The p_x and p_y orbitals, Fig. 3.12, on each nitrogen combine to each form a π set: $\pi_u(2p_x)$, $\pi_g^*(2p_x)$, $\pi_u(2p_y)$, $\pi_g^*(2p_y)$. The form of these orbitals has been shown in Fig. 3.5. The p_x and p_y orbitals are perpendicular (orthogonal) to each other, as are $\pi_g^*(2p_x)$ and $\pi_g^*(2p_y)$, Fig. 3.12. Now, if we refer to our MOED, Fig. 3.9, we see that there are sufficient electrons to occupy completely the $\pi_u(2p_x)$ and $\pi_u(2p_y)$ orbitals, but that the $\pi_g^*(2p_x)$ and $\pi_g^*(2p_y)$, as well as the $\sigma_u^*(2p_z)$, remain unoccupied. The net bonding between the N atoms then comes from the $\sigma_g(2p_z)$ and the $\pi_u(2p_x)$ and $\pi_u(2p_y)$ bonding molecular orbitals, as any bonding contribution from $\sigma_g(1s)$ and $\sigma_g(2s)$ is canceled by the corresponding antibonding contribution from $\sigma_u^*(1s)$ and $\sigma_u^*(2s)$.

3.8   Hybridization

The molecular orbital energy diagram (MOED) displayed in Fig. 3.9 has been shown to be incorrect by a careful analysis of the spectrum of N_2. In fact, the highest occupied MO is a σ_g orbital. In order to understand this fact we must consider the phenomenon of hybridization or mixing of atomic orbitals.

The mixing of AO's in the same atom, in our case the 2s and $2p_z$ orbitals, is a favorable process because the hybrid orbitals which result have much better bonding properties than pure 2s or 2p orbitals. Without hybridization, the combinations of orbitals would be as follows: $2s_A$ and $2s_B$ combine to give:

$$\sigma_g(2s) = N_1(\phi_{2s_A} + \phi_{2s_B})$$

$$\sigma_u^*(2s) = N_2(\phi_{2s_A} - \phi_{2s_B})$$

and the $2p_{z_A}$ and $2p_{z_B}$ combine to give:

$$\sigma_g(2p) = N_3(\phi_{2p_A} + \phi_{2p_B})$$

$$\sigma_u^*(2p) = N_4(\phi_{2p_A} - \phi_{2p_B})$$

With hybridization we mix the 2s and 2p orbitals on each atom before we combine the orbitals of different atoms. Thus the hybrid orbitals are AO's, as they are restricted to a single atom and do not encompass the entire molecule. If we mix the 2s and $2p_z$ orbitals on one N atom we obtain:

$$\phi_{hy_1} = N_5(\phi_{2s} + \lambda\phi_{2p_z}) \quad \text{or} \quad \phi_{hy_1} = \frac{s + \lambda p}{\sqrt{1 + \lambda^2}}$$

$$\phi_{hy_2} = N_6(\lambda\phi_{2s} - \phi_{2p_z}) \quad \text{or} \quad \phi_{hy_2} = \frac{\lambda s - p}{\sqrt{1 + \lambda^2}}$$

where the N's are normalization constants; the λ is a weighting factor, that is, a number larger than 0 but less than 1; and s and p stand for the 2s and $2p_z$ orbitals, respectively. The appearance of λ in the form shown assures orthogonality of the two wave functions and equality of the normalization factors, $N_5 = N_6 = 1/\sqrt{1 + \lambda^2}$. The normalization factor $N_5 = 1/\sqrt{1 + \lambda^2}$ here is exact, since the overlap integral S between s and a p orbital (or between any two AO's) on *any one* atom is exactly 0, that is, the various AO's of any one atom are truly orthogonal. The weighting factor would be unnecessary if the hybrid orbitals were made up of equal parts of s and p, that is, $\lambda = 1$, but one orbital (ϕ_{hy_1}) has more s than p character, and the other (ϕ_{hy_2}) has less, as is seen from the form of the wave functions of the hybrid orbitals ($0 < \lambda < 1$).

A physical picture of the hybridization is given in Fig. 3.13. It will be noted that in this representation the angular part of the 2p orbital is shown by the two touching spheres. The angular part of the wave function is used because, although the 2s and 2p orbitals have different radial distributions near the nucleus, at bonding distances the radial parts are similar and hence are factored out. Thus we have obtained on each N atom, two sp hybrid orbitals, which are oriented at 180° from one another. The p part of these orbitals gives them strong directional character, and hence the sp hybrids overlap other orbitals much more effectively than either the s or the p orbital alone.

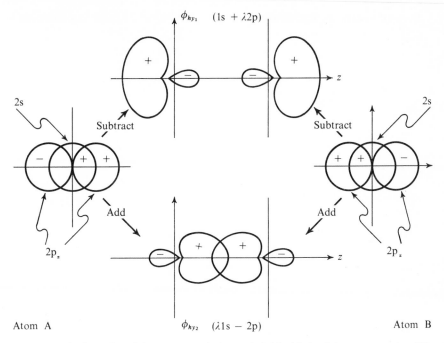

Fig. 3.13   The formation of the two nonequivalent sp hybrid orbitals of nitrogen atom A and B.

In combining the sp hybrids of two different nitrogen atoms (A and B), we want to use that set of hybrids which will give the best bond: in this case the hybrids (ϕ_{hy_2}) with the greater p character. Combining the ϕ_{hy_2} of the two N atoms to obtain MO's gives:

$$\psi\sigma_g = N_7(\phi_{hy2A} + \phi_{hy2B})$$
$$\psi\sigma_u^* = N_8(\phi_{hy2A} - \phi_{hy2B})$$

This process is represented in Fig. 3.14. We are left with one hybrid orbital (ϕ_{hy_1}) on each N atom. These hybrids, however, are oriented so that the large lobes are pointing away from each other, and hence their interaction is quite small. Combining the ϕ_{hy_1} of the two different N atoms, we obtain:

$$\psi\sigma_g = N_9(\phi_{hy1A} + \phi_{hy1B})$$
$$\psi\sigma_u^* = N_{10}(\phi_{hy1A} + \phi_{hy1B})$$

These combinations are shown pictorially in Fig. 3.15. If overlap is not neglected, N_7 and N_9 will have the form $1/\sqrt{2 + 2S}$ and N_8 and N_{10} $1/\sqrt{2 - 2S}$. However, they are essentially the lone pair non-bonding orbitals, see p. 37.

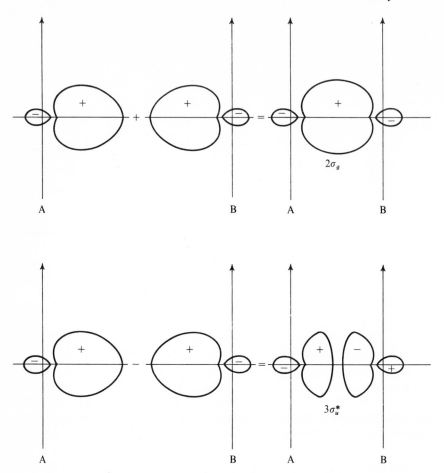

Fig. 3.14   Formation of σ_g and σ_u^* orbitals from the ϕ_{hy_2} sp hybrid orbitals.

When any two orbitals are of the same type (say σ_g), they can interact with one another. As a result of the quantum-mechanical treatment of the problem, this interaction *always* depresses the lower one and raises the higher one of the pair. This process of mixing may be thought of as being analogous to the process of hybridization. However, there is one major difference. In hybridization the AO's are orthonormal, whereas $2\sigma_g$ and $3\sigma_g$ (as well as $2\sigma_u$ and $3\sigma_u$) are not orthogonal since nothing has been done to make them so. As a result the hybrid orbitals are intermediate in energy between the AO's from which they are mixed, whereas the interaction of the MO's spreads their energies.

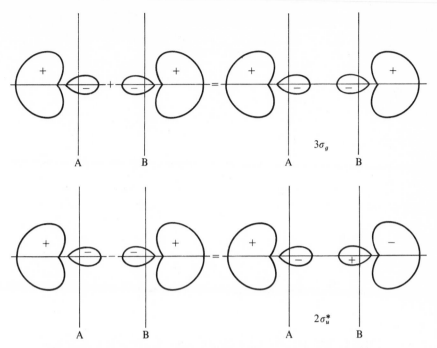

Fig. 3.15   Formation of σ_g and σ_u^* orbitals from the ϕ_{hy_1} sp hybrid orbitals.

Accordingly, this mixing depresses $2\sigma_g$ and $2\sigma_u^*$ and raises $3\sigma_g$ and $3\sigma_u^*$, as shown in Fig. 3.16. It must then be true that the weakly bonding $3\sigma_g$ has been raised sufficiently to end up above $\pi_u(2p)$, so that it becomes the highest occupied orbital, and the correct MOED is shown in Fig. 3.17.

Hybridization of AO's before combining them into MO's is actually an artificial process. In describing the molecular wave function, for example, $2\sigma_g$, we can simply include the appropriate contribution from the 2s and 2p orbital of each atom:

$$\psi 2\sigma_g = N\left[(c_1\phi_{2s_A} + c_2\phi_{2p_A}) + (c_1\phi_{2s_B} + c_2\phi_{2p_B})\right]$$

The addition of some p to the s character places extra electron density between the atoms and stabilizes the bonding orbital to a very great extent. This stabilization of the hybrid orbital is accompanied by a large destabilization of the other bonding orbital of similar symmetry, here the $3\sigma_g$ orbital. Electron density is removed from between the atoms, and the $3\sigma_g$ orbital is raised above the π level.

In Fig. 3.17 we have introduced a very common and useful notation, the symbols $2\sigma_g$, $3\sigma_g$, $2\sigma_u^*$, and $3\sigma_u^*$ for the new hybrid MO's. Since we mixed s and p_z orbitals to obtain the hybrids, the resulting MO's can no longer be

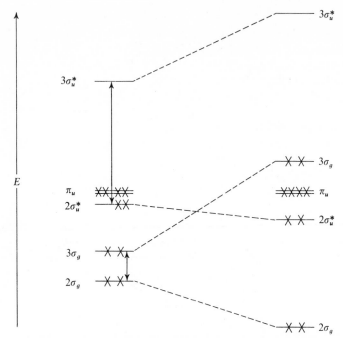

Fig. 3.16 Spreading of energies of hybrid orbitals of the same type.

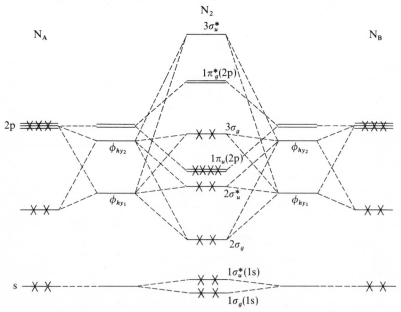

Fig. 3.17 The correct molecular orbital energy diagram for N_2.

44

identified with specific AO's and some notation is needed to indicate the difference. The numbers preceding the σ's are order numbers, or pseudo-quantum numbers, simply ranking the MO's of the same type in the order of increasing energy (i.e., $1\sigma_g$ is of lower energy than $2\sigma_g$, etc.). For clarity and consistency, we extend the notation to all of the MO's of the diagram, even if they can be identified with a specific AO.

3.9   The Structure of Oxygen

The valence electrons of oxygen are in the 2s and 2p orbitals with the configuration $2s^2 2p^4$. The MOED would be essentially that constructed for nitrogen (Fig. 3.17), as we are using the same set of AO's in our combinations (the same basis set). However, it is now necessary that the MO's accommodate sixteen electrons, of which twelve (all 2s and 2p electrons) are referred to as valence electrons. If we follow the principles used for the periodic classification, the two electrons which must be added to the nitrogen system (ten valence electrons) must go separately into the $1\pi_g^*(2p_x)$ and $1\pi_g^*(2p_y)$ orbitals, with spins parallel (Hund's rule). The two unpaired electrons in the π^* orbitals give rise to the paramagnetic properties of molecular oxygen. (Molecules or atoms with one or more unpaired electrons are attracted by an external magnetic field and hence are said to be *paramagnetic*.) The diradical character and accompanying paramagnetism of oxygen constitute perhaps its outstanding property.

The occupation of antibonding orbitals by one or more electrons cancels some of the bonding attraction between the atoms. In the O_2 example, we have two π bonding orbitals, each doubly occupied, and a σ bonding orbital, doubly occupied, or a total of three bonding orbitals. However, each of the two electrons in an antibonding orbital cancels the bonding effect of an electron in a bonding orbital, and so the net bonding in oxygen can be considered to result from a double bond. (Since two electrons constitute a bond and there are six bonding electrons and two antibonding electrons, for a net of four bonding electrons, we have two net bonds.) Evidence for the effect of occupation of the antibonding orbitals comes from bond distances. In the ground (or normal) state, the bond distance between oxygen atoms is 1.2074 A. However, when O_2 is ionized by loss of an electron from one of the π^* antibonding orbitals, the resulting O_2^+ species should have a stronger interaction between the two O atoms. In fact, the bond distance of O_2^+ is 1.1227 A, a considerable decrease, indicative of stronger bonding in the ion.

The O_2^+ species is very interesting. It not only carries a positive charge but also has one unpaired electron; it is properly called a radical-cation. The bond lengths of O_2^- (a radical anion) and O_2^{2-} are 1.26 and 1.49 A, respectively, confirming the fact that electrons have been added to antibonding orbitals.

3.10   The Structure of Fluorine

The MOED of F_2, formed from two F atoms with the valence configuration of $2s^2 2p^5$, is basically that of the other second-row homonuclear diatomics and differs from O_2 in that two additional electrons must be accommodated. These two electrons will go into the $\pi_g^*(2p_x)$ and $\pi_g^*(2p_y)$ orbitals in Fig. 3.17. Now the two antibonding π^* orbitals are completely filled, and this antibonding character cancels the bonding generated from the two π bonding orbitals. The net bonding thus results from the σ bond-p_z interaction between the atoms. This fact is apparent in the conventional valence bond structure of F_2: $:\ddot{F}:\ddot{F}:$. The three lone-pair electrons on each F atom are associated with the canceling effect of $\sigma_g(2s) - \sigma_u^*(2s)$, $\pi_u(2p_x) - \pi_g^*(2p_x)$, and $\pi_u(2p_y) - \pi_g^*(2p_y)$. Since the antibonding cancels the bonding, the net result in each interaction is to have a lone pair on each atom corresponding to the three bonding-antibonding interactions.

The effect of increasing occupancy of antibonding orbitals is dramatically illustrated by the decreasing bond energies of N-N, O-O, and F-F, which are 225, 118, and 36 kcal/mole, respectively; thus it is about six times more difficult to dissociate N_2 than F_2 to its atoms. Furthermore, the bond distances increase in the expected order: 1.10 A (N_2), 1.21 A (O_2), and 1.42 A (F_2). Although the increasing nuclear charge of the atoms as we move across the periodic table is partly responsible for the decreased bonding between them, the contribution of antibonding orbital occupation is frequently overlooked.

3.11   The Structure of Carbon Monoxide

In considering the structure of N_2, O_2, and F_2, we combine the orbitals on two identical atoms to form the homonuclear diatomic molecules discussed above. We shall now consider the combination of two different atoms in the first-row elements. The general problem is to consider the effects of the difference in electronegativity of the two atoms. We know that the nuclear charge increases with increasing atomic number. In the first row of the periodic system all of the valence electrons occupy the same shell ($n = 2$) and hence are at about the same distance from the nucleus. Consequently, the greater the atomic number the more difficult it is to remove the electrons (the higher the ionization potential), that is, the electrons in the 2s and 2p orbitals decrease in energy as the nuclear charge increases, and they are thereby stabilized. In constructing the MOED's, we should place the 2s and 2p orbitals of the more electronegative element (the one with the higher nuclear charge) lower than the corresponding orbitals of its partner.

In constructing the MOED of carbon monoxide (Fig. 3.18), we see that the levels of the isolated oxygen atom are below those of the isolated carbon atom. In making the combination of σ bonding orbitals, we involve the same general concepts of hybridization as we employed in the case of N_2. However,

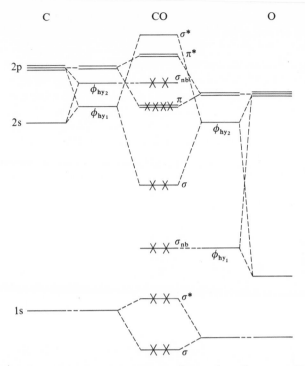

Fig. 3.18 The molecular orbital energy diagram for carbon monoxide.

the hybrid orbitals on each N atom were equal in energy, whereas in the CO case the situation is different. The two hybrid orbitals on oxygen do not match the energies of those on carbon; as a matter of fact, the lower-energy sp hybrid on carbon matches the energy of the higher-energy hybrid on oxygen, with the result that the two lone pairs of electrons—the electrons in nonbonding orbitals σ_{nb}—are on the lower-energy hybrid of oxygen and the higher-energy hybrid of carbon. These relationships are shown in Fig. 3.18. (Note that no g and u subscripts occur, since heteronuclear diatomic molecules have no center of inversion, and hence the g-u classification is not appropriate.)

There are thus two lone pairs of electrons on CO, neither involved substantially in bonding. One pair, as explained above, is on oxygen in a hybrid orbital, mostly s in character, and thus in a very stable (unreactive) orbital. The other lone pair is on the C atom and has a high degree of p character. This orbital extends away from the C—O bond and is a high-energy orbital in which the electrons are loosely bound. The reactivity of electrons in this orbital accounts for the well-known donor properties of CO. The bonding between carbon and oxygen results from the σ bond, mostly on the oxygen, and the set of π orbitals π_x, π_y, also mostly on the O atom. The energy diagram shows the π^*'s lying much closer to the carbon p's than to the oxygen p's.

The lone-pair orbitals on carbon and oxygen and a π and a π^* orbital are illustrated in Fig. 3.19. The π orbitals are mostly on the O atom, since they receive a greater contribution from this atom than from the C atom. The other π bonding MO is at right angles to the one shown, that is, in front and behind the paper. The antibonding orbital, π_x^*, in Fig. 3.19, shows that the carbon makes a larger contribution to this orbital than does oxygen. There is another π^* orbital, the π_y^*, at right angles to the π_x^* orbital. The fact that the π^* orbitals are close in energy to the carbon p orbital and are concentrated on the C atoms is of very great importance in considering the acceptor properties of CO in a variety of carbonyl complexes.

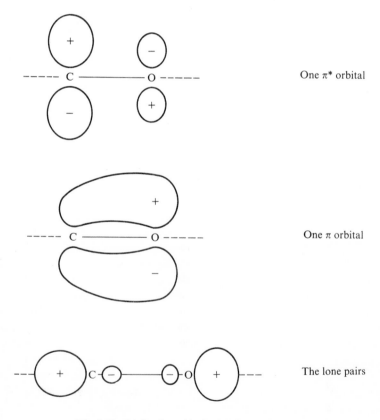

One π^* orbital

One π orbital

The lone pairs

Fig. 3.19   Molecular orbitals of carbon monoxide.

3.12   The π-Electron Structure of 1,3-Butadiene

In order to obtain a qualitative picture of the π molecular orbitals of 1,3-butadiene, we can consider the molecule as generated from two ethylenes. For this purpose, then, we would make linear combinations of the two ethylene π bonding orbitals and linear combinations of the two ethylene π^* orbitals. Such combinations are indicated in Fig. 3.20. In this treatment we are assuming that butadiene is a linear molecule, which of course it is not, since it actually exists as an equilibrium mixture of the *s-trans* (*a*) and *s-cis* (*b*) forms:

$$
\text{(a)} \quad\rightleftharpoons\quad \text{(b)}
$$

(a)   (b)

The equilibrium lies to the left, and the energy difference between the two conformations is about 3–4 kcal/mole.

The resonance structures of butadiene are written conventionally as

$$
\overset{1}{C}H_2{=}\overset{2}{C}H{-}\overset{3}{C}H{=}\overset{4}{C}H_2 \;\leftrightarrow\; CH_2{-}CH{=}CH{-}CH_2 \;\leftrightarrow\; \overset{\mp}{C}H_2{-}CH{=}CH{-}\overset{\pm}{C}H_2
$$

(c)   (d)   (e)

Structures *d* and *e* involve a double bond between atoms 2 and 3, and structure *d* a "long" bond between 1 and 4; in structure *e* there are formal charges on atoms 1 and 4. The resonance structures tell us that there is some double-bond character in the 2,3 bond in the ground state, albeit small, since structures *d* and *e* make substantially less contribution to the ground state than does structure *c*. It is this double-bond character, however, which provides the energy barrier between the *s-cis* and *s-trans* conformations.

It is frequently possible to draw qualitative conclusions about double-bond character from resonance structures. For the important resonance structures of benzene, for example, we would write the two equivalent structures:

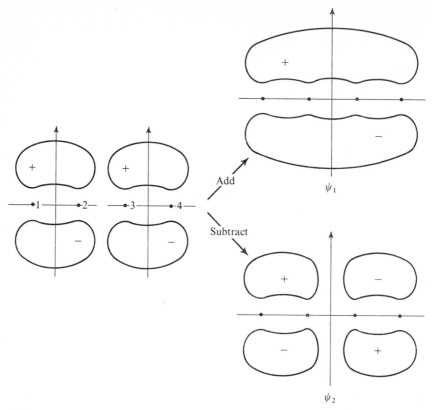

Fig. 3.20a The molecular orbitals of $CH_2{=}CH{-}CH{=}CH_2$ made by the linear combination of ethylene π orbitals.

These two structures are the principal ones that need to be considered in benzene, and we see that each should contribute equally and that each bond is approximately one half of a double bond. In butadiene, the 2,3 bond is called an *essential single bond*, and the 1,2 and 3,4 bonds are called *essential double bonds* because there is only one principal resonance structure (*c*) and in this structure the double and single bonds are fixed as shown. In contrast, there are no essential single or double bonds in benzene.

Now let us consider the kind of information the MO picture of butadiene gives us. The four MO's have different energies, and these increase with the increasing number of nodes. Since all π systems must necessarily be planar in order for the $p\pi$ orbitals to be parallel and overlap maximally, the nodal plane in the plane of the molecule is usually neglected in counting nodes. We see then (Figs. 3.20a and *b*) that ψ_1, ψ_2, ψ_3, and ψ_4 have 0, 1, 2, and 3 nodes, respectively. In ψ_1 there is a bonding interaction between C_1 and C_2,

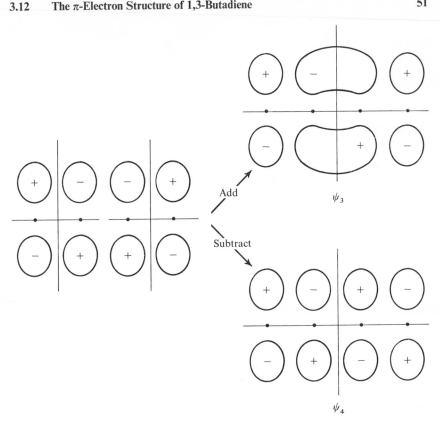

Fig. 3.20b   The molecular orbitals of $CH_2{=}CH{-}CH{=}CH_2$ made by the linear combination of ethylene π^* orbitals.

between C_2 and C_3, and between C_3 and C_4, or a total of three bonding interactions. This orbital is strongly bonding. A bonding interaction exists when the sign of the orbital between bonded (adjacent) atoms does not change, that is, where there is not a node between them. In ψ_2, there are two bonding and one antibonding interactions; hence this orbital is a bonding MO. In ψ_3 there are two antibonding and one bonding interactions; hence this orbital is antibonding. As for ψ_4, the three antibonding interactions make this a strongly antibonding orbital. The MOED is given in Fig. 3.21. This diagram also shows that the energy required for the $\pi \rightarrow \pi^*$ transition in ethylene (185 nm) is considerably greater than that required (217 nm) for the lowest-energy $\pi \rightarrow \pi^*$ transition in butadiene.

The form of the MO's for butadiene is:

$$\psi_j = c_{j1}\phi_1 + c_{j2}\phi_2 + c_{j3}\phi_3 + c_{j4}\phi_4$$

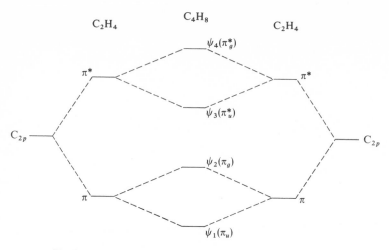

Fig. 3.21   The molecular orbital energy diagram for butadiene.

or in general for any molecular wave function:

$$\psi_j = \sum_{r=1}^{n} c_{jr}\phi_r$$

where c_{jr} is the coefficient of the rth atomic orbital in the jth molecular orbital, and the summation is over all n atomic orbitals. The evaluation of the coefficients in these MO's and of the associated orbital energies will be discussed at length in Chapters 10 and 12. However, a qualitative treatment is in place here. We had thought of the butadiene MO's as generated from the ethylene MO's by linear combinations. Let the ethylene MO's be:

$$\pi_1 = 1/\sqrt{2}\,(\phi_1 + \phi_2) \quad \pi_2 = 1/\sqrt{2}\,(\phi_3 + \phi_4)$$
$$\pi_1^* = 1/\sqrt{2}\,(\phi_1 - \phi_2) \quad \pi_2^* = 1/\sqrt{2}\,(\phi_3 - \phi_4)$$

where π_1 and π_1^* are the bonding and antibonding MO's of ethylene 1 (made up of carbons 1 and 2 of butadiene), respectively, and π_2 and π_2^* are those of ethylene 2 (made up of carbons 3 and 4 of the butadiene molecule). Figure 3.21 shows the combining of π_1 and π_2 to form ψ_1 and ψ_2:

$$\psi_1 = N'[1/\sqrt{2}\,(\phi_1 + \phi_2) + 1/\sqrt{2}\,(\phi_3 + \phi_4)]$$
$$= N\,(\phi_1 + \phi_2 + \phi_3 + \phi_4)$$
$$\psi_2 = N'[1/\sqrt{2}\,(\phi_1 + \phi_2) - 1/\sqrt{2}\,(\phi_3 + \phi_4)]$$
$$= N\,(\phi_1 + \phi_2 - \phi_3 - \phi_4)$$

Here we have absorbed the $1/\sqrt{2}$ factors into N. Similarly, we combine π_1^* and π_2^*:

$$\psi_3 = N'[1/\sqrt{2}\,(\phi_1 - \phi_2) - 1/\sqrt{2}\,(\phi_3 - \phi_4)]$$
$$= N\,(\phi_1 - \phi_2 - \phi_3 + \phi_4)$$
$$\psi_4 = N'[1/\sqrt{2}\,(\phi_1 - \phi_3) + 1/\sqrt{2}\,(\phi_3 - \phi_4)]$$
$$= N\,(\phi_1 - \phi_2 + \phi_3 - \phi_4)$$

Although this procedure is not quite valid, because it neglects all interactions between π's and π*'s and therefore makes all c_{jr}'s numerically equal (although distinct as to sign), it produces the correct nodal and symmetry properties of the MO's. These MO's can be normalized as follows, using ψ_1 as an example:

$$\psi_1 = N(\phi_1 + \phi_2 + \phi_3 + \phi_4)$$
$$\int N^2(\phi_1{}^2 + \phi_2{}^2 + \phi_3{}^2 + \phi_4{}^2)\,d\tau = 1$$
$$4N^2 = 1, \quad \text{and} \quad N = \pm\tfrac{1}{2}$$

The wave functions would then be:

$$\psi_1 = \tfrac{1}{2}(\phi_1 + \phi_2 + \phi_3 + \phi_4)$$
$$\psi_2 = \tfrac{1}{2}(\phi_1 + \phi_2 - \phi_3 - \phi_4)$$
$$\psi_3 = \tfrac{1}{2}(\phi_1 - \phi_2 - \phi_3 + \phi_4)$$
$$\psi_4 = \tfrac{1}{2}(\phi_1 - \phi_2 + \phi_3 - \phi_4)$$

These wave functions are all orthogonal to each other, that is,

$$\int \psi_1\psi_2\,d\tau = \tfrac{1}{2}\cdot\tfrac{1}{2}\int [(\phi_1 + \phi_2 + \phi_3 + \phi_4)\cdot(\phi_1 + \phi_2 - \phi_3 - \phi_4)]\,d\tau$$
$$= \tfrac{1}{4}(\int \phi_1{}^2\,d\tau + \int \phi_2{}^2\,d\tau - \int \phi_3{}^2\,d\tau - \int \phi_4{}^2)\,d\tau$$
$$= \tfrac{1}{4}(1 + 1 - 1 - 1) = 0$$

and hence they are orthonormal wave functions.

Now, if we ask ourselves how the MO picture permits us to deduce the extent of double-bond character in the 2,3 bond of butadiene, we make the following observations. In ψ_1 there is a bonding interaction between C_2 and C_3, but in ψ_2 there is an antibonding interaction between these two atoms. These are the only MO's we need consider because they are the only ones occupied in the ground state. Since we have assumed that C_2 and C_3 contribute equally to both ψ_1 and ψ_2, and since both ψ_1 and ψ_2 are equally (doubly) occupied, the bonding interaction between these atoms in ψ_1 is exactly canceled by the antibonding interaction between them in ψ_2, and hence there should be no net π bonding between C_2 and C_3 in the ground state. However, as we shall see, the coefficients on C_2 and C_3 in ψ_1 are in fact greater than they are on these atoms in ψ_2, and hence there is some net π bonding between these two atoms in the ground state; see section 4.3. These considerations illustrate how important it is to consider the detailed form of the wave function, and we shall accordingly turn our attention in Chapter 4 to a method of obtaining a form of the wave function that is particularly illustrative and simple.

PROBLEMS

3.1 Draw the MOED for (*a*) B_2 and (*b*) Be_2.

3.2 Assume that in CO_2 each of the three atoms has sp hybridization. (*a*) Draw the molecular orbital energy diagram, and (*b*) sketch the $p\pi$ molecular orbitals.

3.3 Form the (*a*) sp, (*b*) sp^2, and (*c*) sp^3 hybrid orbitals and show that each set of these hybrid orbitals forms an orthonormal set. Show that there are other sets of orthonormal hybrid orbitals for sp^2 and sp^3.

3.4 Given two unnormalized, orthogonal hybrid orbitals:

$$\psi_1 = s + \lambda p$$

$$\psi_2 = s + \delta p$$

Normalize both orbitals and then find values for λ and δ such that the two hybrid orbitals are equivalent and point in opposite directions.

3.5 Given the wave functions for 1,3-butadiene:

$$\psi_1 = 0.373\phi_1 + 0.600\phi_2 + 0.600\phi_3 + 0.373\phi_4$$

$$\psi_2 = 0.600\phi_1 + 0.373\phi_2 - 0.373\phi_3 - 0.600\phi_4$$

$$\psi_3 = 0.600\phi_1 - 0.373\phi_2 - 0.373\phi_3 + 0.600\phi_4$$

$$\psi_4 = 0.373\phi_1 - 0.600\phi_2 + 0.600\phi_3 - 0.373\phi_4$$

Show that these four wave functions form an orthonormal set.

3.6 When the overlap integral, S, is neglected, the bonding MO in ethylene is:

$$\psi_b = \frac{1}{\sqrt{2}}(\phi_A + \phi_B).$$

If the value for S in ethylene were 0.3, calculate the normalization constant including overlap. What is the numerical difference in N when S is 0.3 and when it is equal to zero?

GENERAL REFERENCES

1. C. A. Coulson, *Valence*, 2nd Ed., Oxford University Press, London, 1961.
2. A. Streitwieser, Jr., *Molecular Orbital Theory for Organic Chemists*, John Wiley and Sons, New York, 1961.
3. H. B. Gray, *Electrons and Chemical Bonding*, W. A. Benjamin, New York, 1964.
4. A. L. Companion, *Chemical Bonding*, McGraw-Hill Book Company, New York, 1964.
5. M. Orchin and H. H. Jaffé, *The Importance of Antibonding Orbitals*, Houghton Mifflin Company, 1967.

4 Linear- and cross-conjugated molecules; the free-electron method and the calculation of spectra

4.1 Introduction

The *free-electron method* (FEM) is an approximate quantum-mechanical method of calculating the wave functions of π electron systems of conjugated compounds. In such a system, the electrons are not restricted to specific nuclei, but are free to move without hindrance over the entire system. In its simplest form, the FEM assumes that there is a region in space in which the potential energy of an electron is finite and constant and that outside this region the energy is infinite. With this assumption the behavior of the electron becomes the problem of the particle or the electron in a box.

The box or potential well should be three-dimensional, but in π electron systems the only motion of interest is along the conjugated chain. The potential energy along this path should be lowest near the carbon nuclei and higher between the atoms, and at some distance beyond the end of the conjugated system the energy should rise steeply to a high value but not to infinity. Such a potential energy function is approximated by Fig. 4.1a, but

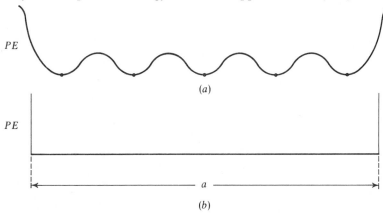

Fig. 4.1 Potential energy: (*a*) as a function of distance along the conjugated chain, and (*b*) the approximate potential commonly assumed.

55

in most applications of the FEM this potential is approximated by the one shown in Fig. 4.1b. The most satisfactory position for the well wall or the end of the box is one bond distance beyond the end atoms of the conjugated system. The total length of the box is usually denoted as a.

4.2   Calculation of the Wave Functions by the Free-Electron Method

The electrons in the π system, according to the FEM, behave approximately as though the potential they experience is a simple square-well function; that is, in the region $0 \le x \le a$ the potential energy function is $U(x) = 0$, where x is the distance along the well between 0 and a. In such a situation the Schrödinger equation implies a wave function of only one coordinate:

$$-\frac{h^2}{8\pi^2 m}\frac{d^2\psi}{dx^2} = E\psi \quad 0 \le x \le a \tag{4.1}$$

The function ψ must be zero outside the well, since there the potential is infinitely high. To be well-behaved and prevent a discontinuity in ψ, the function ψ in the region between 0 and a must be such that it equals zero at $x = 0$ and at $x = a$. Functions which are solutions of the differential equation 4.1 and also satisfy these boundary conditions can be seen by inspection to be:

$$\psi = A \sin\frac{n\pi x}{a} \quad \text{with } n = 1, 2, 3, \text{ etc.} \tag{4.2}$$

To verify that equation 4.2 is a proper solution of equation 4.1, we make the proper substitution:

$$-\frac{h^2}{8\pi^2 m}\frac{d^2(A \sin n\pi x/a)}{dx^2} = EA \sin\frac{n\pi x}{a}$$

The first derivative of $A \sin (n\pi x/a)$ is

$$\frac{d\psi}{dx} = A\frac{n\pi}{a}\cos\frac{n\pi x}{a}$$

and the second derivative is

$$\frac{d^2\psi}{dx^2} = -A\frac{n^2\pi^2}{a^2}\sin\frac{n\pi x}{a}$$

Hence

$$\left(-\frac{h^2}{8\pi^2 m}\right)\left(-\frac{An^2\pi^2}{a^2}\sin\frac{n\pi x}{a}\right) = EA \sin\frac{n\pi x}{a}$$

$$\frac{An^2 h^2}{8ma^2}\sin\frac{n\pi x}{a} = EA \sin\frac{n\pi x}{a}$$

Canceling $A \sin (n\pi x/a)$, we obtain:

$$E = \frac{n^2 h^2}{8ma^2} \tag{4.3}$$

Now, $\sin (n\pi x/a)$ is zero at $x = 0$ for any n, and zero at $x = a$ whenever n is an integer. Consequently $A \sin (n\pi x/a)$ is a solution of equation 4.1 and for any integral values of n obeys the boundary conditions. The energies, wave functions, and probability functions for the four MO's in butadiene are shown in Fig. 4.2.

If we are to use equation 4.2, we need to evaluate the constant A in the equation, that is, normalize ψ. Since $\int \psi^2 \, d\tau = 1$ is required of any wave function.

$$\int N^2 \sin^2 \frac{n\pi x}{a} \, d\tau = 1$$

$$= N^2 \int_0^a \sin^2 \frac{n\pi x}{a} \, dx$$

$$= N^2 \frac{a}{n\pi} \int_0^a \sin^2 \frac{n\pi a}{a} \, d\left(\frac{n\pi x}{a}\right)$$

If we let $n\pi x/a = y$, this becomes:

$$N^2 \sin^2 \frac{n\pi x}{a} \, d\tau = N^2 \frac{a}{n\pi} \int_0^{n\pi} \sin^2 y \, dy = N^2 \frac{a}{n\pi} \cdot \frac{n\pi}{2} = N^2 \frac{a}{2} = 1$$

$$\therefore N = \sqrt{\frac{2}{a}}$$

and

$$\psi = \sqrt{\frac{2}{a}} \sin \frac{n\pi x}{a} \tag{4.4}$$

It is of considerable interest to calculate the amplitude (the value of ψ) of these FEM functions at the position of the various carbon atoms since these values bear a particular relation to Hückel MO's, to be discussed in Chapter 10. The solution of equation 4.4 provides this answer only if we can evaluate the normalization constant, $\sqrt{2/a}$. For the present purpose it is sufficient to use as a unit of length for the measurement of a, the length of one bond in our compound, for example, in butadiene $a = 5$ of these dimensionless units. Using the wave function as defined,

$$\psi = \sqrt{\frac{2}{5}} \sin \frac{n\pi x}{a} \tag{4.5}$$

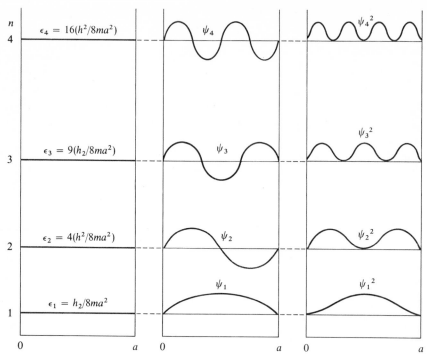

Fig. 4.2   The energies, ψ, and ψ^2 for the four molecular orbitals of butadiene.

$$\psi_1(C_1) = \psi_1(C_4) = \sqrt{\frac{2}{5}} \sin \frac{\pi}{5} = \sqrt{\frac{2}{5}} \sin \frac{4\pi}{5} = 0.632 \sin 36° = 0.372 = c_{11}$$

$$\psi_1(C_2) = \psi_1(C_3) = \sqrt{\frac{2}{5}} \sin \frac{2\pi}{5} = \sqrt{\frac{2}{5}} \sin \frac{3\pi}{5} = 0.632 \sin 72° = 0.602 = c_{12}$$

we may now make the calculations. In butadiene, x at carbon atom 1 is $a/5$. For the lowest-energy orbital, $n = 1$, and for C_1

$$\psi = \sqrt{\frac{2}{5}} \sin \frac{1\pi(a/5)}{a} = 0.632 \sin \frac{\pi}{5} = 0.632 \sin 36° = 0.632 \times 0.588$$

$$= 0.372$$

For C_2, x is $2a/5$, whence

$$\psi = \sqrt{\frac{2}{5}} \sin \frac{2\pi}{5} = 0.632 \sin 72° = 0.632 \times 0.951 = 0.602$$

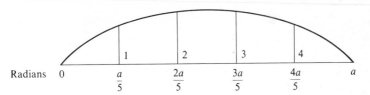

Fig. 4.3   The unnormalized coefficients in ψ_1 of butadiene. (The c's are the coefficients, before normalization, of the corresponding Hückel molecular orbital.)

Since by symmetry carbon 1 is equivalent to 4 and 2 is equivalent to 3, we have the value of the amplitude at each of the four C atoms, as shown in Fig. 4.3.

These values of ψ are exactly the same values as are obtained by the use of the Hückel MO treatment described in Chapter 10. Actually this relation is quite general. If an FEM function of a *linear polyene* is evaluated as shown, and if the FEM box is taken to extend just one bond length beyond each terminal atom, then the amplitude of the FEM orbital evaluated at the position of each atom is equal to the coefficient of the AO of this atom in the corresponding MO (as evaluated by the Hückel method).

The amplitude of the sine curves at each atom in ψ_2, ψ_3, and ψ_4 can be determined in an analogous manner, as shown in Fig. 4.4. Thus, by employing the electron-in-the-box or FEM model, we obtain the wave functions:

$$\psi_n = \sqrt{\frac{2}{a}}\, \sin \frac{n\pi x}{a}, \quad n = 1, 2, 3, \dots$$

and the energies of these solutions have been shown in equation 4.3. When $n = 0$, $\sin (0\pi x/a)$ is zero for all values of x and thus such a function has no meaning as a wave function.

In the general case, we are interested in evaluating ψ at various distances x along the length of the box. Suppose we wish to apply the method just described to hexatriene. In the lowest-energy molecular orbital, ψ_1, $n = 1$. From $x = 0$ to $x = a$, $\sin (1\pi x/a)$ goes from 0 through 1 to 0, always positive (cf. Fig. 4.5), and thus the wave function has no node. At carbon atom 1, the value of x/a is $\frac{1}{7}$ and hence the value of ψ is $\sqrt{\frac{2}{7}} \sin (1\pi\frac{1}{7}) = 0.535 \sin 25.7° = 0.23$, which is also the value of ψ at carbon 6, where $x/a = \frac{6}{7}$. [This value of ψ is equal to the coefficient c_{jr} (here c_{11} and c_{16}) occurring in the molecular orbital ψ_1 obtained by the lowest approximation of the Hückel method.] The value of ψ_1 at carbon 2 (and carbon 5) is $\sqrt{\frac{2}{7}} \sin (1\pi\frac{2}{7}) = 0.535 \sin 51.4° = 0.42$, and at carbon 3 (as well as carbon 4) is $\sqrt{\frac{2}{7}} \sin (1\pi\frac{3}{7}) = 0.535 \sin 77.1° = 0.52$.

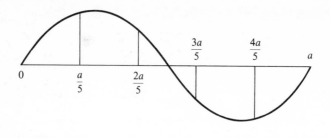

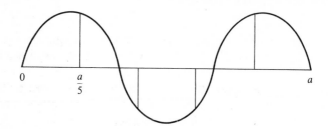

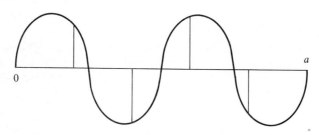

Fig. 4.4 The wave functions ψ_2, ψ_3, and ψ_4 of butadiene.

$$n = 2 \quad \psi_2(1) = \sqrt{\frac{2}{5}} \sin \frac{2\pi}{5} = 0.632 \sin 72° = 0.602 = -\psi_2(4)$$

$$\psi_2(2) = \sqrt{\frac{2}{5}} \sin \frac{4\pi}{5} = 0.632 \sin 144° = 0.372 = -\psi_2(3)$$

$$n = 3 \quad \psi_3(1) = \sqrt{\frac{2}{5}} \sin \frac{3\pi}{5} = 0.632 \sin 108° = 0.602 = \psi_3(4)$$

$$\psi_3(2) = \sqrt{\frac{2}{5}} \sin \frac{6\pi}{5} = 0.632 \sin 216° = -0.372 = \psi_3(3)$$

$$n = 4 \quad \psi_4(1) = \sqrt{\frac{2}{5}} \sin \frac{4\pi}{5} = 0.632 \sin 144° = 0.372 = -\psi_4(4)$$

$$\psi_4(2) = \sqrt{\frac{2}{5}} \sin \frac{8\pi}{5} = 0.632 \sin 288° = -0.602 = -\psi_4(3)$$

In $\psi_2 = \sqrt{\tfrac{2}{7}}\,\sin(2\pi x/a)$, the value of $2x/a$ varies from 0 to 2 as x goes from 0 to a; hence the sine goes from 0 through 1 (at $\pi/2$), 0 (at π), and -1 (at $3\pi/2$) to 0 (at 2π) (cf. Fig. 4.5), and thus we have a single node at $x = a/2$. The value of $\sqrt{\tfrac{2}{7}}\,\sin(2\pi x/a)$ at carbon 1 $(x/a = \tfrac{1}{7})$ is $0.535\,\sin(2\pi/7) = 0.42$. The value at carbon 6 is -0.42; at carbon 2, $\sqrt{\tfrac{2}{7}}\,\sin(4\pi/7) = 0.52$ (-0.52 at C_5); and at carbon 3, $\psi_2 = \sqrt{\tfrac{2}{7}}\,\sin(6\pi/7) = 0.23$ (and -0.23 at C_4); etc. The wave functions for the other MO's can be similarly determined; the shapes of the ψ's are shown in Fig. 4.5.

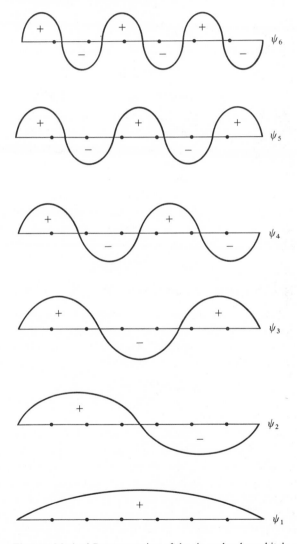

Fig. 4.5   Free Electron-Method Representation of the six molecular orbitals of hexatriene.

4.3   The Importance of the Amplitude Values

Let us now return to the question of the amount of double-bond character between atoms 2 and 3 in butadiene, discussed in the last paragraph of Chapter 3. The wave functions for ψ_1 and ψ_2, the orbitals occupied in the ground state, were calculated by the FEM method (Figs. 4.3 and 4.4) to be:

$$\psi_1 = \sqrt{\frac{2}{5}} \sin \frac{\pi x}{a}$$

$$\psi_2 = \sqrt{\frac{2}{5}} \sin \frac{2\pi x}{a}$$

Evaluation of the amplitudes of these functions at carbon atoms 1–4 gave:

$$\psi_1: \quad 0.372, 0.602, 0.602, 0.372$$

$$\psi_2: \quad 0.602, 0.372, -0.372, -0.602$$

We are particularly interested in the magnitude of these values at C_2 and C_3. We see that in ψ_1, which is bonding between C_2 and C_3, the absolute values of the amplitudes at these atoms are considerably larger than the corresponding ones in ψ_2, which is antibonding between the same atoms. In other words, the bonding interaction between C_2 and C_3 due to ψ_1 is greater than the antibonding interaction between these atoms contributed by ψ_2, and hence there is some net π bonding in the ground state between these atoms. In particular the amount of π bonding between adjacent C atoms is related to the amplitudes at these atoms by the formula:

$$p_{rs} = \sum_j n_j \psi_j(r)\psi_j(s)$$

(or, in the Hückel treatment see Chapter 10), $p_{rs} = \sum_j n_j c_{jr} c_{js}$), where p_{rs} is called the mobile double-bond order, n_j is the number of electrons on the jth molecular orbital, $\psi_j(r)$ and $\psi_j(s)$ are the amplitudes of adjacent atoms r and s, and the summation is taken over all orbitals. Thus, the mobile double-bond order between C_2 and C_3 in the ground state is:

$$
\begin{aligned}
p_{23} &= 2[\psi_1(2)\psi_1(3)] + 2[\psi_2(2)\psi_2(3)] \\
&= 2(0.602 \times 0.602) - 2(0.372 \times 0.372) \\
&= 0.447
\end{aligned}
$$

It is interesting to note that in the lowest excited state of butadiene, where ψ_1 is doubly occupied and ψ_2 and ψ_3 are singly occupied, the double-bond character in the 2,3 bond is considerably greater; the antibonding due to ψ_2 is eliminated since ψ_3 is bonding between these atoms, and hence p_{23} is 0.72. The increased double bond order would, of course, increase the rotational barrier between s-cis and s-trans forms in the excited state. The different electron distribution in the excited state as compared to the ground state leads to different chemistry in the two states, and thus the chemistry of excited butadiene (i.e., the photochemistry of butadiene) is quite different from the ground-state chemistry of this compound.

Although we will discuss later the so-called Woodward-Hoffmann rules for the cyclization reactions of conjugated systems (cf. Chapter 11), one aspect of these rules may be mentioned here. In applying the rules, it is particularly important to ascertain the signs of the wave functions of the highest occupied and lowest unoccupied orbitals at the atoms which become bonded in the cyclization. Accordingly, it is useful to have short-cut methods for this purpose, since only the signs are of interest. Let us consider the problem of ascertaining the signs of the wave function at atoms 1 and 6 in ψ_4 of hexatriene, the lowest unoccupied orbital of this molecule. First we remember that in the six-carbon system we deal with a chain seven bonds in length. In ψ_4 there are three nodes (there are always $n - 1$ nodes in ψ_n in a linear polyene), equally spaced, and the odd number of nodes requires that the signs at carbons 1 and 6 be opposite to each other. The necessary information is thus obtained quickly without any calculation whatsoever.

4.3a Cross Conjugation

The importance of the amplitudes is also effectively illustrated by a consideration of the *cross-conjugated* systems. Examples of such compounds are encountered very early in the study of organic chemistry, and, as reference to examples **1–8** shows, there is a large variety of compounds of this type:

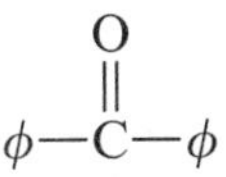

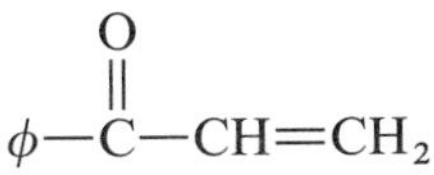

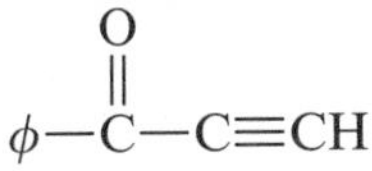

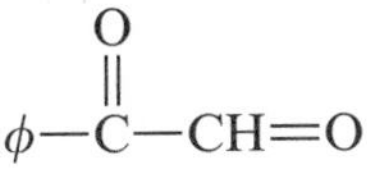

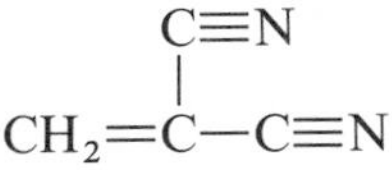

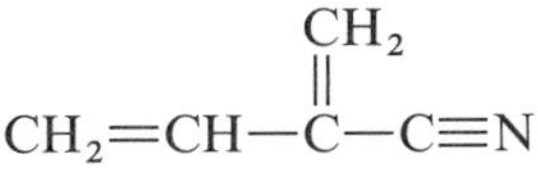

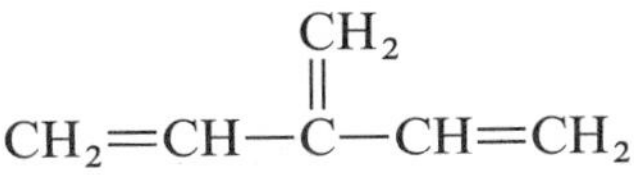

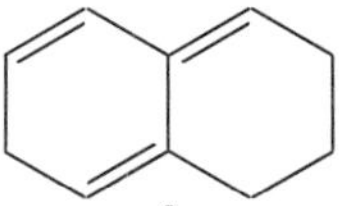

A cross-conjugated compound may be defined as a compound possessing three unsaturated groups, two of which, although conjugated to a third unsaturated center, are not conjugated to each other. The word conjugated is defined here in the classical sense of denoting a system of alternating single and double bonds. Accordingly, the two centers separately conjugated to the third are themselves separated by two essential single bonds. Each unsaturated center possesses $2n$ pπ electrons, where n is an integer. Accordingly, the lone pair of electrons in a nonbonding p orbital of a singly bonded nitrogen or oxygen atom is an unsaturated center isoelectronic with a vinyl group, and compounds **9–13** may also be considered to be cross-conjugated:

$$\phi-\overset{\overset{\displaystyle O}{\|}}{C}-\ddot{N}H_2 \qquad CH_2=CH-\ddot{O}-CH=CH_2 \qquad \phi-\ddot{N}H-\overset{\overset{\displaystyle O}{\|}}{C}OEt$$

$$\qquad\quad 9 \qquad\qquad\qquad\quad 10 \qquad\qquad\qquad\quad 11$$

$$R-\overset{\overset{\displaystyle O}{\|}}{C}-\ddot{O}-\overset{\overset{\displaystyle O}{\|}}{C}-R \qquad CH_2=CH-\ddot{O}-\overset{\overset{\displaystyle O}{\|}}{C}-CH_3$$

$$\qquad\qquad 12 \qquad\qquad\qquad\quad 13$$

Indeed urea, NH_2CONH_2, the first organic compound to be synthesized, is a legitimate example of a cross-conjugated compound.

Let us consider in detail the bonding in the cross-conjugated compound 3-methylene-1,4-pentadiene, **7**, for which the following resonance structures may be written:

$$\underset{1}{CH_2}=\underset{2}{CH}-\underset{3}{\overset{\overset{\displaystyle \overset{6}{C}H_2}{\|}}{C}}-\underset{4}{CH}=\underset{5}{CH_2} \qquad\qquad CH_2=CH-\overset{\overset{\displaystyle \overset{\mp}{C}H_2}{|}}{C}=CH-\overset{\pm}{C}H_2$$

$$(a) \qquad\qquad\qquad\qquad\qquad (b)$$

$$\overset{\pm}{C}H_2-CH\doteq\overset{\overset{\displaystyle \overset{\mp}{C}H_2}{|}}{C}-CH=CH_2 \qquad\qquad \overset{\mp}{C}H_2-CH-\overset{\overset{\displaystyle \overset{\pm}{C}H_2}{|}}{C}=CH-CH_2$$

$$(c) \qquad\qquad\qquad\qquad\qquad (d)$$

$$CH_2\text{---}CH\text{---}\overset{\overset{\displaystyle CH_2}{\|}}{C}\text{---}CH\text{---}CH_2$$

$$(e)$$

We see from representations a–c that each terminal vinyl group is written as though it conjugates separately with the central methylene, leaving one terminal vinyl group intact. The two terminal vinyls cannot be represented as interacting directly with each other except by writing higher multiply charged species or charged species simultaneously involving long bonds, such as *d*. Such interactions are inferred by structure *e*. We would correctly infer from these resonance structures that, because the central methylene must share its double bond with two terminal vinyls, the 3,6 bond in **7** would have substantially less double-bond character than the 1,2 (and 3,4) bonds in butadiene. We would also expect that the 2,3 and 3,4 bonds in **7** would have somewhat less double bond character than the 2,3 bond of butadiene. With the knowledge of the values of the amplitudes at the C atoms (coefficients in the HMO treatment) we can quantify these predictions about double-bond character.

TABLE 4.1
The Wave Functions (MO's) of 3-Methylene-1,4-pentadiene, 7

j	$\psi_j(1)$	$\psi_j(2)$	$\psi_j(3)$	$\psi_j(4)$	$\psi_j(5)$	$\psi_j(6)$	Energy, eV
1	+0.247	+0.435	+0.615	+0.435	+0.247	+0.349	−20.16
2	+0.500	+0.500	0.000	−0.500	−0.500	0.000	−16.26
3	+0.435	+0.247	−0.349	+0.247	+0.435	−0.615	−14.05
4	+0.435	−0.247	−0.349	−0.247	+0.435	+0.615	−8.27
5	+0.500	−0.500	0.000	+0.500	−0.500	0.000	−6.06
6	+0.247	−0.435	+0.615	−0.435	+0.427	−0.349	−2.16

(f) (g)

The results of MO calculations for 3-methylene-1,4-pentadiene, **7**, are given in Table 4.1. Also shown are two of the many possible conformers, the all *s-trans* (f) and the all *s-cis* (g). The calculations were performed according to a known method and will not be discussed here. The MO's of Table 4.1 are represented pictorially in Fig. 4.6, and the MOED is shown in Fig. 4.7[a]. The calculated bond orders for 3-methylene-1,4-pentadiene are listed in Table 4.2.

---

[a] For consistency with later portions, these energies are taken from HMO calculations, not from FEM work (cf. Chapter 10).

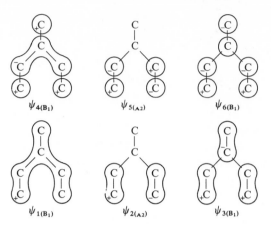

Fig. 4.6 The six molecular orbitals of 3-methylene-1,4-pentadiene. (The symbols in parentheses represent the symmetry of the orbitals; cf. Chapter 5.) From N. F. Phelan and M. Orchin, *J. Chem. Ed.*, **45**, 633 (1968), with permission.

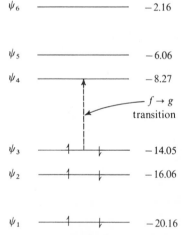

Fig. 4.7 The molecular orbital energy diagram for 3-methylene-1,4-pentadiene (energy in electron volts). (The $f \to g$ transition is explained in Section 4.8.) From N. F. Phelan and M. Orchin, *J. Chem. Ed.*, **45**, 633 (1968), with permission.

In calculating the amount of double-bond character in the 2,3 and 3,4 bonds, we need to examine only ψ_1 and ψ_3, since the amplitude at atom 3 in ψ_2 is zero. The amplitudes at atoms 2, 3, and 4 in ψ_1, the MO which is bonding everywhere, are larger than the amplitudes at these atoms in ψ_3, which is antibonding between 2 and 3 and between 3 and 4. Hence, there is a net π bonding interaction between 2 and 3 and between 3 and 4, a result which could also be predicted qualitatively from the resonance formulations of structures 7*b* and 7*c*.

TABLE 4.2
Bond Orders of 3-Methylene-1,4-pentadiene, 7

$$
\begin{array}{c}
C_6 \\
|\\
C_3 \\
\diagup \quad \diagdown \\
C_2 \qquad C_4 \\
| \qquad\quad | \\
C_1 \qquad C_5
\end{array}
\qquad
\begin{array}{l}
p_{12} = p_{45} = 0.930 \\[8pt]
p_{23} = p_{34} = 0.363 \\[8pt]
p_{36} = 0.859
\end{array}
$$

(For butadiene, $\overset{1}{C}=\overset{2}{C}-\overset{3}{C}=\overset{4}{C}$, $p_{12} = p_{34} = 0.921$ and $p_{23} = 0.388$.)

4.3*b*   Competition between the Vinyl Groups and Other Groups for Conjugation with Carbonyl

In order to evaluate the relative conjugating ability of individual unsaturated groupings, Z, the double-bond orders in a variety of unsaturated aldehydes of the formula

$$
\begin{array}{c}
O_1 \\
\| \\
Z_3-C_2-H
\end{array}
$$

have been calculated, and these are listed in Table 4.3. The more effectively Z conjugates with the carbonyl, the smaller will be the expected value of p_{12} and the larger the value of p_{23}. With the value of p_{12} as the criterion, eight selected Z groups are listed in Table 4.3 on a scale of decreasing order of conjugating ability.[b]

We now turn our attention to the series of compounds

$$
\begin{array}{c}
O_1 \\
\| \\
Z_3-C_2-C=C,
\end{array}
$$

[b] It will be noted that, although p_{12} for 4-aminophenyl is the smallest in the series, p_{23} is not the largest, but is the third largest, in the series, being exceeded by 1,3,5-hexatrienyl and 1,3-butadienyl. However, the changes in p_{12} and p_{23} of 4-aminophenyl, as compared to the parent phenyl, are larger than the changes in p_{12} and p_{23} of either 1,3,5-hexatrienyl or 1,3-butadienyl, as compared to the parent vinyl. Accordingly, on this rather arbitrary basis, we consider 4-aminophenyl to be the best conjugating group in the series.

TABLE 4.3
Effect of Cross Conjugation

Aldehyde Series: $Z_3 - \overset{\overset{\displaystyle O_1}{\|}}{C_2} - H$ Vinyl Ketone Series: $Z_3 - \overset{\overset{\displaystyle O_1}{\|}}{C_2} - C_4 = C_5$

Z	p_{12}	p_{23}	$f \rightarrow g$ eV	p_{12}	p_{23}	p_{24}	p_{45}	$f \rightarrow g$ eV
4-Aminophenyl	0.710	0.476	3.27	0.632	0.444	0.408	0.904	2.61
1,3,5-Hexatrienyl	0.716	0.495	4.24	0.643	0.457	0.403	0.905	3.97
1,3-Butadienyl	0.725	0.480	5.29	0.653	0.445	0.404	0.904	4.91
Vinyl	0.750	0.439	7.29	0.673	0.407	0.407	0.903	6.52
3-Nitrophenyl	0.764	0.400	6.93	0.682	0.374	0.411	0.901	6.41
Phenyl	0.766	0.400	6.88	0.682	0.372	0.411	0.901	6.37
3-Aminophenyl	0.766	0.399	2.61	0.684	0.371	0.411	0.901	2.08
4-Nitrophenyl	0.770	0.393	6.55	0.685	0.366	0.412	0.900	6.18
Hydrogen	...	...	...	0.750	...	0.439	0.983	7.29

in which the vinyl group competes with a variety of Z groups for conjugation with the carbonyl. The ability of Z to conjugate with the carbonyl should follow the decreasing order previously ascertained in the series

$$Z - \overset{\overset{\displaystyle O}{\|}}{C} - H,$$

and reference to the right-hand side of Table 4.3 shows this to be the case. What is perhaps unexpected is the very small effect the nature of Z has on the conjugation of the vinyl to the carbonyl. We might have expected that the better conjugating vinyl groups would conjugate more effectively with the carbonyl when Z is a poor conjugating group. Although there is a slight trend in this direction, by and large the extent of conjugation of the vinyl has not been appreciably influenced by the Z group. All of this interesting information is derived from the data relating to the numerical value of the amplitudes.

4.4 Calculation of the Energy for the Longest-Wavelength Transition in Polyenes

We indicated earlier that the energy of any MO of a molecule treated by the FEM can be calculated from equation 4.3:

$$E_n = \frac{n^2 h^2}{8ma^2} \tag{4.3}$$

To calculate the energy of the longest-wavelength transition, we need to know the energy of the highest occupied and the lowest unoccupied MO. In a polyene with k double bonds, the highest occupied level will be the $n = k$ level and the lowest unoccupied MO, frequently called a *virtual orbital*, will be the $n = k + 1$ level. The difference in energy will then be:

$$\Delta E = E_{k+1} - E_k = \frac{(k + 1)^2 h^2}{8ma^2} - \frac{k^2 h^2}{8ma^2} = \frac{h^2(2k + 1)}{8ma^2} \qquad (4.6)$$

According to the Bohr condition,

$$\Delta E = h\nu = \frac{hc}{\lambda} \quad \text{and} \quad \lambda = \frac{hc}{\Delta E} \qquad (4.7)$$

Substitution of ΔE from 4.6 into equation 4.7 gives:

$$\lambda = \frac{8ma^2 c}{h(2k + 1)} \qquad (4.8)$$

We can evaluate the constants $8mc/h$ so that λ can be expressed in nanometers and a in angstrom units, whereupon we obtain:

$$\lambda \text{ (nm)} \approx \frac{33a^2}{2k + 1} \qquad (4.9)$$

If we wish to calculate, for example, the longest-wavelength transition in octatetraene, the denominator in equation 4.9 becomes 9 and we must then assign a value in angstrom units to a, the length of the chain or box. We know that in the polyenes both the double bonds and the single bonds are essential, and hence we have reason to assume that the average C—C distance is about the mean of a single (1.54) and a double bond (1.33), namely, about 1.4 A. If we include one bond length at each end, we obtain $9 \times 1.4 = 12.6$ A $= a$, and substitution equation 4.9 gives:

$$\lambda = \frac{33 \times 12.6^2}{9} = 582 \text{ nm (49.1 kcal/mole)} \qquad (4.10)$$

The observed value is 286 nm (100 kcal/mole), so obviously one or more assumptions are in error.

In the first place, the use of an average bond length based on the C—C distance in the simple molecules of ethane and ethylene is an oversimplification. Furthermore, multiplying such an average bond length by the number of bonds to calculate the length of the box assumes a linear molecule, and the polyenes are not in fact linear. In the most stable, all *trans-s-trans* configuration, the distance between carbons 1 and 3 is obviously shorter than the sum of two average bond lengths. If the C-C-C bond angle were 120°, then the distance between adjacent carbons should be the average bond length times cos 30°. In addition to the appropriate bond length, there is also the question of how far beyond the carbon chain one should extend the box. The assumption of one bond length at each end has no rigorous justification, although it is just this assumption which leads to the close correspondence between FEM and the Hückel method; in fact, a half bond length at each end beyond the chain would give a better answer in the present series, because a smaller a here results in closer agreement between calculated and experimental values. It is also possible, of course, to calculate a from the experimental measurement of λ_{max} for a polyene of known structure, and this method perhaps can lead to the most acceptable value for bond length. In any case, however, there is a more fundamental argument for the failure to secure rigorous agreement between theory and experiment in the polyene series.

Application of the FEM assumes that the potential along the chain of the conjugated system is constant. We have seen that a conjugated polyene consists of a series of bonds of alternating length. Accordingly, there is a variable rather than a constant potential along the chain. Although, as expected, λ_{max} increases as the length of a polyene chain increases, the proportionality is good only with relatively short polyenes. In fact, as the number of double bonds increases, the wavelength of absorption approaches about 550 nm as a limiting value.

4.5 The Cyanine Dyes

Conjugated acyclic polyenes unquestionably possess adjacent bonds of alternating length. On the other hand, consider the situation with the symmetrical cyanine dyes:

$$R_2\overset{+}{N}{=}CH{-}(CH{=}CH{-})_r\ddot{N}R_2 \leftrightarrow R_2\ddot{N}({-}CH{=}CH)_r{-}CH{=}\overset{+}{N}R_2$$

In such compounds there is no bond alternation, since both structures are equivalent and contribute equally to the real structure of the dye. In such a system, the potential energy in the well is more uniform and constant and hence the FEM is more nearly applicable.

If we look at the structure of the cyanine dye given above, we see that there are $2r$ pπ electrons contributed by the r vinyl groups plus four pπ electrons contributed by the nitrogen lone pair and the methine double bond. Since there is a total of $(2r + 4)$ π electrons, these will occupy $r + 2$ molecular orbitals and the highest occupied orbital will be $r + 2$, while the lowest unoccupied orbital will be $r + 3$. Substitution of these values into equation 4.6 gives:

$$E = \frac{(r + 3)^2 h^2}{8ma^2} - \frac{(r + 2)^2 h^2}{8ma^2} = \frac{(2r + 5)h^2}{8ma^2} \tag{4.11}$$

and substituting this value of E in equation 4.7 gives:

$$\lambda(\text{nm}) = \frac{8mc}{h} \cdot \frac{a^2}{2r + 5} = \frac{33a^2}{2r + 5} \qquad (4.12)$$

The average length of the vinyl chromophore $-C{=}C$ is considered to be 2.48 A, and the additive term to allow for the lengths of the bonds outside this group and to the ends of the box is 5.65 A. The total length of the box is thus considered to be $a = (2.48r + 5.65)$ A. Substituting this value of a into equation 4.12 gives:

$$\lambda(\text{nm}) = \frac{33(2.48r + 5.65)^2}{2r + 5}$$

The excellent agreement between calculated and observed transition energies for the cyanine dyes with end methyl groups is shown in Table 4.4.

TABLE 4.4

Absorption Spectra of Cyanine Dyes

$$Me_2\overset{+}{N}{=}CH-(CH{=}CH-)_r NMe_2$$

Longest-Wavelength Transition

r	Calculated by eq. 4.8	Observed
1	309	309
2	409	409
3	509	511

4.6   Unsymmetrical Cyanine Dyes

As indicated above, in symmetrical cyanine dyes, in which the end groups are identical, the two principal resonance structures contribute equally to the ground state. If the end groups are not identical, however, there is a difference in the tendency of the two nitrogens to become positive by release of their electrons. Thus there will be unequal contribution of the two resonance forms to the ground state; the nitrogen which is more basic will tend to exist in the positive form to a greater extent than the less basic nitrogen at the other end. Thus, for example, the N atom in benzoxazole is more basic than the N atom in pyrrole, and hence, in the unsymmetrical cyanine dye shown in Fig. 4.8, structure a is favored over b. The symmetrical dye, in which the benzoxazole nuclei (A) are end groups (Fig. 4.8c) and the length of the conjugated system is equal to that of the unsymmetrical dye, absorbs at 580.5 nm, whereas the symmetrical dye of equal chromophoric length with pyrrole (B) end groups (Fig. 4.8d) absorbs at 536.5 nm. One might

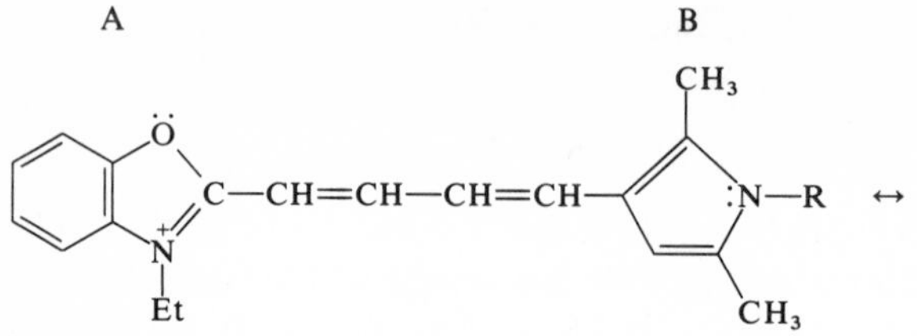

$[\lambda = 516.0$ nm (obs'd) for R $=$ Et; 492.0 nm for R $=$ Ph$]$

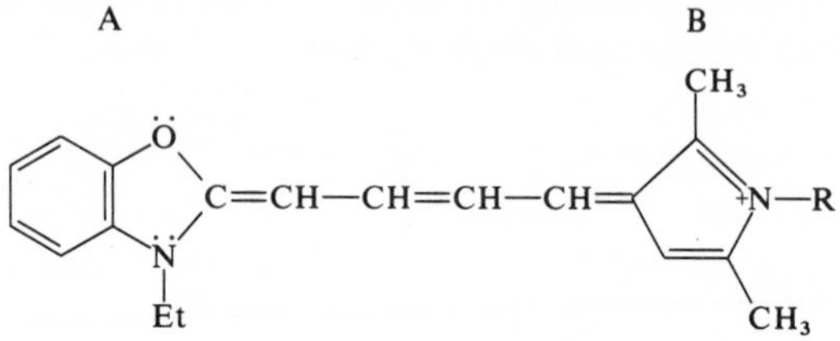

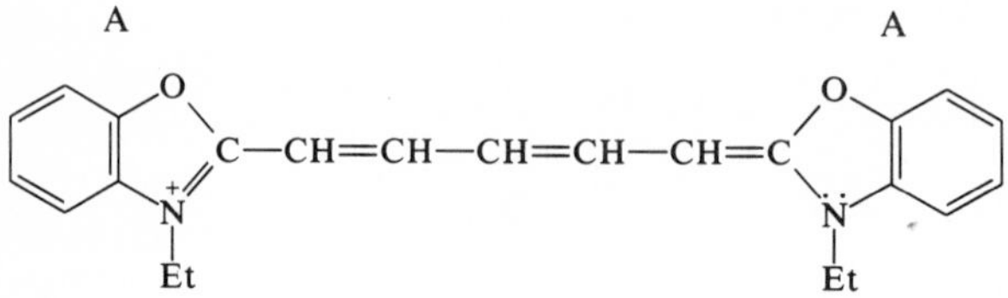

$(\lambda = 580.5$ nm$)$

(R = Et, Ph)

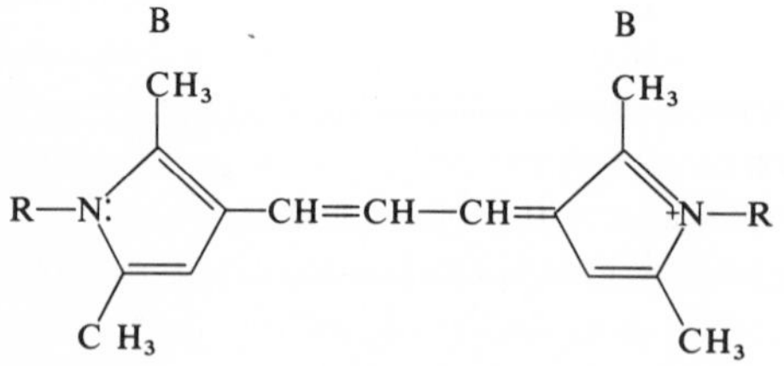

$(\lambda = 536.5$ nm for R $=$ Et; 540.5 nm for R $=$ Ph$)$

Fig. 4.8 (a) and (b) Resonance structures of an unsymmetrical cyanine dye. (c) and (d) The corresponding symmetrical dyes.

72

conclude that in the unsymmetrical dye the expected absorption would be the average of the two symmetrical dyes, that is,

$$\frac{AA + BB}{2} = \frac{580.5 + 536.5}{2} = 558.5 \text{ nm}$$

The observed λ_{max} is 516.0 nm (using MeOH as solvent in all cases), and hence the deviation from the expected value is 42.5 nm. The greater the difference in basicity between the two N atoms, the greater is this deviation, called the *Brooker deviation*. The greater the difference in basicity, the more the bond alternation, and hence the greater is the resulting deviation from the average of the two corresponding symmetrical dyes.

In the above example, the benzoxazole moiety is more basic than the pyrrole moiety. The replacement of the ethyl group on the pyrrole N atom of Fig. 4.8 by phenyl reduces the basicity even further, since now the lone pair is delocalized over the phenyl ring as well as over the pyrrole. In calculating the Brooker deviation, λ for AA is again 580.5 nm and now λ for BB, the symmetrical N-phenylpyrrole compound, is found to be 540.5 nm. The average of these two values is 560.5 nm, whereas the observed λ_{max} of the unsymmetrical AB compound is 492.0 nm for a Brooker deviation of 68.5 nm. This is a considerably larger deviation for the compound of Fig. 4.8a in which R = Ph, as expected, than for the (electronically less) unsymmetrical dye in which R = Et. The increased unsymmetrical nature has enhanced the bond alternation.

A similar treatment can be applied to many neutral dyes, which, because they are in part cyanine structures, are called merocyanine dyes, Fig. 4.9a. In this class of dyes, bond migration in the formalism of resonance requires that a charge-separated (dipolar) species be written, as shown in Fig. 4.9b. Here the two end groups are the benzthiazole and rhodanine nuclei, and, as with all merocyanine dyes, the resonance is of the amidic type:

$$> N-C=O \leftrightarrow > \overset{+}{N}=C-\overset{-}{O}$$

These two structures are the principal structures that can be written for both the ground and excited states. Since the two end groups differ, we expect a Brooker deviation and indeed we find (Figs. 4.9c and d) that $\lambda_{AA} = 557.5$ nm, $\lambda_{BB} = 532.0$ nm, average = 544.8 nm, observed = 524.0 nm, and Brooker deviation = 20.8 nm. Although the observed λ is again lower than the average, there is no *a priori* answer to the question of whether the neutral or the dipolar form is lower in energy and hence which makes the greater contribution to the ground state. This information may be secured as follows.

Let us make the assumption that the neutral form (N) is lower in energy. The energy of the ground-state molecule is lower than the energy of the neutral form because of resonance stabilization. The excited-state energy

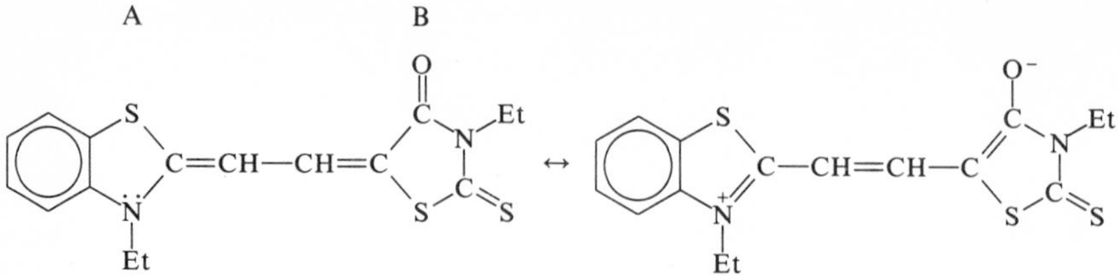

(Obs'd λ_{max} = 524.0 nm)

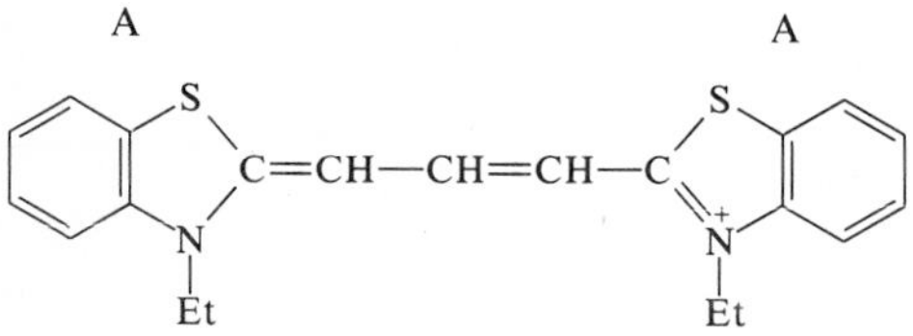

(λ_{AA} = 557.5 nm)

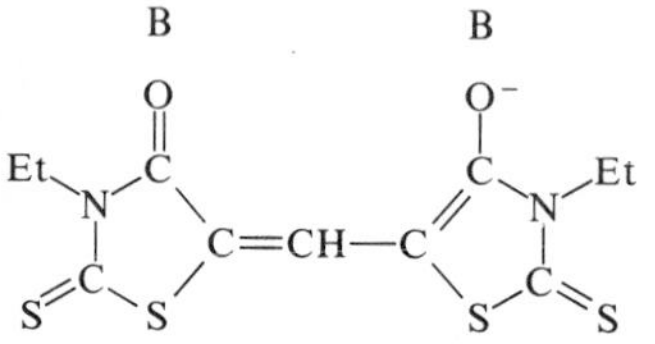

(λ_{BB} = 532.0 nm)

Fig. 4.9   (a) and (b) Resonance structures of an unsymmetrical merocyanine dye.
(c) and (d) The corresponding symmetrical dyes.

of the molecule is probably higher than the energy of the dipolar form, because
if two structures mix to lower a ground state, the corresponding mixed
excited state is raised. Accordingly, the horizontal solid lines in Fig. 4.10a
represent the relative energies of the ground and excited states, and the
distance between them represents the energy difference between the ground
and excited states, and hence is related to the position of λ_{max} in the spectrum.
In a common notation suggested by Mulliken, this essentially $\pi \rightarrow \pi^*$ type
of transition is referred to as the $V \leftarrow N$ (to be read as V from N) transition
(see Chapter 8).

   If now the basicity of the benzthiazole end group is increased by changing
the substituent on the nitrogen to the trimethylene derivative, Fig. 4.11, it

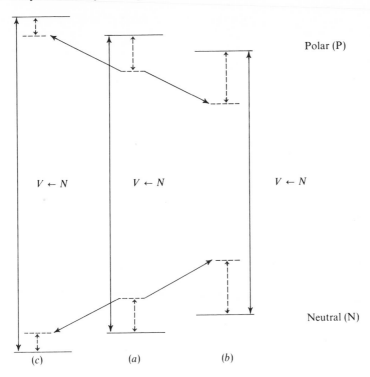

Fig. 4.10   (*a*) Starting compound, for example, Fig. 4.9*a*, with N more stable than P. (*b*) Increasing basicity of nitrogen decreases bond alteration. (*c*) Decreasing basicity increases bond alteration. Broken lines are energies of the N (nonpolar) and P (polar) resonance forms, while solid lines are energies of the ground and excited states. It will be noted that, the closer in energy the two forms, the greater is their splitting (broken arrows).

is found that the Brooker deviation is diminished from 21.0 to 16.0 nm. The decreased Brooker deviation implies that the two forms, neutral and dipolar, have been brought closer together in energy and the bond alternation has been decreased. The closer in energy the two forms, the greater is the interaction between them, with strong lowering of the ground state and strong raising of the excited state. This effect is shown by the length of the vertical broken arrows in Fig. 4.10*b*. On the other hand, when an electron-withdrawing substituent is placed on the nitrogen of the benzthiazole nucleus to reduce its basicity, the Brooker deviation is increased, Fig. 4.10*c*. These facts are consistent with the assumption that the nonpolar form is the more stable form in the ground state; otherwise the electron-releasing and electron-withdrawing effects would have an influence opposite to that observed on the Brooker deviation.

It is possible to prepare compounds in which the polar form is more stable than the neutral form; such a dye is shown in Fig. 4.12. Here the ground state is best represented by the dipolar form. The reason for the greater stability of the dipolar form is, of course, the fact that in this form both rings are aromatic, whereas the neutral form requires quinoid structures in both rings.

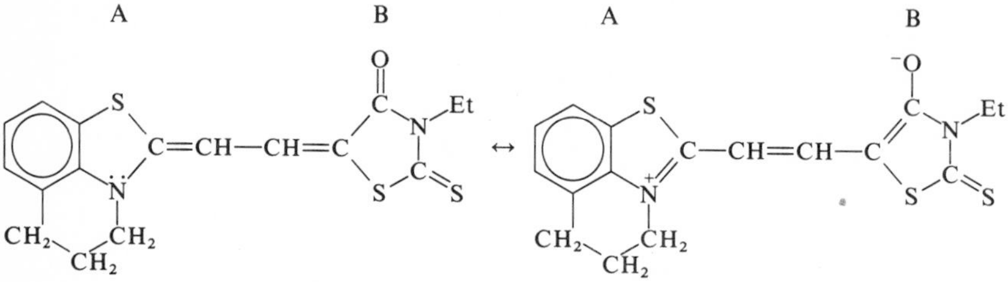

Fig. 4.11   An unsymmetrical merocyanine dye. (λ_{AA} = 556.0 nm, λ_{BB} = 532.0 nm, (AA + BB)/2 = 544.0 nm, observed = 528.0 nm, Brooker deviation = 16.0 nm.)

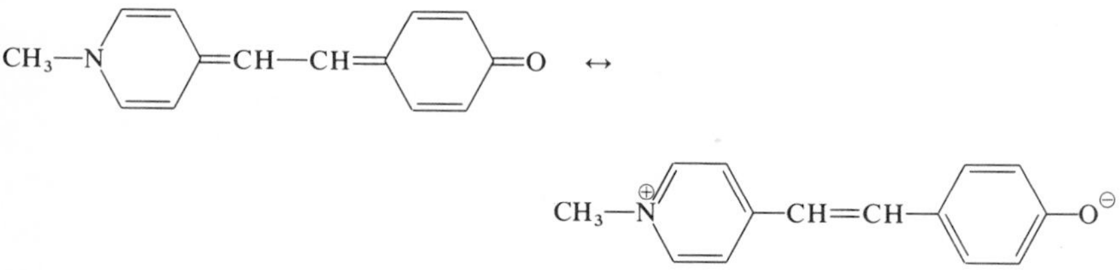

Fig. 4.12   A merocyanine dye in which the polar form is more stable.

We have just seen that, in one case, we can have the neutral form more stable, and by a change of structure have the polar form more stable. It should be theoretically possible to find structures in which the two forms have equal stability. It is also possible to employ solvents that will preferentially raise or lower one of the two forms. Thus, for example, if the polar form is slightly less stable than the neutral form, and if the compound is dissolved in a polar solvent, the polar form will be stabilized to a greater extent than the neutral form and it may be possible, by the proper choice of solvent, to reverse the stabilities.

Dyes in which the polar form is more stable than the neutral form have been prepared. According to the simple resonance interpretation, both the neutral and the polar structures contribute to the ground and to the excited states, with the polar structure making a larger contribution to the ground state and the nonpolar one a larger contribution to the excitation state. Hence the dipole moment of the excited state is less than that of the ground state.

Consequently, in changing from a nonpolar solvent the excited state is less stabilized than the ground state. Accordingly, the excitation energy increases, and a polar solvent produces a hypsochromic shift. Thus, the dye shown in Fig. 4.13 ($n = 1$) has λ_{max} in dry pyridine at 400.0 nm but absorbs in water at 363.5 nm, while when $n = 3$ the change from pyridine to water results in a blue shift of 220.0 nm compared to the usual solvent shifts of 10–20 nm!

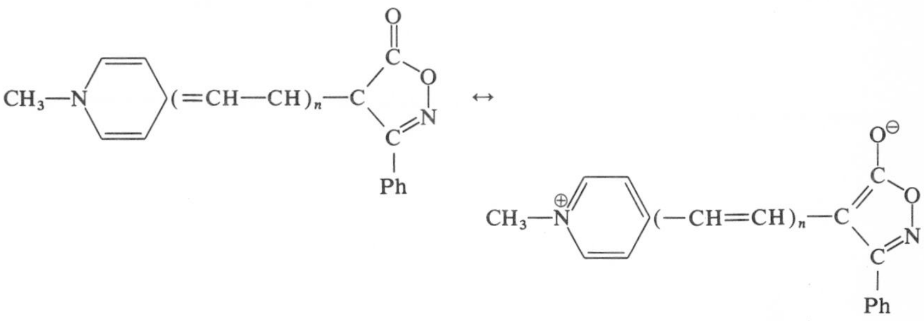

Fig. 4.13   A merocyanine dye showing dramatic changes in color upon addition of traces of water to a solution in dry pyridine.

In practice it has been found that the color of certain dyes in a solvent like pyridine can be remarkably altered by the addition of a drop of water. In fact, such a color change can be used to titrate the amount of water in pyridine.

4.7   Spectra of Polyenyl Cations[c]

The FEM is extremely successful in the treatment of the symmetrical cyanine dyes because, as has been pointed out, in such molecules there is no bond alternation. The same phenomenon applies also to the vinylogs of the allyl system. Consider, for example, the 2,6-dimethylheptadienyl cation, Fig. 4.14, generated by adding a proton to the precursor indicated. In such an ion, again there is no bond alternation because both structures contribute equally to the ground state. Accordingly, one might expect a good correlation between calculated and experimental λ_{max} as a function of the chain length

---

[c] There are two general schemes for naming the class of compounds which possess a carbon atom having an empty orbital, called generically carbonium ions. The derived nomenclature scheme is analogous to the naming of carbinols, where CH_3OH is carbinol, CH_3CH_2OH is methylcarbinol, etc. According to this scheme CH_3^+ is carbonium ion, $CH_3CH_2^+$ is methyl-carbonium ion, and $(CH_3)_3C^+$ is trimethylcarbonium ion. The alternative naming scheme consists of giving the accepted systematic group or radical name to the ion, followed by the word cation. Thus CH_3^+, $CH_3CH_2^+$, and $(CH_3)_3C^+$ are called, respectively, methyl cation, ethyl cation, and tert-butyl cation. The names methyl carbonium ion, ethyl carbonium ion, and tert-butyl carbonium ion are ambiguous and should be avoided.

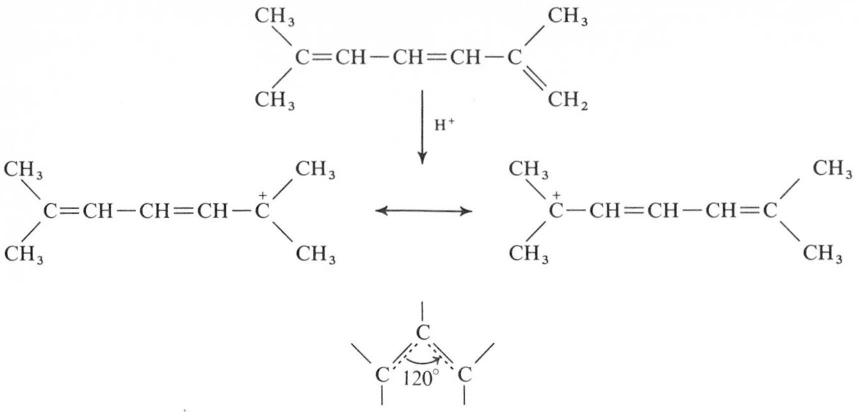

Fig. 4.14   2,6-Dimethylheptadienyl cation and the C-C-C bond angle.

in this series. In one such study, the length of the box a was taken as

$$a = \left[(l_C \cos 30°)(n_C - 1)\right] + 2p$$

where l_C is the average bond length between carbons, set equal to 1.4 A; $\cos 30°$ is the term which takes into account the probably all *trans* configuration, n_C is the number of carbon atoms in the π system, since $n_C - 1$ is the number of bonds (e.g., in allyl $n_C - 1 = 2$); and $2p$ is the penetration of the π system beyond the ends of the box, which is the polyenyl cation. The measured spectrum of one member of the series was used to calculate p, and the value of 1.4 A was thereby assigned to it. The agreement between experimental and calculated values is shown in Fig. 4.15.

4.8   Spectra of Catacondensed Polycyclic Hydrocarbons

The free-electron model has been extended to the calculation and classification of the spectra of catacondensed aromatic hydrocarbons. Aromatic hydrocarbons in which all carbon atoms lie on the periphery of the carbon skeleton are called *catacondensed*. Thus, naphthalene, phenanthrene, anthracene, triphenylene, and chrysene, Fig. 4.16, are catacondensed; no C atom is common to more than two rings.[d] On the other hand, pyrene,

[d] There is an unfortunate misuse of the practice of depicting aromatic rings by the use of the solid circle inscribed in a hexagon. In accordance with Sir Robert Robinson's original suggestion, the circle should represent only a sextet of pπ electrons. Accordingly, drawing naphthalene with two solid circles not only is an aberration from the original suggestion but is misleading as well, since there are ten and not twelve pπ electrons in naphthalene. In this example only one of the two rings should be drawn with a solid circle inside it, but it is immaterial which ring is chosen. The choice, when it exists, as to which of the rings in the polycyclic system should be shown as possessing sextets (solid circles) is not completely arbitrary. Eric Clar, *Polycyclic Hydrocarbons*,

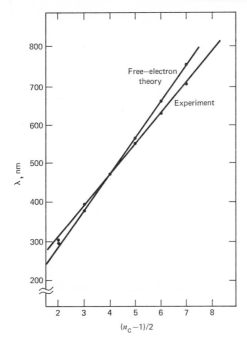

Fig. 4.15 $\lambda_{\max}$ vs. length of polyene chain. For these cations, the value of a was calculated from $a = [(l_C \cos 30°)(n_C - 1)] + 2p$ and the value of $p = 1.4$ A was obtained experimentally from the spectrum of $(CH_3)_2C{=}CH\text{-}CH{=}CH\text{-}\overset{+}{C}(CH_3)_2$, where $n_C = 5$ and $l_C = 1.4$ A. With these values, $a = [(1.4 \cdot \sqrt{3}(2)(4)] + 2 \cdot 1.4 = 7.65$ A. The values of λ (nm) were then calculated from $\lambda = 33a^2/(2k + 1)$, where k is the number of double bonds. From D. Osseen, R. B. Flewwelling, and W. G. Laidlaw, *J. Am. Chem. Soc.*, **90**, 4209 (1968), by permission.

Academic Press, Vols. 1 and 2, 1969, makes skillful use of the placement of the sextets to show chemical and spectral relationships between aromatic systems. His volumes are essential references for anyone interested in the chemistry and spectra of aromatic compounds.

For most purposes, aromatic systems can be drawn in the conventional manner without solid circles, bearing in mind that the double bonds should be drawn in a way that permits the maximum number of rings to have the Kekulé structure (three double bonds). Thus, for example, it is preferable to draw chrysene as in (*a*) rather than (*b*), since in the former all four rings have

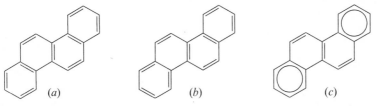

Kekulé structures, but in (*b*) only three rings are Kekulé and one is quinoidal. In this case, the two terminal rings may be shown as sextets as in (*c*).

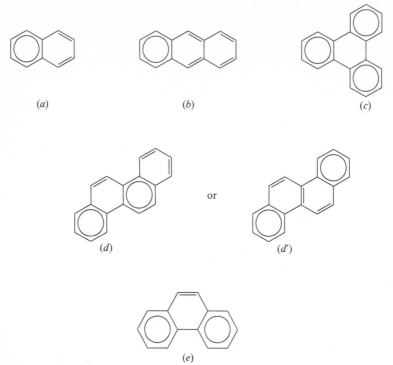

Fig. 4.16 Catacondensed aromatics: (a) naphthalene; (b) anthracene; (c) triphenylene; (d) chrysene; (e) phenanthrene.

perylene, and anthanthrene, Fig. 4.17, are examples of molecules in which all C atoms are not on the periphery and some of the C atoms are common to three rings; such molecules are called *pericondensed*.

A catacondensed aromatic can be considered roughly as a circle of C atoms, and in this crude approximation the pπ electrons can be thought of as traveling on the periphery of the circle. Such a concept is analogous to the particle-in-a-box model of linear conjugated systems except that now the treatment is that of a particle on a ring (rigid rotator). The periphery of the molecule is considered to be a perfect circle; the potential is assumed to be constant on the circle and infinite off it. The particle has motion in only one dimension, as in the box, but here it is along the periphery of the circle. If the electron is moving on a circle of constant radius, the only variable is the angular motion, Fig. 4.18. The Schrödinger equation for the particle in a ring becomes:

$$-\frac{h^2}{8\pi^2 mr^2}\frac{d^2\psi}{d\varphi^2} = E\psi \qquad (4.13)$$

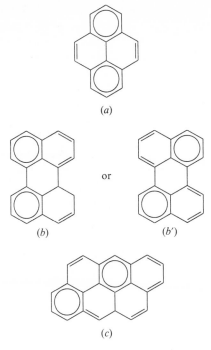

Fig. 4.17   Pericondensed aromatics: (a) pyrene; (b) perylene; (c) anthanthrene.

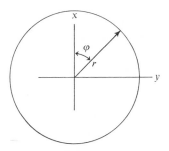

Fig. 4.18   Coordinates for a particle-on-a-ring wave function. From R. L. Flurry, Jr. *Molecular Orbital Theory in Organic Molecules*, Marcel Dekker, Inc, 1968, by permission.

where φ is the angle by which the motion is measured, and r is the radius of the ring.

Now again, just as for the particle in the box (equation 4.1), we must find acceptable solutions of ψ. To be a satisfactory solution, that is, to be continuous and single-valued, the values of ψ and its first derivative at 0° must be equal to the corresponding values at 2π. There are two equally acceptable solutions:

$$\psi = A \sin q\varphi \tag{4.14a}$$

$$\psi = A \cos q\varphi \tag{4.14b}$$

where A is a constant and q is an integer which Platt has called the *orbital ring quantum number*. For $q = 0$, $A \sin q\varphi = A \sin 0$ is zero everywhere and hence not an acceptable wave function, but $\psi = A \cos q\varphi = A \cos 0 = A$ is an acceptable solution. Thus for $q = 0$ there is only one solution, whereas for each other value of q there are two degenerate ones and hence the familiar MOED for aromatic compounds (see, e.g., Fig. 4.20).

Associated with the above ψ are the energies:

$$E = \frac{q^2 h^2}{8\pi^2 m r^2} \tag{4.15}$$

which can be written in more convenient form:

$$E = \frac{q^2 h^2}{2ml^2} = 1{,}21 \times 10^6 \frac{q^2}{l^2} \text{ cm}^{-1} \tag{4.16}$$

where $l = 2\pi r$ is the perimeter or circumference in angstrom units, and h and m are Planck's constant and the mass of the electron, respectively. Since E depends on q^2, the energies are again quadratically spaced, as in the particle in the box. Although there are two independent functions ψ (equations 4.14a and 4.14b), for a given value of q greater than zero, both have the same energies and hence they are *degenerate*. We can rationalize the double degeneracy by recognizing that the angular momentum, if different from zero, may point along either direction of the rotational axis, that is, its z component, m_q, may be equal to $+q$ or $-q$; by a classical analogy we may say that the electron, if rotating ($q \neq 0$), may be moving in a clockwise or counterclockwise direction.

The two degenerate orbitals, represented by equations 4.14a and 4.14b, are orthogonal. The orthogonality becomes readily apparent when we consider the nodal properties of the wave functions. When $q = 0$, there is no node. When $q = 1$, there is one node, Fig. 4.19. If the top of the circle is arbitrarily taken to be $\varphi = 0$ and clockwise rotation is taken as positive, the circles on the far right represent $\psi = A \sin q\varphi$. Thus, for $q = 1$, $\sin q\varphi$ goes from zero at $\varphi = 0$ and is positive until $\varphi = \pi$, where it is zero again. Then, between $\varphi = \pi$ and 2π, $\sin \varphi$ is negative. For $\psi = A \cos q\varphi$, ψ is positive from $\varphi = 0$ to $\varphi = \pi/2$, but negative from $\varphi = \pi/2$ to $\varphi = 3\pi/2$. The product of the two wave functions (orbitals) integrated over all space is obviously zero, and hence the wave functions are orthogonal, as required for MO's

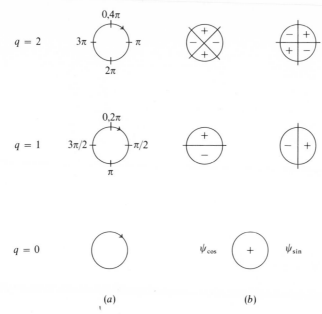

Fig. 4.19 Nodal properties for the particle-on-a-ring model: (*a*) gives the value of $q\varphi$ around the circle; (*b*) gives the sign of the function.

in a degenerate set. When $q = 2$, $\sin 2\varphi$ is positive from $\varphi = 0$ to $\varphi = \pi/2$, negative from $\varphi = \pi/2$ to $\varphi = \pi$, etc., while $\cos 2\varphi$ changes sign at $\varphi = \pi/4$, $3\pi/4$, $5\pi/4$, etc., Fig. 4.19.

Thus far we have dealt with the orbital ring quantum number of individual electrons. The molecular state is characterized by the sum of the individual m_q's, the projection of q on an external z-axis, to give a total ring quantum number, Q, with z components: $M_Q = \Sigma m_q$. In a catacondensed system of n rings there are $4n + 2$ carbon atoms and an equal number of π electrons. In the ground state the $4n + 2$ electrons fill the first, nondegenerate level $(q = 0)$, plus n doubly degenerate levels, and the highest filled level has $q = n$. This highest level, which is completely occupied in the ground state, is called the f level by Platt, and the electrons in this level are f electrons. Electrons in the shell or level below this one are called e electrons, and the electrons in still lower shells are designated successively by the next preceding letters of the alphabet. The first vacant level, $q = n + 1$, is called the g level, and higher levels are given succeeding letters of the alphabet. The notation starting with f and g for the highest filled and the lowest empty levels of the ground state has the advantage that the transition of longest wavelength is always $f \to g$ transition in one-electron notation, independent of the size of the catacondensed hydrocarbon.

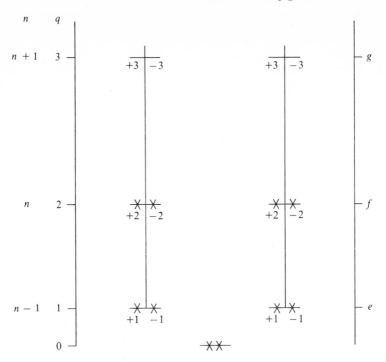

Fig. 4.20　The molecular levels for naphthalene.

In the ground state, the nondegenerate level, $q = 0$, is occupied by two electrons. All other levels are occupied by four electrons, and hence there will be an electron in an $m_q = -q$ orbital for every electron in the corresponding orbital, $m_q = +q$. Hence, in the ground electronic state, the total angular momentum number, Q, will be zero. When $Q = 0$, we have what Platt calls the A state. As we will see below, when $Q = 1, 2, \ldots$, etc., we have B, C, etc., states.

In the excited state, where there are partially occupied levels, the value of Q will not be zero. Let us examine the situation of naphthalene, a cata-condensed hydrocarbon with $n = 2$, and hence $4n + 2 = 10$ π electrons, placed in the quadratically spaced levels, Fig. 4.20. If an electron is promoted from the f to the g level to give an excited state with the $f^3 g$ configuration (or, more explicitly, $d^2 e^4 f^3 g$), we have the following possibilities with respect to Q: an electron with $m_q = +2$ can go to either $m_q = +3$ or -3 to give a total M_Q of $+5$ or -1, or an electron with $m_q = -2$ can go to either $+3$ or -3 to give a total M_Q of $+1$ or -5. Thus, we can have four final M_Q values: ± 1 and ± 5, or two doubly degenerate states with $Q = 1$ or 5, respectively.

If we generalize the situation and excite an electron from the orbital with $q = n$ to one with $q = n + 1$, the possible contributions to M_Q will be $\pm n$ for the electron remaining in the $m_q = \pm n$ orbital and $\pm(n + 1)$ for the electron in the excited orbital. With a jump of an electron from an orbital with $m_q = \pm n$ to one with $m_q = \pm(n + 1)$, states are reached with $M_Q = \pm 1$ or $\pm(2n + 1)$, and $Q = 1$ or $2n + 1$. These values apply for the upper states of the $f \rightarrow g$ transition, but other transitions are possible, such as $e \rightarrow g$, and in the upper state of such a transition, Q will have integral values different from 1 and $2n + 1$.

There are two series of Q values; one goes upward from zero in integral units and the other upward from $2n$. As mentioned previously, Platt assigned alphabetical symbols to these, starting with A for the zero series and K for the $2n$ series. Thus, for $Q = 1$ and $Q = 5$ in the $f^4 \rightarrow f^3g$ transition, we have a B state and an L state, respectively, and thus the one-electron transition gives rise to two states. Each of these states, however, is doubly degenerate. This degeneracy is a consequence of our assumption of a circle; when the circle is not perfect (as is the case in our molecule), each of the two states consists of two substates, and thus there are four transitions from the ground or A state to the states B_a, B_b, L_a, L_b. The subscripts a and b refer to the direction of polarization, the a band being polarized along the short axis in a transverse polarization and the b band along the long or longitudinal axis.

Of the two states, B and L, described above, the state with the highest spin angular momentum is most stable, just as for atoms, the electronic structure giving rise to the state of highest multiplicity is the most stable (Hund's rule). Accordingly, we expect the L state to be more stable than the B state, and the transition at longest wavelength then would correspond to the L band.

In deciding the direction of polarization of the bands, that is, the assignment of the subscripts a and b, Platt suggested that the periphery of the catacondensed carbon system be expanded into a circle, Q nodes drawn through the circle, and the circle reformed into the perimeter of the molecule, retaining the nodes. The signs of the wave functions for the various areas are then indicated by alternating positive and negative signs on each side of the node. Since every Q (except $Q = 0$) involves double degeneracy, two sets of nodes, mutually orthogonal to each other, must be generated for each Q, as illustrated for our example of naphthalene in Fig. 4.21a. Thus, for the L state there are five nodes spaced at $\pi/5$ intervals. The a state is associated with MO's having a node bisecting the cross link and the bonds of the molecule; the b state, with a node including the cross link and the atoms. In naphthalene the cross link joins the 9,10 carbon atoms, and the b states have a node passing through these atoms, while all the states subscripted with a have a node perpendicular to this cross link. It will be noted also that the b states

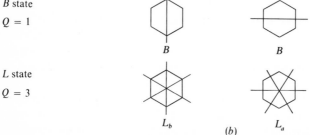

(a)

Fig. 4.21   Polarization diagrams for the B and L states of naphthalene (a) and benzene (b). From R. L. Flurry, Jr. *Molecular Orbital Theory in Organic Molecules*, Marcel Dekker, Inc, 1968, by permission.

are polarized along the long or longitudinal axes and the a states along the short or transverse direction, as indicated by the double-headed arrows either to the right (transverse) or below (longitudinal) the naphthalene structure.

In the $f \to g$ states of benzene, Q takes values of 1 and 3. The B state, corresponding to $Q = 1$, is doubly degenerate, since benzene is rather well approximated by a circle. However, the approximation is not perfect. Because we have six C atoms in the shape of a hexagon, the potential is not

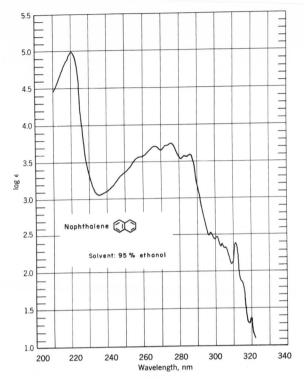

Fig. 4.22 The ultraviolet spectrum of naphthalene. (The α, *para*, and β bands in the Clar nomenclature, corresponding to ^1L_b, and ^1L_a, and ^1B_b bands in the Platt nomenclature are at about 300, 275, and 220 nm, respectively.)

truly constant, and consequently the $Q = 3$ states are split into two separate states, L_a and L_b, the latter of lowest energy. The nodal patterns of these states are given in Fig. 4.21*b*. All three bands are observed in the spectrum of benzene: the ^1L_b band at 256 nm ($\epsilon = 220$), the ^1L_a band at 203 nm ($\epsilon = 6900$), and the doubly degenerate ^1B band at 183 nm ($\epsilon = 46{,}000$). In recent years the particle-on-a-ring model has not been widely employed, although the nomenclature has gained a fair measure of acceptance.

Most aromatic hydrocarbons have three rather well defined band systems in the ultraviolet or visible. The longest wavelength band has the lowest intensity, ϵ, and the shorter-wavelength bands have successively higher intensities. The spectrum of naphthalene, Fig. 4.22, is an example. Clar calls these three bands α, *para*, and β bands, respectively. Very successful correlations between these bands and structures of the polycyclic hydrocarbons have been made, most notably in two series of aromatic compounds. One series is the acene series, in which successive rings are fused linearly to

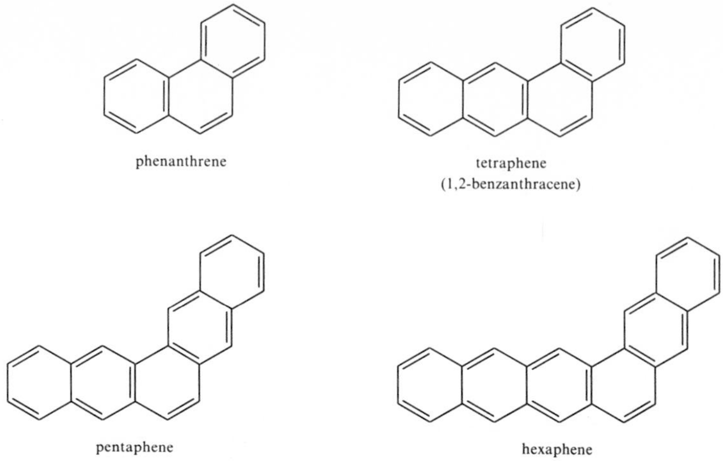

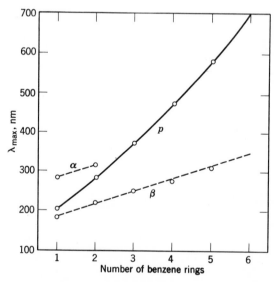

phenanthrene

tetraphene
(1,2-benzanthracene)

pentaphene

hexaphene

Fig. 4.23   The first four members of the phene series.

Fig. 4.24   The α, β, and *para* bands in the acene series. From G. M. Badger, *The Structure and Reactions of Aromatic Compounds*, Cambridge University Press, 1954, with permission.

benzene; thus, benzene, naphthalene, anthracene, and tetracene constitute the first four members of the acenes series. The other series of compounds is the phene series, considered to be derived from phenanthrene by linear

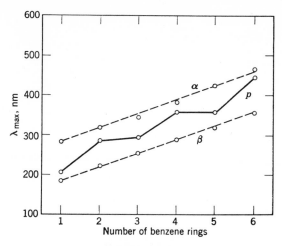

Fig. 4.25 The α, β, and *para* bands in the phene series. From G. M. Badger, *The Structure and Reactions of Aromatic Compounds*, Cambridge University Press, 1954, with permission.

fusion on either of the angular rings of phenanthrene. The first four members of this series are shown in Fig. 4.23. It will be noted that each phene compound consists of two sets of linearly annellated rings fused onto a more or less centrally placed ring so as to make the system angular. The correlations between the α, *para*, and β bands in the two series are shown in Figs. 4.24 and 4.25.

PROBLEMS

4.1   Prove that $\phi(x) = A \sin(n\pi x/a)$ is a solution to the second order differential equation:

$$\frac{-h^2}{8\pi^2 m} \frac{d^2\phi(x)}{dx^2} = E\phi(x)$$

4.2   (*a*) Show that

$$\frac{d^2 f(x)}{dx^2} + \frac{4\pi^2}{\lambda^2} f(x) = 0$$

is equivalent to

$$\frac{-h^2}{8\pi^2 m} \frac{d^2\phi(x)}{dx^2} = E\phi(x)$$

4.3   (*a*) Using the free-electron model, calculate the wavelength (nm) of the longest-wavelength transition (lowest energy) for decapentaene, and the corresponding energy, in kilocalories per mole, for this transition. (*b*) What are the amplitudes of the wavefunction ψ_7 at the various positions and hence the coefficients of Hückel MO's?

4.4   Using Platt's perimeter model, determine which states are possible in anthracene.

4.5   Using the free-electron model, calculate the energy for the longest-wavelength transition for:

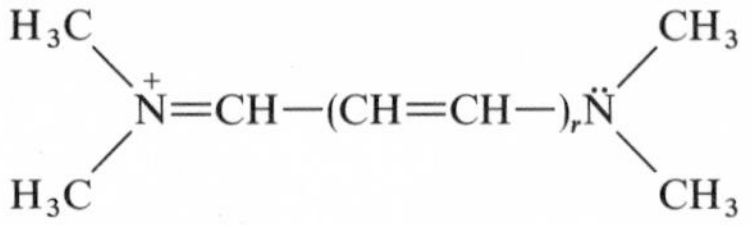

$$
\begin{array}{c}
H_3C \\

\end{array}
\!\!\!\!\!\!\!
\overset{+}{N}\!\!=\!\!CH\!-\!(CH\!=\!CH\!-\!)_r\overset{\cdot\cdot}{N}
\!\!\!\!\!\!
\begin{array}{c}
CH_3 \\

\end{array}
$$

H₃C \ + ⁄CH₃
N=CH—(CH=CH—)ᵣN
H₃C ⁄ \CH₃

where $r = 3$, the $C=C-$ units are 2.48 A, and the box extends a total of 5.65 A (including both sides) beyond the carbon chain.

GENERAL REFERENCES

1. J. R. Platt, *Free Electron Theory of Conjugated Molecules: A Source Book*, John Wiley and Sons, New York, 1964.
2. R. L. Flurry, Jr., *Molecular Orbital Theory and Bonding in Organic Molecules*, M. Dekker, New York, 1968.
3. W. Kauzman, *Quantum Chemistry*, Academic Press, New York, 1957.
4. H. H. Jaffé and M. Orchin, *Theory and Applications of Ultraviolet Spectroscopy*, John Wiley and Sons, New York, 1962.

5 Symmetry, point groups, and character tables

5.1 A Symmetry Operation; Point Symmetry

Although what follows is applicable to all objects, we shall generally confine our discussion to molecules and their properties.

If we partially rotate (rotation through less than 360°) any molecule around an axis which passes through its center and if, after such rotation, the molecule appears to be exactly as it was originally, we have performed a symmetry operation on the molecule. Of course, if we rotate the molecule exactly 360°, it will appear unchanged, but this is a trivial operation, that is, it is applicable to every molecule. Accordingly, here we are interested in

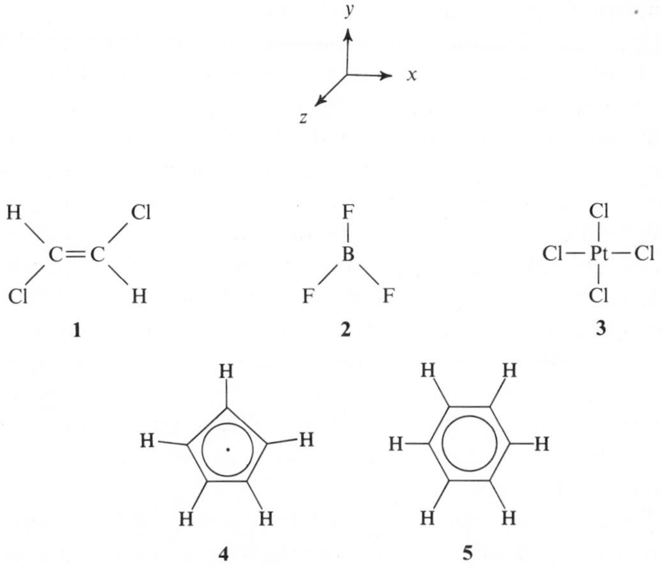

Fig. 5.1 Planar molecules with C_p rotational axes. (The molecules are in the xy-plane and C_p is the z-axis, where $p = 2, 3, 4, 5$, and 6 for **1, 2, 3, 4**, and **5**, respectively.)

rotations by less than 360°. Thus, rotation by 180° around the z-axis in *trans*-1,2-dichloroethylene, **1** (Fig. 5.1), produces a new orientation superimposable on the original. In this molecule there are three *sets* of like atoms; the pair of H, the pair of Cl, and the pair of C atoms. The individual atoms of each set of two cannot be distinguished, and hence the 180° rotation transforms each like atom into its indistinguishable partner. Similarly, rotations around the z-axis by 120 or 240° in **2**; 90, 180, or 270° in **3**; 72, 144, 216, or 288° in **4**; 60, 120, 180, 240, or 300° in **5** all give new orientations which are indistinguishable from (superimposable on) the original; and hence in all cases we have performed a symmetry operation.

In these cases of rotation around an axis, one point in the molecule, the center of gravity, remains unchanged. Symmetry of this kind is therefore called *point symmetry*; it is thus distinguished from other symmetry such as translational symmetry, which characterizes, for example, a picket fence. If we place **1** in a coordinate system and assume that the center of gravity is at the origin (i.e., the point defined as 0, 0, 0) and now perform the 180° rotation around the z-axis, we find that the coordinates which describe like atoms before the operation will also describe these atoms after the operation. Thus, if we consider atoms as points and if Cl atoms are at points 2, 2, 0 and −2, −2, 0 before the operation, Cl atoms will also be found at 2, 2, 0 and −2, −2, 0 after the operation.

5.2   Rotation about an Axis; Rotational Axis, C_p

During a 360° rotation around the z-axis, of the molecules shown in Fig. 5.1, **1** repeats itself twice, **2** three times, **3** four times, **4** five times, and **5** six times; the z-axis in each of these molecules is thus called a two-fold, three-fold, four-fold, five-fold, and six-fold rotation axis, respectively. If the angle through which the molecules must be rotated in order to secure the superimposable image is called θ, the molecules have $360/\theta$-fold rotational axes. The rotational axis is usually denoted as C_p, where C is for cyclic and $p = 360°/\theta$. The subscript p is thus the number of times the superimposable orientation appears during a 360° rotation. It should be understood that the *symmetry operation* we have been discussing is rotation around an axis, while the *symmetry element* is the rotational axis C_p. Frequently the symmetry operation C_p is also called a proper rotation, and the symmetry element C_p a *proper rotational axis*. As we shall see, this notation distinguishes this symmetry operation and element from the ones designated as an improper rotation and an improper rotational axis.

All of the molecules depicted in Fig. 5.1 are planar, and the rotational axis in each case is normal to the molecular plane. The rotational axes of some nonplanar molecules are shown in Fig. 5.2. In *trans*-1,2-dichlorocyclohexane (for our purposes here let us assume that the ring is planar), **6**, there is a C_2

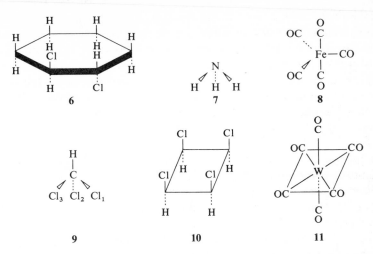

Fig. 5.2   Molecules with C_p axes (C_p = 2, 3, 3, 3, 4, and 4, respectively, for **6, 7, 8, 9, 10,** and **11**).

axis in the plane of the molecule bisecting the bond between the two chlorine-bearing C atoms and the bond opposite; in ammonia, **7**, a C_3 axis passes through the N atom and the center of the triangular base formed by the three H atoms; iron pentacarbonyl, **8**, and chloroform, **9**, have C_3; the all cis-1,2,3,4-tetrachlorocyclobutane, **10**, and tungsten hexacarbonyl, **11**, have C_4. All linear molecules such as $HC\equiv CH$ and CO_2 (and hence also all diatomic molecules) possess C_∞, since rotation about the internuclear axis by any angle gives an orientation identical with the original.

Although the fact appears to be trivial, it is nevertheless important to recognize that all molecules have an infinite number of C_1 axes, since a 360° rotation around any or all axes passing through the center of gravity of the molecule returns it to its original position.

5.3   The Identity Operation

The operation which leaves a molecule unchanged and hence in an orientation identical with the original is called the *identity operation*. It is desirable to distinguish between identical and equivalent orientations. Thus, in performing the C_2 operation on *trans*-dichloroethylene, **1** (Fig. 5.1), we exchange like atoms and the new orientation is not identical with the original; rather, it is equivalent to the original because like atoms cannot be distinguished. Only a second C_2 operation results in the orientation identical to the original. Chloroform, **9**, has a C_3 axis which includes the H-C bond and passes through the center of the triangular base formed by the three Cl atoms. A 120° clockwise rotation takes Cl_1 into Cl_3, Cl_3 into Cl_2, Cl_2 into

Cl_1, and the H and C atoms into themselves. Again, the new orientation is indistinguishable from the original because it is impossible to distinguish among Cl atoms. However, if the first C_3 were followed by a second C_3 and then by a third, we would return the molecule to its original position. The two orientations obtained by C_3 and by $C_3 \times C_3$ are orientations *equivalent* to the original, but the $C_3 \times C_3 \times C_3$ or C_3^3 brings the molecule back to its original and *identical* orientation. Even if the Cl atoms could be distinguished, it would appear that nothing at all had been done to the molecule by C_3^3.

An operation which leaves the molecule identical to the original, that is, an identity operation, is denoted by I [or by E from the German *Einheit*, meaning unit or unity since the transformation matrix (*vide infra*) for this operation is always a unit matrix]; thus $C_3^3 = I$, $C_2^2 = I$, $C_1 = I$. Since all these operations, by definition, leave the molecule unchanged, doing nothing to the molecule is also an identity operation. The necessity of including this operation will become apparent when we consider the properties of a point group.

5.4   Operations of the Same Class

It should be noted that, in the chloroform example, performing $C_3 \times C_3$ or C_3^2 gives an orientation that is identical with the ones obtained by a C_3 operation in the counterclockwise direction, that is, $C_3 \times C_3 = C_3'$, where the prime refers to counterclockwise rotation. Furthermore, C_3 and C_3' give equivalent but not identical orientations; these two operations are said to belong to the same *class*, and we say that chloroform has $2C_3$ axes, where 2 is called the *order* of the class. On the other hand, in a molecule such as *trans*-dichloroethylene, performing either a C_2 or a C_2' operation results in the *identical* orientation, and hence we have only one C_2 in a class in this example.

If we examine a tetrahedral molecule, CH_4, drawn so that the four H atoms are at the corners of a tetrahedron, Fig. 5.3, we see that each of the four faces of the tetrahedron, A, B, C, and D (D is unlabeled in Fig. 5.3; it is the face opposite the H_3 atom and cannot be seen), is an equilateral triangle. The most obvious symmetry feature of such a triangle is its C_3 symmetry. It is clear then that a rotational axis which passes through the center of a triangular face and the corner opposite it is a C_3 axis, since it transforms corners into each other. Thus, the axis passing through the center of face A and H_4 is a three-fold axis, clockwise rotation around which interchanges the atoms on face A, carrying H_3 into H_1, H_1 into H_2, and H_2 into H_3. This axis is also a C_3' axis, and the counterclockwise rotation or C_3' operation carries H_1 into H_3, H_3 into H_2, and H_2 into H_1. Since the tetrahedron has four faces, there is a total of $8C_3$. All of the $8C_3$ are said to belong to the same class. For the time being we can say that all $8C_3$ transform the four hydrogens in an equivalent way.

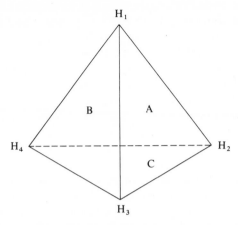

Fig. 5.3   Methane, with the four hydrogen atoms at the corners of a tetrahedron.

5.5   The Inverse of a Symmetry Operation

We saw above that, if two or more operations performed in sequence return the molecule to an orientation identical to the original, the product of the operations is equal to the identity. Another very simple way to return a molecule to its original position after a symmetry operation consists of simply reversing the operation. Thus, the C_3 or clockwise rotation of the molecule shown in Fig. 5.3, 120° around the axis passing through H_1 and the base of the triangle, takes H_2 into H_3, H_3 into H_4, and H_4 into H_2. Now, if this operation were followed by C_3' or C_3^{-1}, the inverse of C_3, the molecule would be returned to its original orientation. The two operations can be considered analogous to the military exercise of "right face" followed by "left face," since the second operation in each case cancels the effect of the first one. For this reason, the second operation is called the *inverse* (or, less desirably, the *reciprocal*) of the first, and the multiplication may be written $C_3^{-1}C_3 = I$ or, more generally, $A^{-1}A = I$, where A is any symmetry operation.

It does not make any difference as to the order in which we perform these two operations, that is, the operations commute. Thus, we can first rotate the molecule in Fig. 5.3 120° around the C_3 axis counterclockwise, followed by a 120° clockwise rotation, and thus return the molecule to its original position. An operation such as C_2 is its own inverse since a second C_2 in either direction gives the identical orientation.

5.6   Reflection at a Plane; the Mirror Plane, σ (Sigma)

If a molecule is bisected by a plane, and each atom in one half of the bisected molecule is reflected (the operation) through the plane and encounters a similar atom in the other half, the molecule is said to possess a mirror plane (the symmetry element). The operation and the element are denoted by σ (probably derived from the first letter in the German word *Spiegel*, meaning a mirror). Every planar molecule, of course, has at least one plane of symmetry, the molecular plane. Thus, the xy-plane is a horizontal mirror plane in all of the molecules shown in Fig. 5.1; BF_3, **2**, has, in addition, three vertical mirror planes, each including one B-F bond and bisecting the angle between the other two B-F bonds. The molecule $PtCl_4$, **3**, has four vertical planes, the two planes along the x- and y-axes (i.e., the xz- and yz-planes) and two diagonal planes which are perpendicular to each other and which bisect the angles between the x- and y-axes. Benzene, **5**, has six vertical planes, a set of three which pass through opposite atoms and which may be designated $3\sigma_v$, and a set of three bisecting opposite C-C bonds, which may be designated $3\sigma_d$. A symmetric linear molecule such as $HC\equiv CH$ has an infinite number of vertical mirror planes, all of which include the internuclear axis, here the z-axis; in addition to $\infty\sigma_v$, there is a horizontal plane, σ_h, which bisects the molecule.

If we place our molecule in a Cartesian coordinate system, reflection in a mirror plane always results in a change in sign of the coordinate normal to this plane and leaves the coordinates parallel to the plane unchanged. Thus, σ_{xy} changes every point x, y, z to x, y, $-z$; σ_{yz} changes every x to $-x$; and σ_{xz} changes every y to $-y$.

5.7   Inversion at a Center of Symmetry (i)

When a straight line is drawn from any atom in a molecule through the center of the molecule and, if continued in the same direction, encounters an equivalent atom equidistant from the center (the operation), and if the same operation can be performed on all atoms, then the molecule possesses a center of symmetry, designated i for inversion. Since each atom is thus reflected through the center into an equivalent atom, atoms must occur in pairs (with the exception of any atom lying on the center itself), with the members of the pair equidistant but in opposite directions from the center. Thus, in Fig. 5.1 molecules **1**, **3**, and **5** have i, but **2** and **4** do not, while in Fig. 5.2 only **11** has i.

The operation of inversion changes the sign of each of the three coordinates which define the position of an atom. Actually an atom should be considered a point, but for our purposes we may consider the atoms as having some volume. Thus, a point on the upper part of the upper right Cl atom in *trans*-dichloroethylene, **1** in Fig. 5.1, may be described as being at x, y, z in the coordinate system shown. On inversion, this point becomes $-x$, $-y$, $-z$, and is found on the lower part of the equivalent lower left Cl atom. Since such an equivalent point was found on inversion, and every other point on every atom can be similarly transformed, the molecule has a center of symmetry.

5.8   Rotation about an Axis, Followed by Reflection at a Plane Normal to This Axis; Rotation-Reflection Axis, S_p

If a molecule is rotated around an axis and the resulting orientation is reflected in a plane perpendicular to this axis (the operation), and if the resulting orientation is superimposable on the original, the molecule is said to possess a *rotation-reflection axis* (the element). The axis around which the rotation was performed is the rotation-reflection axis, and it is designated as S_p, where p, as usual, is the order. This axis is also called an *alternating* or *improper axis* and is thus distinguished from a rotational or proper axis, C_p. In *trans*-dichloroethylene, **1** in Fig. 5.1, the x-axis is an S_2 axis, since, for example, the lower right H atom, on a $180°$ rotation around the x-axis, followed by reflection in the yz-plane (which is normal to the x-axis), is placed on top of the upper left H atom in the original. The other atoms are similarly transposed. It should be noted that in this example the S_2 axis, which coincides with the x-axis, is not a C_2 axis. It should also be noted that the S_2 operation achieves exactly the same result as i. Thus, if we focus on a point in front of the molecule and on the right-hand Cl atom in **1**, Fig. 5.1, and rotate the molecule $180°$ around the x-axis, the initial point x, y, z goes to x, $-y$, $-z$, and reflection across σ_{yz} gives $-x$, $-y$, $-z$, so that S_2 reverses the signs of all three coordinates of each point. But this is precisely what i does, and so S_2 is equivalent to i. Thus, if a molecule possesses i, the S_2 need not be separately specified because it is equivalent and implied. Actually, any axis through a molecule having a center of symmetry is an S_2 axis.

Some molecules have alternating axes of order twice that of the highest rotational axis, S_{2p} and C_p. Thus, in the chair form of cyclohexane, **12** (Fig. 5.4), the vertical axis which passes through the center of the molecule is a C_3 axis (e.g., clockwise $120°$ rotation transforms hydrogen 1 into 5, 5 into 10, and 10 into 1, as well as 2 into 6, 6 into 9, and 9 into 2), but this axis is also an S_6 axis (e.g., clockwise $60°$ rotation followed by reflection in the normal plane takes 1 into 4, 4 into 5, 5 into 8, etc.). A vertical axis through the carbon of methane, **13** (Fig. 5.4), is a S_4 axis, taking 1 into 2, 2 into 3, 3 into 4, and 4 into 1. Inscribing the tetrahedron of CH_4 in the cube helps to illustrate other symmetry properties of the tetrahedral structure. Thus, the S_4 axis described above passes through the center of opposite faces of the cube, and since there are six faces there are three S_4 axes. And because for each S_4 (clockwise) there is a S_4' (counterclockwise), there is thus a total of $6S_4$ operations in a

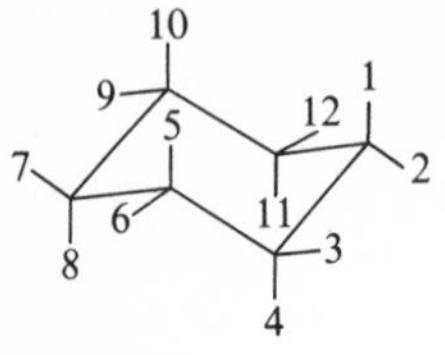

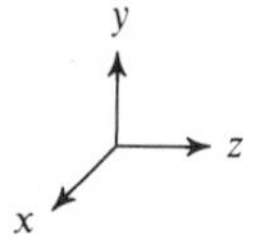

Fig. 5.4 Molecules with S_6 (**12**), S_4 (**13**), and S_1 (**14**) rotation-reflection (alternating or improper) axes.

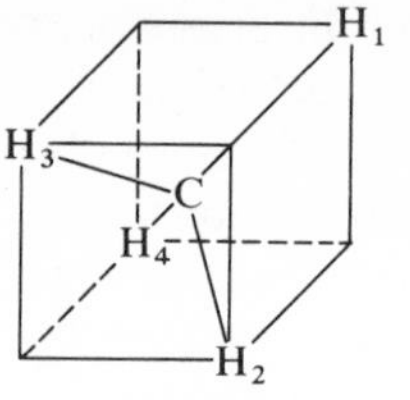

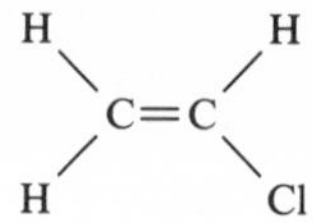

tetrahedral molecule, and all of these S_4 transform in a similar manner and hence belong to the same class. Note again that these S_4 axes are not C_4 axes; they coincide with C_2 axes.

The highest-fold axis implies a great deal about the symmetry properties of a molecule. Thus, the outstanding symmetry property of the chair form of cyclohexane is its six-fold alternating axis, S_6, which transforms all axial hydrogens into each other, as well as all equatorial hydrogens into each other; thus the twelve hydrogens in C_6H_{12} belong to two sets of six chemically equivalent H atoms. The S_4 operation on CH_4 transforms all H atoms into each other, and hence the four H atoms in CH_4 constitute one set of chemically equivalent atoms. The S axis is called an *alternating axis* because the equivalent atoms transformed by the operation lie alternately on one side and then on the other side of the plane of reflection.

It should also be pointed out that not only is i equivalent to S_2 but also σ is equivalent to S_1. Usually it is easier to find the plane of symmetry than the S_1 axis, but the S_1 axis will be any axis perpendicular to the plane of symmetry. Thus, in chloroethylene, **14** (Fig. 5.4), the x-axis is an S_1 axis; rotation of 360° around x, followed by reflection in the yz-plane, gives a molecule identical with the original. Traditionally, a student is taught to look for a plane or center of symmetry in a molecule because this is more easily recognized, but it should now be clear that these elements are equivalent to S_1 and S_2, respectively.

The point symmetry operations may be summarized as follows. If a molecule can be rotated or reflected around its center in such a way that after the operation the molecule is identical with, or superimposable on, the original, the molecule possesses point symmetry of some kind. There are only two fundamental point symmetry operations: simple rotation around an axis, C_p; and rotation-reflection about an axis, S_p. Furthermore, C_1 is equivalent to doing nothing, the identity operation, I; reflection in a mirror plane, σ, is equivalent to S_1; and inversion, i, is equivalent to S_2.

5.9   Symmetry and Optical Activity

The classical criterion for optical activity of a molecule involves the test of superimposability of its mirror image; if the mirror image can be super-imposed on the original, the two are identical and the molecule is inactive. If the original and the mirror image cannot be superimposed, the molecule is active and the compound is resolvable. Thus, in the 1,2-dichlorocyclopropanes in Fig. 5.5 we see that the *cis* isomer, **15**, and its mirror image, **16**, are directly

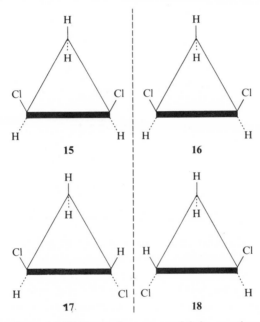

Fig. 5.5   *cis*- and *trans*-1,2-Dichlorocyclopropane and their respective mirror images.

superimposable and hence identical. On the contrary, the *trans* isomer, **17**, and its mirror image, **18**, are not superimposable; accordingly, *trans*-1,2-dichlorocyclopropane is optically active and resolvable into the enantio-morphs, **17** and **18**.

It is always relatively easy to draw the mirror image of a structure, but to test whether or not the mirror image is superimposable on the original can be a tricky exercise. The problem of comparison is made more complicated by the practice of using projection formulas. If a molecule is represented by a three-dimensional drawing, there are no restrictions on the manipulation of the drawing; it can be rotated out of the plane of the paper or in the plane of the paper in any manner desired for comparison because the spatial relationship of each atom to every other atom is specified in the three-dimensional drawing. However, there are restrictions on the way that plane projection formulas can be manipulated because there is no way of knowing (except by convention) which atoms were projected forward and which atoms projected back. In Fig. 5.6 the enantiomorphs shown three dimensionally

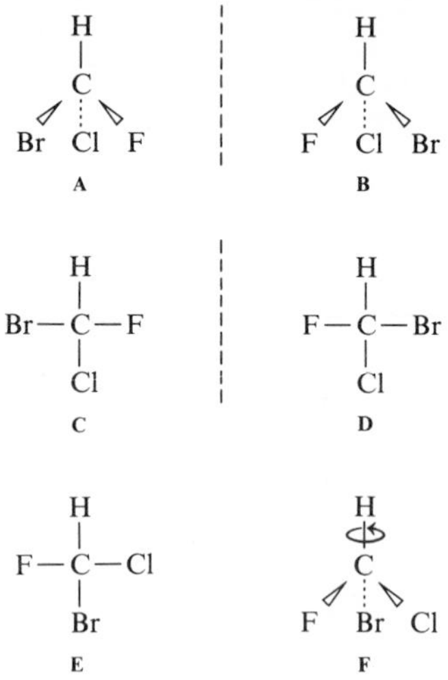

Fig. 5.6   Plane projection formulas of an optically active compound.

as **A** and **B** are drawn as plane projections in **C** and **D** respectively. The latter appear to be superimposable by, for example, rotating **D** 180° out of the plane of the paper around a vertical axis in the plane of the paper. Such an operation on a plane projection formula is not permitted, however, for it results in changing the orientation of the atoms with respect to each other. The enantiomorph of **C** in plane projection can be obtained by exchanging one pair of atoms; two exchanges, for example, **C** → **E** gives **C** (equal to **A**) back again. That **E** is identical to **A** can perhaps be seen more readily by drawing **E** in three dimensions as in **F** and then rotating **F** 120° around the H-C bond in the direction shown.

The test of superimposability is readily performed by visualization when simple molecules are involved, but when complicated molecules are examined such visualization is difficult. Of course, a molecular model and its mirror image may be constructed to determine whether the members of the pair are identical or enantiomorphic, but this procedure is cumbersome and time-consuming. Accordingly, it is desirable to determine from the symmetry properties of the molecule whether or not it is optically active.

It can be shown mathematically that the requirement for optical activity, the lack of superimposability of molecule and mirror image, is equivalent to absence of an alternating axis of any order, S_p. Similarly, existence of an S_p is sufficient to indicate that molecule and mirror image are superimposable, and optical activity is absent. All molecules which lack a rotation-reflection axis of any order (fold), S_p, are said to be dissymmetric and are optically active, that is, the molecule and its mirror image cannot be made to coincide in space by any kind of rotational or translational motion of the whole molecule.[a] It will be recalled from our earlier discussion that a mirror plane is equivalent to S_1 and a center of inversion is equivalent to S_2, and hence molecules possessing these symmetry elements have mirror images that are superimposable on the original and therefore are inactive. If we examine *trans*-1,2-dichlorocyclopropane, **17** (Fig. 5.5), we see that there is a C_2 axis in the plane of the three-membered ring passing through the bond between the chlorine-bearing C atoms and the methylene C atom. Since the molecule does not possess an S_p axis, however, it is active. On the other hand, because **17** and **18** have a C_2 axis, they do not lack symmetry, that is, they are *dissymmetric* but not *asymmetric*. The *only* symmetry that dissymmetric (optically active) compounds possess is one or more C_p axes; although many common dissymmetric compounds have C_2 axes, most optically active compounds are asymmetric as well as dissymmetric and hence have only C_1.[b]

The spiro compound shown in Fig. 5.7 has an S_4 axis, which bisects both rings and the N atom, but no planes of symmetry (S_2 axes). It has been synthesized and found to be unresolvable. The S_4 axis is also necessarily a C_2 axis, but, as pointed out above, if C_2 were the only symmetry element present the molecule would be active. In summary, if a molecule possesses only C_p, it is dissymmetric and optically active; if $p = 1$ the molecule is asymmetric as well as dissymmetric, and if $p > 1$ the molecule is dissymmetric; if a molecule possesses S_p with any p, it cannot be optically active. All optically active molecules are necessarily dissymmetric, but not all dissymmetric molecules are asymmetric.

---

[a] An interesting example of a molecule which is dissymmetric "on paper" but is experimentally inactive is the compound $(+)$-menthyl$(-)$-menthyl-2,2′,6,6′-dinitro-4,4′-diphenate, which in any single conformation lacks an S_p axis but is inactive because of the free rotation around the 4,4′ bonds.

[b] One useful classification of stereoisomers (isomers that have the same number and kinds of atoms attached to each other but differ because of different fixed geometry) is a division into two classes—stereoisomers that are enantiomeric and those that are not; the latter are called *diastereomers*. Thus, *cis*- and *trans*-2-butene are diastereomers by this definition. The enantiomeric relationship is essentially the relationship between right- and left-hand helices or of handedness in the screw sense of being right- or left-handed. The handedness of a given enantiomer is called its *chirality* (pronounced "kirality," from the Greek *cheir*, meaning hand).

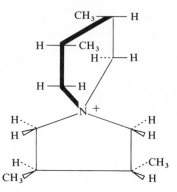

Fig. 5.7   A molecule with an S_4 axis as the only S axis.

It is important to recognize that practically all molecules exist in some conformation possessing only C_p symmetry, and hence if all the molecules of a compound could be frozen in that conformation, the compound would be optically active. Take as an example 1,2-dichloroethane, Fig. 5.8; a is the

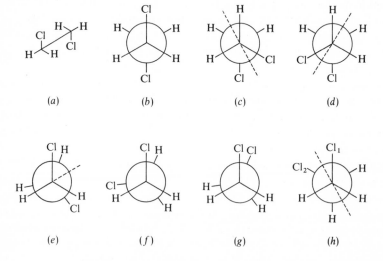

Fig. 5.8   The conformations of 1,2-dichloroethane.

sawhorse representation of the staggered form (or conformer) with chlorine atoms *trans*. This form is also drawn by means of the Newman projection formula in Fig. 5.8*b*. The other two staggered forms (*gauche*) are shown in Figs. 8*c* and *d*. Conformation *a* (and its equivalent *b*) have $i = S_2$ as well as $\sigma = S_1$ and C_2; *c* and *d* have only C_2, the broken line which bisects the C-C bond. Conformations e, f, and g are the three eclipsed conformations (drawn slightly askew for clarity); e and f have C_2 only, and g also has $2\sigma = 2S_1$. There are an infinite number of conformations resembling h which are neither completely eclipsed nor completely staggered; all of these have C_2 symmetry. However, the barrier to rotation around the C-C bond is small, and at all reasonably accessible temperatures the compound exists as a mixture of all conformations. Thus, statistically there are as many molecules in conformation c as in d, and hence any optical activity due to c is exactly canceled by its mirror image, d. The statistical probability of finding equal populations of enantiomorphs at all times holds for all conformations, and hence the molecule is inactive because the aggregate of molecules is at all times a mixture of racemic mixtures. More simply stated, on the average, for every optically active molecule with a left-handed configuration there is another with a right-handed configuration. As a practical matter, it is sufficient to determine whether a molecule in *any* readily accessible conformation has an S_p axis.

5.10   Chemical Equivalence and Isomers

In molecules such as methane, ethane, benzene, and ethylene all the hydrogen atoms are *chemically equivalent* and in each of these molecules all the H atoms belong to one *set*. In propane there are six equivalent methyl hydrogens constituting one set, and two methylene hydrogens constituting a second set. We now ask how exactly we can demonstrate that the hydrogens discussed above are members of a set and are chemically equivalent.

If H atoms are chemically equivalent and are members of a set, replacement of one of them by another atom A will give a molecule identical with that obtained by replacement of any other H atom in the set by A. In order for a set of atoms to be chemically equivalent they must meet the atom replacement test. We can predict the results of such a test by symmetry considerations. Thus if we refer to Fig. 5.4 and examine methane, **13**, shown inscribed in a cube, we see that the C_2 axis, passing through the center of the top and bottom faces, carries H_1 into H_3 and vice versa, and H_2 into H_4 and vice versa. The C_2 axis passing through the center of the side faces carries H_1 into H_2 and vice versa, and H_3 into H_4 and vice versa, and since H_1 can thus be carried into all the other H atoms by C_2 operations, all the hydrogens are equivalent.

We could alternatively show the equivalence by means of the C_3 axes. Thus, rotation around the C_3 axis passing through the corner occupied by H_1, the center C atom, and the corner opposite takes H_2 into H_3, H_3 into H_4, and H_4 into H_2. Rotation around the C_3 axis passing through H_3, C, and the corner opposite exchanges H atoms 1, 2, and 4. All four hydrogens are thus equivalent and can be transformed into each other by appropriate

C_3 operations; and there is only one chloromethane. Similarly, there is only one monosubstitution product of ethane, benzene, and ethylene. With respect to symmetry requirements, atoms or groups (methyl group, e.g.), are equivalent if they can be exchanged by rotation about a symmetry axis, $C_p(\infty > p > 1)$, of the molecule or group.

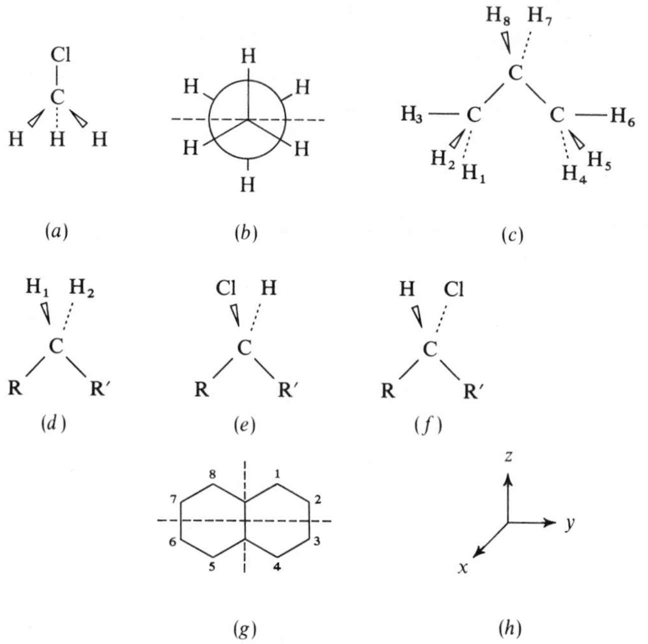

(a)   (b)   (c)

(d)   (e)   (f)

(g)   (h)

Fig. 5.9   Equivalent and enantiotopic hydrogens.

Examination of CH_3Cl, Fig. 5.9a, shows a C_3 axis passing through the Cl and C atoms and the center of the triangular base formed by the three H atoms. Rotation around this axis transforms the hydrogens into each other, and hence the three hydrogens are equivalent; substitution of any H atom, by, for example, chlorine, gives only one product, methylene chloride, CH_2Cl_2. Analogous arguments for the numbers of isomers formed as a result of substitution of hydrogens by other atoms or groups can be used for all molecules. Consider ethane, Fig. 5.9b, in the staggered conformation. Rotation around the C_3 axis perpendicular to the paper which passes through the two C atoms carries the hydrogens of each individual methyl group into each other, and the C_2 operation around the axis (broken line) in the plane of the paper and perpendicular to the C_3 axis carries the top front H atom into the bottom rear H atom and vice versa, etc. Hence, all H atoms are equivalent and only one monosubstitution product is possible. The hydrogens are equivalent not only in this conformation but also in all the infinite number of possible conformations, since they all have the C_3 and C_2 axes described above.

The chemical equivalence of hydrogen atoms in *planar* compounds can be ascertained by exchange of hydrogens on reflection in mirror planes. Such mirror planes in planar molecules always include a C_2 axis. Thus, if we examine naphthalene, Fig. 5.9g, we see that H_1 and H_4, H_5 and H_8, H_2 and H_3, H_6 and H_7 are exchanged by reflection through the xy-plane (or by C_2^y), while reflection through the xz vertical plane (or by C_2^z) exchanges H_1 and H_8, H_4 and H_5, H_2 and H_7, H_3 and H_6. Thus there are two sets of hydrogens: H_1, H_8, H_4, H_5, and H_2, H_3, H_6, H_7, and hence only two possible monosubstituted naphthalenes.

5.11   Enantiotopic Hydrogens and Nuclear Magnetic Resonance Spectroscopy

If a molecule has a C_p axis, the atoms exchanged by the C_p operation are chemically equivalent and substitution of any of these equivalent atoms leads to only one compound. Now let us examine the situation in which similar atoms are exchanged only by reflection through a mirror plane or, more generally, by an S_p operation. In a general case, RCH_2R_1, Fig. 5.9d, the molecule does not possess $C_p, p > 1$, but does possess a mirror plane bisecting the H_1CH_2 angle. Reflection through this plane exchanges H_1 and H_2. Now, if we use the atom substitution test for equivalence, we see that substitution of H_1 by another atom leads to the enantiomorph of the compound obtained by substitution of H_2 by the same atom. If substitution is by a Cl atom, the enantiomorphs are Figs. 5.9e and f. In this situation the H atoms of conformation d bear a special relationship to each other and may be said to be *pseudo-equivalent*. In recently suggested nomenclature, such atoms are called *enantiotopic* atoms.

Now let us examine a molecule such as RCH_2R, in which the R groups are equivalent; the simplest example in the hydrocarbon series is propane, where $R = CH_3$. Let us assume that propane has the staggered conformation shown in Fig. 5.9c. The C_2 axis, which bisects the HCH angle of the methylene group, takes H_7 into H_8, H_3 into H_6, H_1 into H_5, and H_2 into H_4. But NMR evidence, as well as the fact that only two different monochloropropanes are known, tells us that there are only two and not four sets of hydrogens, one set with two and the other with six atoms. Logically, the first set is assigned to H_7 and H_8, the two secondary H atoms, and the other six must all belong to a single set. How can we justify this?

The answer lies in two facts. (1) If R in CH_3R is an atom, for example, chlorine, the molecule has a three-fold axis. But we cannot consider

CH_3CH_2—, the actual R group in propane, as an atom. However, (2) rotation around the C—R axis is unhindered, so that this axis is a pseudo-symmetry axis, or an *axis of local symmetry*. The free rotation about this axis makes H atoms 1, 2, and 3 equivalent, as well as H_4, H_5, and H_6, and since rotation around the real C_2 axis takes H_1 into H_5, all six H atoms 1–6 are equivalent and members of a set.

Chemically equivalent protons give only one signal in NMR spectroscopy. Thus, we would expect one proton signal in methane, ethane, ethylene, benzene, chloroform, and methylene chloride and two different signals in propane. Although H atoms that can be exchanged *only* by reflection through a mirror plane are not chemically equivalent, such protons still give only one proton signal. The NMR cannot distinguish enantiotopic hydrogen atoms, except perhaps when the spectrum is determined in an optically active solvent. All hydrogen atoms that can be exchanged by *any* symmetry operation give a single nmr signal and such atoms may be called symmetry equivalent atoms; all symmetry equivalent atoms are not necessarily chemically equivalent. If either R or R′ in $RCH_2R′$ is optically active, the methylene hydrogens no longer give the same signal. A mixture of the (+) and (−) forms of the enantiomorphs, of, say, lactic acid $CH_3CHOHCO_2H$, in a solvent that is not not optically active gives a single signal for the proton on the asymmetric C atoms since the NMR is a probe which cannot distinguish enantiomorphs. Because there are thousands of compounds of the basic formula $RCH_2R′$, this problem is of general importance. It is quite likely that the optical activity of biological compounds is induced in nonactive compounds by the interaction of the latter with optically active enzymes. Thus it is conceivable that enzymatic oxidation of one of the enantiotopic methylene hydrogens in propionic acid, $CH_3CH_2CO_2H$, could lead to optically active lactic acid.

In the compound $RCH_2R′$, the C atom is said to be a *prochiral* atom because exchange of one of the H atoms, as pointed out above, for a group or atom other than H, R, or R′ leads to an optically active compound and a chiral C atom. Replacement of one of the hydrogens would lead to the enantiomorph with the S configuration, while replacement of the other H atom would lead to the R configuration. The H atoms can therefore be designated as *pro*-S and *pro*-R hydrogen atoms. For example, if we consider ethylbenzene and intend to replace a hydrogen by a chlorine atom, the H atoms are designated as follows:

5.12   Dipole Moments

A dipole moment is a vector property, that is, it has both magnitude and direction, and it results from the unequal sharing of electrons between atoms. Although a vector property, the dipole moment is a stationary and not a dynamic property of a molecule. Stationary properties must remain unchanged by every symmetry operation of a molecule, and in order to remain unchanged the dipole moment vector must lie on each of the symmetry elements. Molecules with a center of symmetry cannot have a dipole moment because a vector cannot lie on a point. Also, molecules with more than one C_p axis ($p > 1$) cannot have a dipole moment because a vector cannot be coincident with two different axes. Thus, molecules with one C_p ($p > 1$) only or with only σ and no C_p ($p > 1$) may have dipole moments. In addition, molecules that have C_p ($p > 1$) and planes of symmetry that include the C_p also may have dipole moments. Such molecules are H_2O and NH_3 (Fig. 5.10); the former has a C_2 axis and two σ, which intersect at the C_2, while pyramidal NH_3 has 3 σ, which intersect at the C_3 axis.

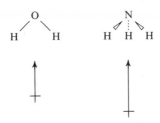

Fig. 5.10   Dipole moments.

In these and all other cases the dipole moment vector must lie in all symmetry elements of the molecule. The direction of the dipole moment is thus determined; however, the magnitude and the positive and negative ends of the dipole cannot be determined from symmetry arguments only.

5.13   Conformers and Isomers

It is important to realize that in what follows we consider the molecule in some fixed geometry. Different conformations of the same molecule, that is, the different geometries obtained by rotations around a single bond, have different symmetry properties. Thus, for example, the symmetry of staggered ethane is different from that of eclipsed ethane and different from the infinite number of other conformations that are less symmetrical than either of the above; in fact, all conformations which are neither eclipsed nor staggered are dissymmetric and hence theoretically resolvable.

Whether or not two conformations of the same molecule are isomers is a matter of definition. By some definitions *cis* and *trans* isomers such as *cis*- and *trans*-dichloroethylene are conformers and not isomers, and by some definitions staggered and eclipsed 1,2-dichloroethane are isomers and not conformers. A highly substituted ethane has been shown to have independent existence in two conformations, and hence such conformers can be considered to be isomers. Thus, 1,1,2,2-tetrachloro-1,2-difluoroethane gives, at 31.5°, a single ^{19}F nuclear magnetic resonance signal which broadens at $-70°$ and splits into two signals at $-120°$. The barrier to rotation around the C-C bond is sufficiently large so that, at low temperatures, the two conformers, one in which the two fluorines are skew or *gauche*, and the other in which they are *trans*, have independent existence. These conformers can legitimately be called isomers at $-120°$, and this example illustrates the difficulty of making an unambiguous distinction between isomers and conformers.

5.14   Conventions Regarding Coordinate Systems and Axes

In the water and phenanthrene examples discussed above, we identified the symmetry elements by the use of subscripts and superscripts related to the coordinate system shown in Fig. 5.11c. It is desirable to place every molecule in a coordinate system in a systematic manner so as to minimize ambiguities and to enhance clear communication. The following conventions or rules are recommended for this purpose.

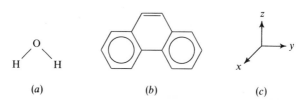

(a)   (b)   (c)

Fig. 5.11   Coordinate axes for H_2O and phenanthrene.

1. *Place the origin of the coordinate system at the center of gravity of the molecule.*

2. *Assign the z-axis as follows:*

(a) *If there is only one rotational axis, take this axis as the z-axis.* Thus, in H_2O and phenanthrene, Fig. 5.11, the C_2 axis becomes the z-axis. *The z-axis is always considered to be vertical.* If *trans*-dichloroethylene, a planar molecule, is written with all the atoms in the plane of the paper, the z-axis is always the axis perpendicular to the plane of the paper, Fig. 5.12b. The designation of planes of symmetry as vertical and horizontal is determined

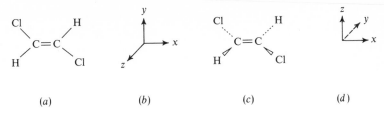

Fig. 5.12   Assignment of axes in *trans*-dichloroethylene.

in relation to the principal or z-axis, always taken as vertical. If Fig. 5.12a were drawn on the blackboard, the xy-plane would appear as a vertical plane, but it is correctly designated as horizontal, σ_h, because it is perpendicular to the C_2 axis. If there is a personal preference for writing the z-axis in an actual vertical position, the molecule can be redrawn as in Fig. 5.12c, so that on the blackboard the plane of the molecule is perpendicular to the plane of the board, whereupon the z-axis will actually be vertical, Fig. 5.12d, and then σ_h is actually horizontal.

(b) *If the molecule has several rotational axes, the one of highest order is taken as the vertical and z-axis.* Thus, $[\mathrm{PtCl_4}]^{2-}$ has a four-fold axis and four two-fold axes, two of them coincident with the coordinate axes directed at corners, and the other two bisecting opposite sides of the square, Fig. 5.13a. Accordingly, the four-fold axis in the plane of the paper is the z-axis,

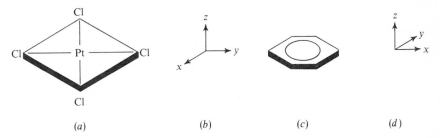

Fig. 5.13   Assignment of coordinate axes in $[\mathrm{PtCl_4}]^{2-}$ and $\mathrm{C_6H_6}$.

Fig. 5.13b, since the molecule is shown oriented in such a way that its molecular plane is perpendicular to the plane of the paper. The molecular plane is perpendicular to the vertical C_4 axis and hence is horizontal, σ_h. Benzene has a six-fold axis, a three-fold and a two-fold axis coincident with it, and six two-fold axes, three bisecting opposite bonds and three passing through opposite atoms. If the benzene molecule is placed in the plane of the paper, the z-axis is perpendicular to it; if it is drawn perpendicular to the plane of the paper, Fig. 5.13c, the z-axis is in the plane of the paper, Fig. 5.13d. The molecular plane is σ_h in either case.

(c) *If there are several rotational axes of the highest order, the axis which passes through the greatest number of atoms is taken as the z-axis.* Thus, in ethylene, Fig. 5.14a, where there are three equal two-fold axes, the z-axis, the vertical axis, is taken as the one which is coincident with the C-C bond, Fig. 5.14b. If one wishes to have the z-axis actually vertical, Fig. 5.14d, then the molecule is rotated 90° around the x-axis, Fig. 5.14c. Naphthalene, Fig. 5.14e, has three C_2 axes; the one passing through the bridging carbon atoms is considered to be the z-axis, Fig. 5.14f. The three two-fold axes in both naphthalene and ethylene are coincident with the x-, y-, and z-axes.[c]

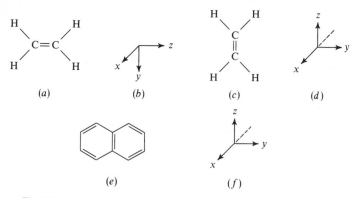

Fig. 5.14　Assignment of coordinate axes in ethylene and naphthalene.

Some molecules which have a C_p axis also have an $S_{2p}(p > 1)$ coincident with the C_p. In such cases, there is no problem. Thus, chair cyclohexane, Fig. 5.15a, has C_3 and S_6 coincident, and this is taken as the z-axis. However, in tetrahedral molecules, for example, besides the C_2 and S_4 axes, there are also C_3 axes, the axes which include the C-H bonds, Fig. 5.15b. In such cases two different representations serve different purposes. Figure 5.15b, with one C-H bond, that is, one C_3 axis vertical, emphasizes this feature; here the molecule sits on one of its faces. On the other hand, Fig. 5.15c, in which the molecule is inscribed in a cube, serves to emphasize the three S_4 (and the coincident C_2) axes by placing one of them vertical and the other two horizontal; this is the preferred assignment.

3. *Assign the x-axis as follows:*

(a) *If the molecule is planar and the z-axis lies in this plane, the x-axis is chosen to be normal to this plane.* This is the situation in water and phenanthrene, Fig. 5.11, and in ethylene and naphthalene, Fig. 5.14. Hence, in these

[c] In the point group D_{2d} (*vide infra*), where there are three two-fold rotational axes and two planes of symmetry, the axis in which the two planes of symmetry intersect is chosen as the z-axis.

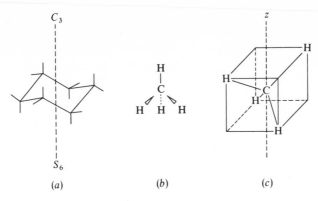

Fig. 5.15   Molecules with C_p and S_{2p}.

molecules if the z-axis is placed vertical and the molecule is drawn in the plane of the paper, the x-axis is the axis perpendicular to the paper.

(b) *If the molecule is planar and the z-axis is perpendicular to this plane, the x-axis* (which, together with the y-axis, must lie in the molecular plane) *is chosen to pass through the largest number of atoms.* Thus, in *trans*-dichloroethylene, the x-axis passes through the C atoms, Fig. 5.12, and in benzene, Fig. 5.13, it passes through opposite C atoms. With $[PtCl_4]^{2-}$ the choices are equivalent, and hence the assignment is arbitrary.

Finally, a word must be said about the assignment of the positive direction of the axes. The convention usually adopted is the so-called right-hand rule. The thumb, index, and middle fingers of the right hand are extended in three mutually perpendicular directions. The directions in which the thumb, index, and middle fingers are pointing then become, respectively, the positive x, y, and z direction. The coordinate systems shown in Figs. 5.11, 5.12, 5.13, and 5.14 obey the right-hand-rule; the head of the arrow indicates the positive direction.

Although the rules given above are widely accepted, conformance still is not universal. Consequently it is essential for an author to indicate explicitly his assignment of the coordinate system for the molecule under discussion.

5.15   Point Groups

Every molecule can be characterized by the symmetry operations that can be performed on it. If precisely the same operations can be performed on two molecules, these molecules, no matter how different chemically, are symmetry-related and must be classified together.

The molecules H_2O and phenanthrene possess the same elements of symmetry, and we can perform the following symmetry operations, and only these, on both of them (Figs. 5.11*a* and *b*): $C_2{}^z$, σ_{xz}, and σ_{yz}. In addition, the identity operation, I, can be performed on every molecule, and hence there is a total of four symmetry operations that can be performed. The four operations together constitute a *group*, and because each of the operations leaves the center of gravity unchanged, the group is called a *point group*. The four operations or elements constituting the group must satisfy the mathematical properties of a group.

5.16   Properties of a Group

For our purposes, the most important requirement for a set of symmetry operations to properly constitute a group is that, if two operations in the group are multiplied together, the product must also be an element (operation) in the group. Let us test this requirement against the four symmetry operations constituting the group to which water and phenanthrene belong, using the coordinate system shown in Fig. 5.11*c*. We first note that the operation $C_2{}^z$ changes a point x, y, z to $-x$, $-y$, z; σ_{yz} changes a point x, y, z to $-x$, y, z; σ_{xz} changes a point x, y, z to x, $-y$, z; and, finally, the operation I changes a point x, y, z to x, y, z, that is, leaves the point unchanged.

Now, if I, $C_2{}^z$, σ_{xz}, and σ_{yz} constitute the elements in the group, multiplication of any two must give a third. In performing multiplications of symmetry operations it should be borne in mind that each operation is a mathematical operator, not a quantity. The operation is an instruction to do something: $C_2{}^z$ says "rotate 180° around the z-axis"; σ_{xz} says "reflect every point in the molecule through the vertical plane *xz*"; etc. The operation requires us to operate on an operand, which is written to the right of the operation. Thus, when forming a product, the elements are taken from right to left; $C_2{}^z \times \sigma_{xz}$ means first to reflect on σ_{xz} and then to rotate through 180° around the z-axis.

Let us now take any point, x, y, z; σ_{xz} gives x, $-y$, z, and the $C_2{}^z$ converts x, $-y$, z to $-x$, y, z. Now, the conversion of x, y, z to $-x$, y, z by $C_2{}^z \times \sigma_{xz}$ is exactly equivalent to a third operation in the group, σ_{yz}, since σ_{yz} also takes the original x, y, z into $-x$, y, z; hence, $C_2{}^z \times \sigma_{xz} = \sigma_{yz}$. If we wish to make use of our atoms in the water molecule, Fig. 5.11*a*, we can focus on a point on the front part of the left H atom: σ_{xz} takes the point to the front of the right H atom; $C_2{}^z$ takes the point to the rear of the left H atom, where we expect to find the point if the original point on the front of the left H atom is reflected through σ_{yz}. The necessity for including the operation I in a point group is now clear. In the present point group the square of every element gives the original: $C_2{}^z \times C_2{}^z = I$, etc.

We indicated above that there are other formal mathematical requirements for a set of elements to constitute a group. Some of these other requirements are as follows:

(a) There is an element I such that $I \cdot X = X \cdot I = X$. Here I is the identity operation, and X is any operation in the set.

(b) The associative law holds, that is, $XYZ = (XY)(Z) = X(YZ)$. (The commutative law, $XY = YX$, does not always hold in groups. In the case of water, however, the four elements do commute, and such a group is said to be *Abelian*.) All point groups that do not have an axis higher than two-fold are Abelian.

(c) For each element, X, there must be another element which is its inverse, $Y = X^{-1}$, such that $YX = XY = I$. The inverse operation was described earlier. In the present example of water, each element is its own inverse; for example, $180°$ rotation around the z-axis in either direction followed by $180°$ rotation in the opposite direction returns the molecule to an orientation identical with the original.

5.17 Classification of Point Groups

In our classification scheme we will start with the point groups[d] that have the least symmetry and finish with the point groups that characterize molecules of the highest symmetry. For such classification it is convenient to use the rotational axis (axes) as the principal criterion.

TYPE I: NO ROTATIONAL AXIS GREATER THAN ONE-FOLD: POINT GROUPS C_1, C_s, AND C_i (see Fig. 5.16).

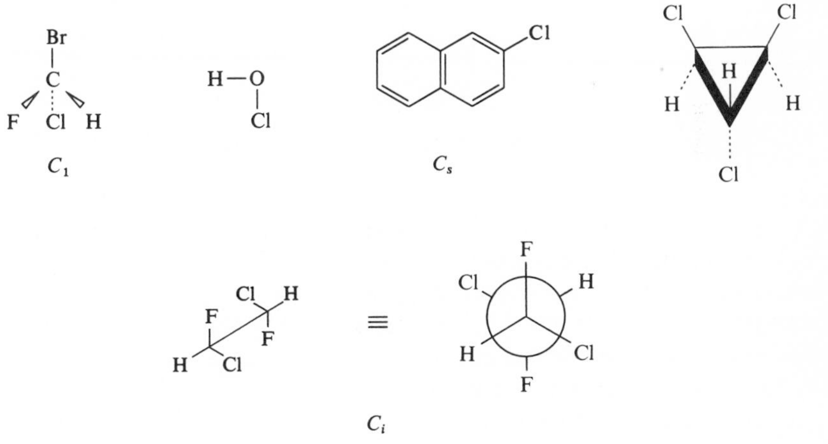

Fig. 5.16   Molecules belonging to Type I: C_1, C_s, C_i.

[d] The symbols used here for the symmetry point groups are called *Schoenflies* symbols, after their inventor. The alternative notation, which is used to describe space symmetry groups, is the *Hermann-Mauguin* notation, named after the two men who independently developed it.

(a) C_1. This point group has no elements of symmetry, and hence all compounds which belong to it are asymmetric

(b) C_s. This group has *only* a single plane of symmetry, σ (which, as we showed earlier, is equivalent to S_1).

(c) C_i. This group has *only* a center of inversion (which is equivalent to S_2).

TYPE II : ONLY ONE AXIS OF ROTATION GREATER THAN ONE-FOLD : POINT GROUPS C_p, S_p, C_{pv}, AND C_{ph}.

(a) C_p. Molecules in these point groups have *only* a single rotational axis, and all such molecules are necessarily dissymmetric and hence optically active. If 1,2-dichloroethane were fixed in the conformation shown, Fig. 5.17, it would belong to this point group.

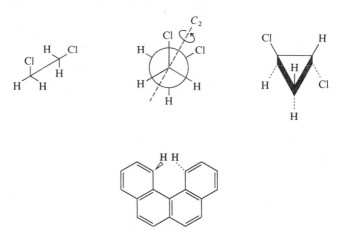

Fig. 5.17   Molecules belonging to point group C_2.

(b) S_p. Only a few molecules are known to belong to these point groups. A molecule with S_4 has also C_2 and one with S_6 has C_3, but if these are the only symmetry elements present, the point groups are S_4 and S_6. Of special interest is the molecule, Fig. 5.7, which belongs to S_4 and which was synthesized explicitly to test for optical activity; the molecule is inactive because the S_4 axis assures that its mirror image is superimposable.

(c) C_{pv}. Molecules belonging to these point groups have the symmetry elements C_p and p vertical planes (σ_v) intersecting in the axis. We have already described in detail the molecules H_2O and phenanthrene, which belong to C_{2v}. Molecules in the point groups C_{pv} are very common; some are shown in Fig. 5.18. A special case is $p = \infty$, which is the point group of all linear molecules lacking a center of symmetry, for example, heteronuclear diatomic molecules.

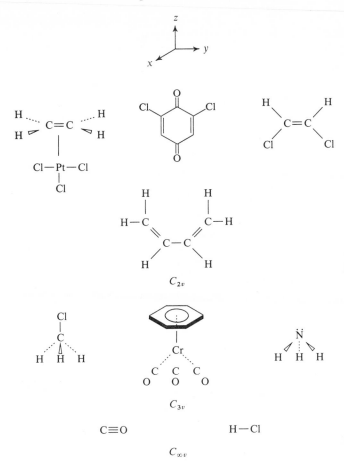

Fig. 5.18   Molecules belonging to C_{pv}.

(d) C_{ph}. Molecules belonging to such point groups have the symmetry element C_p, and at right angles to this vertical axis a mirror or horizontal plane σ_h. Since $C_p \times \sigma_h = S_p$, these molecules also have S_p. When p is an even number, the presence of σ_h implies i. Some examples are shown in Fig. 5.19.

TYPE III: ONE p-FOLD AXIS AND p TWO-FOLD AXES PERPENDICULAR TO THE PRINCIPAL AXIS: POINT GROUPS D_p, D_{ph}, AND D_{pd}.

(a) D_p. If, to the principal axis C_p, are added p two-fold axes as the only symmetry elements, the molecule belongs to the point group D_p. This is not a common point group, but D_2 is encountered more frequently than other D_p point groups. Molecules belonging to D_2 normally have two equivalent

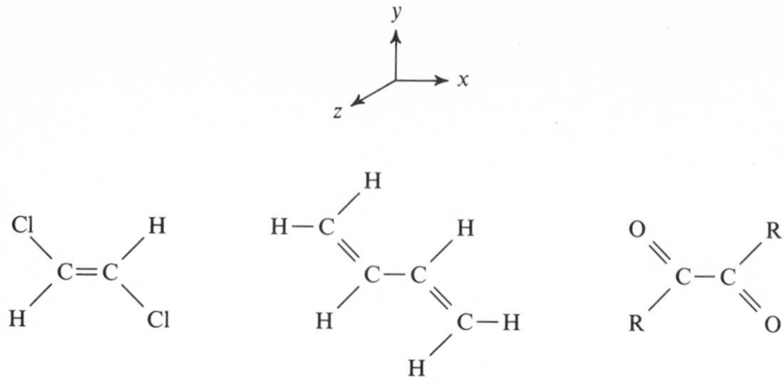

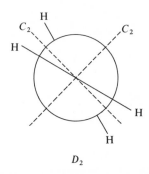

Fig. 5.19 Molecules belonging to C_{2h}.

halves twisted with respect to each other at an angle greater than zero and less than 90°. Thus, twisted ethylene and twisted biphenyl belong to D_2. The Newman projection of twisted ethylene is shown in Fig. 5.20. The vertical or z-axis is the C-C bond axis, and the two two-fold axes are in a plane perpendicular to this axis and are perpendicular to each other. They are shown as broken lines, Fig. 5.20.

Fig. 5.20 Twisted ethylene $(0° < \theta < 90°)$, D_2.

(b) D_{ph}. If σ_h is added to D_p, the molecules having C_p, p two-fold axes, and a mirror plane perpendicular to the principal axis (C_p) belong to D_{ph}; these symmetry elements imply, in addition, p vertical planes, $p\sigma_v$. Many molecules belong to this point group, especially to D_{2h}. In D_{2h} we find one two-fold axis and perpendicular to it a mirror plane which also contains two two-fold axes. In D_{2h} all three axes are C_2, so no one axis stands out as the principal axis. The vertical or z-axis is arbitrarily chosen as the one which passes through the largest number of atoms. Examples of D_{ph} molecules are shown in Fig. 5.21. Thus, benzene has a C_6 axis (the vertical axis), and in the

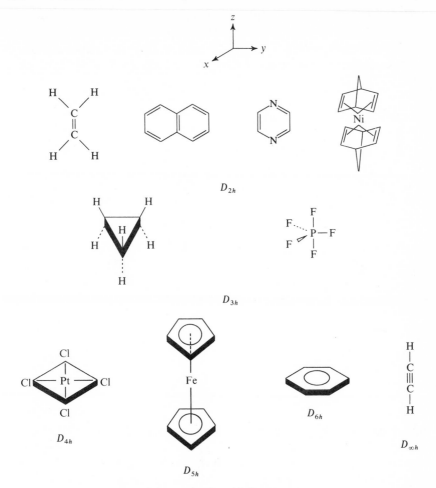

Fig. 5.21 Molecules with D_{ph} symmetry.

symmetry plane, which is the molecular plane and is at right angles to the C_6 axis, there are six two-fold axes: three which pass through opposite C atoms and three which bisect opposite bonds. If p is even, i is present. A special case is $p = \infty$, giving $D_{\infty h}$, the point group to which all linear molecules with a center of symmetry belong: homonuclear diatomics, carbon dioxide, acetylene, etc.

(c) D_{pd}. Molecules in these point groups, in addition to having the axes defining D_p, have p diagonal planes, σ_d, which bisect the angles between successive two-fold axes. Because of the σ_d and the C_2 axes, there is also an S_{2p}, and, if p is odd, i.

In D_{2d} there are normally two equivalent halves of a molecule twisted exactly 90° with respect to one another. The spiran shown in Fig. 5.22a, allene (5.22b), and biphenyl twisted exactly 90° (5.22c) are examples. Actually, in these molecules it is usually easier to recognize the planes of symmetry

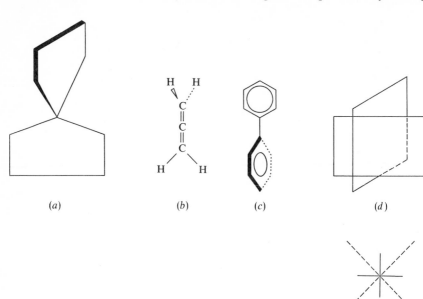

(a)   (b)   (c)   (d)

Fig. 5.22   Molecules belonging to D_{2d}.   (e)

than the C_2 axes perpendicular to the vertical axes. Each of the molecules in Fig. 5.22 has two planes of symmetry, one in the plane of the paper and another perpendicular to the plane of the paper, Fig. 5.22d. The principal C_2 axis coincides with the intersection of these two planes. The two C_2 axes, which together with the principal C_2 axis define D_2, are in a plane horizontal to the major axis and are perpendicular to each other. These two two-fold axes bisect the angles between the planes of symmetry; they are difficult to see without a model but are shown as broken lines in Fig. 5.22e, a view looking down on 5.22d. Now, since it is the C_2 axes which are the frame of reference, the planes are considered to bisect the angles between the axes, and the C_2 are the coordinate axes. Thus, the planes of symmetry become diagonal with reference to the C_2 axes, and hence the designation D_{2d}, where d stands for diagonal planes of symmetry.

Completely staggered ethane, Fig. 5.23a, belongs to D_{3d}. Here the principal axis is C_3 and includes the C-C bond. At right angles to this axis are three C_2 axes, shown as broken lines in the Newman projection formula. The three planes of symmetry, σ_d, bisect the angles between successive two-fold axes. Chair cyclohexane, Fig. 5.23b, has a vertical C_3 axis, three C_2 axes at right angles which bisect opposite C-C bonds, and three σ_d which pass through opposite C atoms.

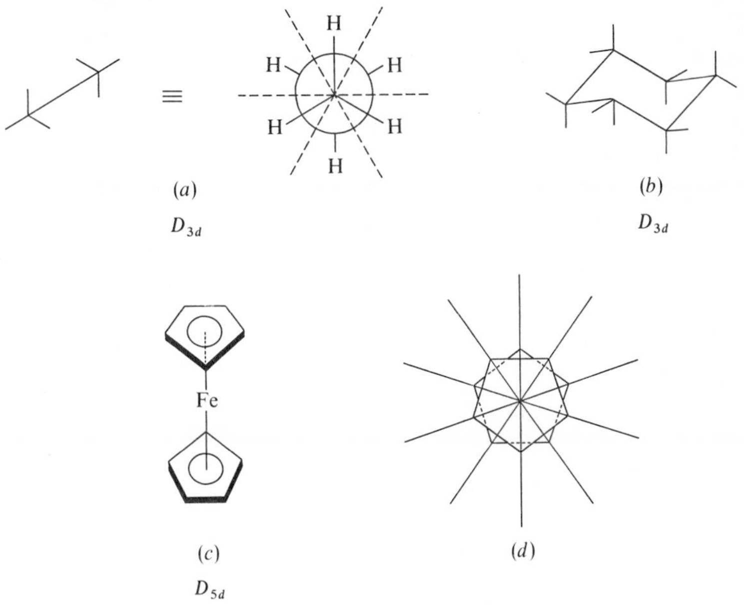

(a)

D_{3d}

(b)

D_{3d}

(c)

D_{5d}

(d)

Fig. 5.23   Molecules with D_{pd} symmetry.

Ferrocene, in the staggered conformation, Fig. 5.23c, belongs to point group D_{5d}, characterized by one five-fold principal and vertical axis, five two-fold axes at right angles to it, and five diagonal planes of symmetry. The view looking down on the molecule is shown in Fig. 5.23d. The heavy lines indicate the five σ_d; the five two-fold axes bisect the angles between these planes. This view is particularly favorable for recognizing S_{10}, which must be present because in D_{5d} there must be S_{2p}.

TYPE IV: MORE THAN ONE AXIS HIGHER THAN TWO-FOLD: POINT GROUPS T_d AND O_h AND ALSO T, T_h, O, I, I_h, K, AND K_h.

(a) T_d. Tetrahedral molecules with identical atoms or groups of atoms around a central atom belong to this point group. Some common molecules are methane, carbon tetrachloride, and nickel tetracarbonyl. The symmetry properties of point group T_d are more readily recognized when the molecule is inscribed in a cube such as that shown for methane in Fig. 5.4. The symmetry elements present are $4C_3$, $3C_2$ (which coincide with $3S_4$), and 6σ. The axis through a corner and opposite face is a C_3 axis, and since there are four corners there are $4C_3$. These $4C_3$ symmetry axes permit $8C_3$ operations, since in the case of a C_3 axis the clockwise and counterclockwise rotations give equivalent but not identical orientations. The $3C_2$ symmetry axes and the $3S_4$ axes coincident with them give rise to $3C_2$ and $6S_4$ operations. The axes through the midpoints of opposite edges are the C_2 axes, and since there are six edges there are $3C_2$. Each edge lies in one mirror plane, and since there are six edges there are 6σ. Organic molecules of the high symmetry required for T_d are not common (adamantane is one), but there are many examples of inorganic molecules (especially of silicon and tetracoordinated aluminum) with this symmetry.

(b) O_h. All symmetrical octahedral compounds belong to this point group, as do the cube and the octahedron. Octahedral molecules are fairly common in inorganic chemistry; some examples are shown in Fig. 5.24. The symmetry

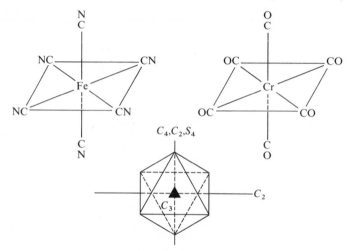

Fig. 5.24   Molecules with O_h symmetry.

elements which characterize this point group and which are perhaps most readily recognized in the octahedron itself, Fig. 5.24, are $6C_4$ (operations $3C_4$ and $3C_4'$), $8C_3$ (operations $4C_3$ and $4C_3'$), $9C_2$ (of which three are coincident with C_4), i, and 9σ. The C_4 axes pass through the opposite corners of the octahedron (six corners, hence $3C_4$ axes). The C_3 axes pass through pairs of opposite faces and the center of the octahedron (eight faces, hence $4C_3$ axes). The C_2 axis bisects pairs of opposite edges (twelve edges, hence $6C_2$; the three C_4 axes are also C_2 axes). There are two classes of mirror planes —one class goes through the centers of four faces and two edges (one set of six σ_d)—and the other class consists of three σ_h, each of which is defined by four edges, for a total of nine σ. The other point groups belonging to type IV are of less importance because they represent only a few molecules. A boron compound with the symmetry of an icosahedron (I_h) has been reported. Such a geometry involves 20 faces that are equilateral triangles, and the point group I_h, to which it belongs, has 120 symmetry operations.

(c) K_h. This point group characterizes the sphere, the geometry possessed by all free atoms. All possible symmetry elements belong to this point group.

5.18   Procedures for Classification into Point Groups

We shall now describe a quick method for the classification of molecules into point groups on the basis of their symmetry properties. Look for the highest rotational axis.

(a) If this axis is C_∞, that is, if the molecule is linear, see whether it has a center of symmetry i.

(a_1). If it has a center, it also has a plane σ_h and belongs to $D_{\infty h}$.

(a_2). If it has *no* center of symmetry, it belongs to $C_{\infty v}$.

(b) If the highest axis is three-, four-, or five-fold, look for other axes of the same order; if there are additional axes of the same order, there are three possibilities.

(b_1). Six five-fold axes. If the molecule also has fifteen planes, it belongs to I_h; otherwise, to I. Both are very rare.

(b_2). Three four-fold axes. If the molecule also has nine planes of symmetry, it belongs to O_h; otherwise, to O.

(b_3). Four three-fold axes, but no four- or five-fold axes. If the molecule has neither a center of symmetry nor any planes of symmetry, it belongs to T; if it has planes but no center, to T_d; and if it has a center, to T_h.

(c) If there is only one axis with $p > 2$, or if the highest axis is two-fold, look for p two-fold axes at right angles to the highest axis. If these exist, look for planes of symmetry.

(c_1). No planes of symmetry: point group D_p.

(c_2). p vertical planes (relative to the C_p axis) but no horizontal plane: point group D_{pd}.

(c_3). p vertical planes *and* a horizontal plane: point group D_{ph}.

(c_4). If the axis of highest order is two-fold, and there are two more two-fold axes, the distinction between D_{2d} and D_{2h} is more difficult. D_{2h} has a center of symmetry; D_{2d} does not.

(d) If only a single p-fold axis exists, check for an S_{2p}.

(d_1). If an S_{2p} exists, the point group is S_{2p}.

(d_2). If no S_{2p} exists, check for planes.

(d_{2a}). No planes: point group C_p.

(d_{2b}). p vertical planes but no horizontal planes: point group C_{pv}.

(d_{2c}). A horizontal plane but no vertical planes: point group C_{ph}.

(*e*) If there are no axes at all, check for a plane and a center.

(*e*₁). One plane: point group C_s.

(*e*₂). A center: point group C_i.

(*e*₃). No element of symmetry: C_1.

Thus the system requires finding the axis of highest order, finding multiple axes of this order, and finding additional two-fold axes as prime steps. No possibilities other than those listed exist, so any finding not in agreement with one of the preceding cases is, of necessity, in error. Note that the axes in D_{pd} are notoriously difficult to find, since they bisect the angles between the planes. Point group D_{2d}, the most common of this type, is readily recognized because there are always two equivalent halves of the molecule twisted exactly 90° with respect to each other.

5.19   Symmetric and Antisymmetric Behavior

In preceding sections we have discussed the symmetry properties of molecules in terms of their geometry and have analyzed the dipole moments of molecules. The geometry and the dipole moment are static properties of the molecule. In order to analyze the dynamic properties of molecules, we must consider whether the particular property is symmetric (unchanged) or antisymmetric under the symmetry operations appropriate to the molecule. All properties must be (or must be decomposable into components which are) either *symmetric* or *antisymmetric* with respect to each symmetry operation that can be performed.[e] In order to determine whether a property is symmetric or antisymmetric, we must have a clear understanding of these terms. We will first examine motions, specifically translations, of objects or molecules.

If we stand in front of a mirror and throw a ball up in the air parallel with the mirror, we see that, when the ball moves upward, so does its reflection in the mirror, and with the same speed. When the ball reaches its apogee and starts to descend, the image likewise changes its direction and descends with the same speed as the ball. The up and down motion of the ball (molecule) is said to be *symmetric* with respect to reflection in the mirror parallel to the motion because the actual motion and its reflection always travel in the same direction with the same speed (or magnitude).

Now let us throw the ball directly at the mirror and analyze this motion, which is perpendicular to the mirror. The reflection or image travels at the same speed as the ball, but the ball is moving toward the mirror and away from the thrower, whereas the reflection is moving in the opposite direction,

---

[e] Except for the properties of molecules with C_p or S_p with $p > 2$ (i.e., molecules belonging to degenerate point groups), which are treated in Section 5.22.

so that the ball and its image collide at the mirror. The motion of the reflection or image has the same magnitude but is now opposite in direction or sign to the motion of the real object (ball). Thus, motion (or translation) perpendicular to the mirror is said to be *antisymmetric* with respect to reflection in the mirror. If we place these motions in a coordinate system with the z coordinate vertical and in the plane of the paper, the x coordinate perpendicular to the paper, and the y coordinate horizontal and in the plane of the paper, we see that, with respect to reflection in a vertical plane, translational motion in the z and y directions, which are parallel with the mirror plane, is symmetric, whereas translational motion in the x direction, perpendicular to the mirror plane, is antisymmetric.

When a molecule has a center of symmetry and we wish to discuss the symmetry behavior of some dynamic property of the molecule with respect to the center, we use the special terms *gerade* (German for even) and *ungerade* (German for odd) for symmetric and antisymmetric behavior, respectively (cf. Section 3.4). With respect to motion in either the x, y, or z direction, reflection through the center of symmetry always reverses the direction of the motion, and hence such motion is always ungerade. Or let us consider a slightly more complicated motion such as a vibration of the hydrogen molecule. In this vibration, when the two H atoms are simultaneously moving away from the center, Fig. 5.25a, the motion of the one on the right can be

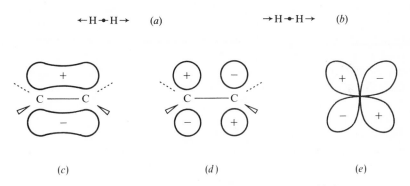

Fig. 5.25 Gerade and ungerade motion and g and u orbitals.

described as an arrow with the arrowhead at the far right end. The motion of the H atom on the left can be similarly described with the arrowhead on the far left end of the arrow. Now, if we reflect the arrowhead on the right through the center and continue an equal distance, we encounter the arrowhead on the left. Such vibrational motion is thus gerade or g. The subsequent in-phase motion of the two H atoms, in which they are moving toward each other, would reverse the two arrows, but the motion is still g.

Consider now the motion whereby the H atoms are moving simultaneously in the same direction, Fig. 5.25b, a translation. Now, on reflection of the arrowhead on the right through the center and continued an equal distance, we encounter the tail of the arrow on the left. Such translational motion is thus ungerade or u.

The g and u notation can be applied also to the orbitals of molecules with a center of symmetry, but here the concept is somewhat more difficult to describe. Although an orbital (or one-electron wave function) describes the "motion" of the electron quantum mechanically, there is no motion which we can follow. The wave function is just what it says, a function which has a certain value (related to the probability of finding the electron there) at each point in space. It is just the values of this function that behave according to symmetry rules, and antisymmetric behavior means that the function changes sign (positive to negative or vice versa) between symmetry-related points. Thus, the π orbital in ethylene, Fig. 5.25c, is ungerade, π_u, whereas the π^* orbital in Fig. 5.25d is gerade, π_g^*, as are all d atomic orbitals, Fig. 5.25e.

5.20 Character Tables

In studying the properties of a molecule we are frequently interested in the motion of the molecule itself and in the motion of the atoms relative to one another (vibrations, infrared absorption), as well as the motion of the electrons in the molecule (molecular orbital theory and electronic spectra). When we are dealing with a molecule like water which belongs to point group C_{2v} (no degeneracies), all motions of the molecule must be either symmetric or antisymmetric with respect to each of the four symmetry operations of the group. For our purposes here we may agree to characterize symmetric behavior as $+1$ and antisymmetric behavior as -1 and to call the $+1$ and -1 the *character* of the motion with respect to the symmetry operation.

For example, let us examine the behavior of the p_y orbital[f] in water, Fig. 5.26, under some of the symmetry operations of point group C_{2v}.

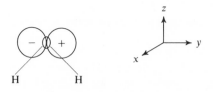

Fig. 5.26   The p_y orbital in water.

[f] When the p orbitals are shown as two circles of opposite sign touching at a point, the angular dependence of the wave function is being illustrated; cf. Section 2.9.

Rotation around the z-axis changes the signs of the two lobes; hence under C_2^z the p_y orbital is antisymmetric and has the character -1. Similarly, reflection on σ_{xz} gives a change in sign and hence is -1, while reflection in σ_{yz} transforms the orbital into itself and hence has the character $+1$. The other symmetry operation in point group C_{2v}, besides C_2^z, σ_{xz}, and σ_{yz}, is the trivial identity operation, I, which, since it leaves the molecule unchanged, obviously has the character $+1$. These four symmetry operations characterize the group, and the total number of symmetry operations in a point group is called the *order* of the point group; in C_{2v} the order is four.

We stated earlier that one of the properties of a group is that the product of any two of its elements is also an element of the group. In C_{2v} we have three nontrivial elements, and since the third is the product of the other two, we need be concerned only with the character of two elements or operations. Since we have two possible independent operations and each can have the character of ± 1, we have a total of four possible combinations. Let us arbitrarily consider C_2^z and σ_{xz} as the two independent operations; both can be $+1$, or they can be $+1$, -1; -1, $+1$; or, finally both -1. We can put this information into a table called a *character table*, and for point group C_{2v} Table 5.1 is such a table. It shows that under the identity operation, I, every property is symmetric, as expected, since the I operation does nothing. The four possible combinations of C_2^z and σ_{xz} are shown, and it is readily verified that in each case the character of σ_{yz} is the product of $C_2^z \times \sigma_{xz}$.

TABLE 5.1
Point Group C_{2v}

C_{2v}	I	C_2^z	σ_v^{xz}	σ_v^{yz}	
A_1	$+1$	$+1$	$+1$	$+1$	z; α_{xx}, α_{yy}, α_{zz}
A_2	$+1$	$+1$	-1	-1	R_z; α_{xy}
B_1	$+1$	-1	$+1$	-1	x, R_y; α_{xz}
B_2	$+1$	-1	-1	$+1$	y, R_x; α_{yz}

5.21   Symmetry Species or Irreducible Representations

We earlier examined the behavior of the p_y orbital in the water molecule, Fig. 5.26, and showed that under the symmetry operations of point group C_{2v}, Table 5.1, the orbital is $+1$ with respect to I, -1 with respect to C_2^z, -1 with respect to σ_{xz}, and $+1$ with respect to σ_{yz}. This $+1$, -1, -1, $+1$ behavior is one of the four possible ways in which every property of the molecule can be described. These four distinct behavior patterns are called *symmetry species* or *irreducible representations*, and their number is equal to the order of the group.

TABLE 5.2
Point Group C_{2h}

C_{2h}	I	$C_2{}^z$	$\sigma_h{}^{xy}$	i	
A_g	$+1$	$+1$	$+1$	$+1$	R_z; α_{xx}, α_{yy}, α_{zz}, α_{xy}
A_u	$+1$	$+1$	-1	-1	z
B_g	$+1$	-1	-1	$+1$	R_x, R_y; α_{xz}, α_{yz}
B_u	$+1$	-1	$+1$	-1	x, y

The symmetry species have, for convenience, been given shorthand symbols, which are shown for C_{2v} in the first column of Table 5.1. All symmetry species that are symmetric with respect to the highest rotational axis are designated by A, and those antisymmetric by B. In the case of point groups like D_{2h}, where there are three two-fold axes and therefore no rotational axis that is of highest order, only the symmetry species that are symmetric with respect to all three C_2 axes are designated A. When more than one species or representation is symmetric with respect to the highest rotational axis, as in C_{2v}, they are distinguished by subscripts (or sometimes by primes), and the totally symmetric species, that is, the one which, as in C_{2v}, is plus with respect to every operation, is always the A_1 species. Subscripting of the B species is more arbitrary. For molecules belonging to C_{2v}, the rules for orientation given earlier are usually unambiguous, and after setting up the coordinate system the B species that is symmetric to σ_{xz} is called B_1.

In all point groups with a center of symmetry, the subscripts u and g are used for all species; the assignment depends on the behavior with respect to the center of inversion, i. Thus, consider the character table of point group C_{2h}, Table 5.2. In this point group there are four operations: I, $C_2{}^z$, σ_h (or σ_{xy}), and i, and again four irreducible representations. The two A species are here designated A_g and A_u; the former is totally symmetric, that is, symmetric with respect to all operations.

Let us examine some property of a molecule in point group C_{2h}. We choose s-trans-butadiene and analyze the lowest-energy π bonding molecular orbital, Fig. 5.27. (The molecule is shown in the plane of the paper, and the $p\pi$ system is perpendicular to the paper.) Under the symmetry operations I, $C_2{}^z$, σ_h, i, this orbital transforms as $+1$, $+1$, -1, -1 and hence belongs to symmetry species A_u.

In a point group like D_{2h}, Table 5.3, where there are six B species, both numbers and the g and u letter designations are required.

Let us examine two possible infrared vibrations of ethylene, a molecule which belongs to point group D_{2h}. An out-of-plane deformation mode, shown in Fig. 5.28a, transforms as species B_{2g}. If the molecule is placed in a plane normal to the plane of the paper, see Fig. 5.28c, C_1 is moving up. C_2 down, H_1

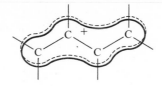

Fig. 5.27 The most bonding molecular orbital in *s-trans*-butadiene, C_{2h}.

TABLE 5.3
Point Group $D_{2h} \equiv V_h$

$D_{2h}{\equiv}V_h$	I	σ_{xy}	σ_{xz}	σ_{yz}	i	$C_2{}^z$	$C_2{}^y$	$C_2{}^x$	
A_g	$+1$	$+1$	$+1$	$+1$	$+1$	$+1$	$+1$	$+1$	$\ldots; \alpha_{xx}, \alpha_{yy}, \alpha_{zz}$
A_u	$+1$	-1	-1	-1	-1	$+1$	$+1$	$+1$	$\ldots$
B_{1g}	$+1$	$+1$	-1	-1	$+1$	$+1$	-1	-1	R_z
B_{1u}	$+1$	-1	$+1$	$+1$	-1	$+1$	-1	-1	$z; \alpha_{xy}$
B_{2g}	$+1$	-1	$+1$	-1	$+1$	-1	$+1$	-1	R_y
B_{2u}	$+1$	$+1$	-1	$+1$	-1	-1	$+1$	-1	$y; \alpha_{xz}$
B_{3g}	$+1$	-1	-1	$+1$	$+1$	-1	-1	$+1$	R_x
B_{3u}	$+1$	$+1$	$+1$	-1	-1	-1	-1	$+1$	$x; \alpha_{yz}$

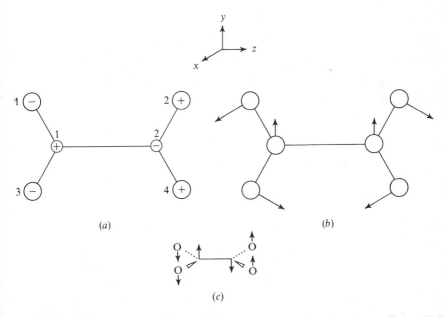

Fig. 5.28 Symmetry species of two vibrational modes in ethylene, D_{2h}: (*a*) B_{2g}; (*b*) B_{2u}; (*c*) B_{2g} at 90° to (*a*).

127

down, H_2 up, H_3 down, and H_4 up. Reflection of these motions through the xy-plane gives motions at each of the atoms which are exactly the reverse of the original motion at that atom, and hence this out-of-plane bending mode is antisymmetric with respect to reflection on σ_{xy} and has the character of -1. Reflection of these motions at the center of symmetry (which is also, of course, the center of gravity and is at the center of the C_1-C_2 bond) converts the upward arrow at C_1 into a downward arrow at C_2, which is exactly the same as the motion of C_2. Similarly, it transforms the down arrows at H_1 and H_3 into up arrows at H_4 and H_2, respectively, which again are the motions of the atoms in this vibration. Thus, the inversion transforms the vibration into itself, the character of i is $+1$, and the vibration is called gerade (g). Testing this bending mode under each of the operations listed in Table 5.3 shows that it transforms as species B_{2g}. The vibrational mode shown in Fig. 5.28b can be similarly demonstrated to belong to species B_{2u}.

5.22 Degenerate Point Groups

The discussion of character tables so far can be applied to point groups C_1, C_s, C_i, C_2, C_{2v}, C_{2h}, D_2, and D_{2h}. None of these point groups involves a symmetry axis C or S greater than two-fold (D_{2d} has an S_4 axis). As soon as an axis greater than two-fold arises, the problems of symmetry species become much more difficult.

Let us use, as an example, the ion $[PtCl_4]^{2-}$ and examine how the three p orbitals of the Pt atom transform under the symmetry operations appropriate to the point group D_{4h}, Table 5.4, to which this square planar ion, Fig. 5.29a, belongs. The x- and y-axes (coordinate systems shown in Fig. 5.29b) are C_2 axes; the C_2 axes which bisect the angles between bonds are designated C_2' axes. The two vertical planes of symmetry that include the

TABLE 5.4
Point Group D_{4h}

D_{4h}	I	$2C_4{}^z$	$C_4{}^2{=}C_2''$	$2C_2$	$2C_2'$	σ_h	$2\sigma_v$	$2\sigma_d$	$2S_4$	$S_2{=}i$	
A_{1g}	$+1$	$+1$	$+1$	$+1$	$+1$	$+1$	$+1$	$+1$	$+1$	$+1$	$\alpha_{xx}+\alpha_{yy};\alpha_{zz}$
A_{1u}	$+1$	$+1$	$+1$	$+1$	$+1$	-1	-1	-1	-1	-1	...
A_{2g}	$+1$	$+1$	$+1$	-1	-1	$+1$	-1	-1	$+1$	$+1$	R_z
A_{2u}	$+1$	$+1$	$+1$	-1	-1	-1	$+1$	$+1$	-1	-1	z
B_{1g}	$+1$	-1	$+1$	$+1$	-1	$+1$	$+1$	-1	-1	$+1$	$\alpha_{xx}-\alpha_{yy}$
B_{1u}	$+1$	-1	$+1$	$+1$	-1	-1	-1	$+1$	$+1$	-1	...
B_{2g}	$+1$	-1	$+1$	-1	$+1$	$+1$	-1	$+1$	-1	$+1$	α_{xy}
B_{2u}	$+1$	-1	$+1$	-1	$+1$	-1	$+1$	-1	$+1$	-1	...
E_g	$+2$	0	-2	0	0	-2	0	0	0	$+2$	$R_x,R_y;\alpha_{xz},\alpha_{yz}$
E_u	$+2$	0	-2	0	0	$+2$	0	0	0	-2	x,y

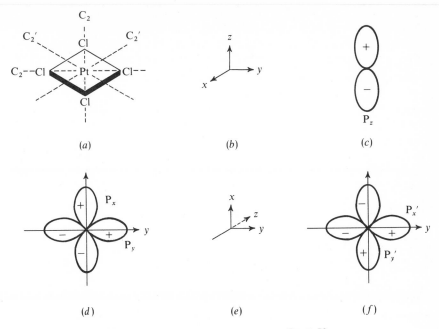

Fig. 5.29   The p orbitals of platinum in $[PtCl_4]^{2-}$.

x- and y-axes, respectively, are σ_v, and those that include the C_2' axes are called σ_d in the character table. The p_z orbital, Fig. 29c, presents no particular problem. If we apply all the symmetry operations on the p_z orbital in turn as listed in the character table, we obtain, starting with I, $+1$, $+1$, $+1$, -1, -1, -1, $+1$, $+1$, -1 and, for i, -1, which tells us that p_z belongs to species A_{2u}.

Now let us examine the behavior of the p_x orbital, Fig. 5.29d. (For convenience, the coordinate system, Fig. 5.29b, has been rotated 90° around the y-axis to give configuration e, which places the xy-plane in the plane of the paper.) If we perform a 90° clockwise rotation, C_4^z, we see that p_x is transformed into p_y and hence p_x is neither symmetric nor antisymmetric under the operation. However, the C_4^z operation simultaneously transforms p_y into $-p_x$; obviously, the transformations are related and the two orbitals transform together into the orbitals shown in Fig. 5.29f.

Although the above transformations can be discussed in general terms in vector notation, we shall develop them by using the familiar orbitals. When we perform a clockwise rotation of 90° on the p_x orbital, we obtain a new orbital which has none of the old one in it and is exactly equal to the original p_y orbital. If we call the new orbital p_x', we may state the fact mathematically:

$$p_x' = 0p_x + 1p_y \qquad (5.1)$$

Similarly, the transformation of the old p_y before the C_4^z into the new p_y' may be written:

$$p_y' = -1p_x + 0p_y \qquad (5.2)$$

These equations are written in matrix notation as:

$$\begin{pmatrix} p_x' \\ p_y' \end{pmatrix} = \begin{pmatrix} 0 & +1 \\ -1 & 0 \end{pmatrix} \begin{pmatrix} p_x \\ p_y \end{pmatrix} \qquad (5.3)$$

The set of numbers in the central parentheses is called the *transformation matrix*. This matrix transforms the old set of p_x, p_y orbitals on a 90° rotation to the new set of p_x', p_y' orbitals. Multiplication of the 2 × 2 matrix

$$\begin{pmatrix} 0 & +1 \\ -1 & 0 \end{pmatrix}$$

by the 2 × 1 matrix or column vector

$$\begin{pmatrix} p_x \\ p_y \end{pmatrix},$$

carried out according to the rules of matrix multiplication, gives equations 5.1 and 5.2. The sum of the numbers appearing in the diagonal from upper left to lower right of the transformation matrix is called the *trace* (in German the *spur*) of the matrix, and the actual number found by the addition is termed the *character* of the transformation matrix. In the present example the character is zero. Accordingly, if we refer to the character table of D_{4h}, we must look for a symmetry species which under the C_4 operation has a character of zero.

Here we can use a short cut to determine the correct symmetry species. We are dealing with a pair of orbitals of equal energy that transform together, which we call a *degenerate* set. A set of two degenerate orbitals always belongs to an E species (a set of three degenerate orbitals to a T species). Hence, our orbitals belong to one of the E species in D_{4h}, and it remains only to specify the behavior under the operation i. A p orbital is always anti-symmetric to a center of symmetry, and hence our p_x, p_y orbitals together belong to species E_u in point group D_{4h}, the characters of which are found in the last row of the character table. As an exercise let us determine the transformation matrix of the p_x, p_y orbitals under operation i to confirm that the character is -2, as shown in the character table. Referring to Fig. 5.29f, we see that under i:

$$p_x' = -1p_x + 0p_y$$
$$p_y' = 0p_x - 1p_y$$

giving the transformation matrix $\begin{pmatrix} -1 & 0 \\ 0 & -1 \end{pmatrix}$, for which the character is -2.

Let us, instead of rotating the p_x, p_y orbitals 90° in the clockwise direction, rotate them in the counterclockwise direction, that is, perform C_4' instead of C_4. The transformations may be written:

$$p_x' = 0p_x - 1p_y \qquad p_y' = 1p_x + 0p_y$$ transformation matrix $\begin{pmatrix} 0 & -1 \\ 1 & 0 \end{pmatrix}$

The character of the transformation matrix is again zero, even though the off-diagonal elements are different from those in the transformation matrix obtained from the C_4 operation. We now see why C_4 and C_4' belong to the same *class*: they have the same character. In the mathematics of group theory, a class includes all the group elements which are *conjugate* to each other. Any two elements X and Y of a group are conjugate to each other if there exists some element of the group, say Z, such that they satisfy the relationship:

$$X = ZYZ^{-1}$$

In our sample C_4 and C_4' are conjugate to each other and belong to the same class; hence there are $2C_4$ because C_4 and C_4' both have the same character.

The C_4 and C_4' or 90° clockwise and counterclockwise rotations give, respectively, the transformation matrices:

$$\begin{pmatrix} 0 & 1 \\ -1 & 0 \end{pmatrix} \text{ and } \begin{pmatrix} 0 & -1 \\ 1 & 0 \end{pmatrix}$$

These numbers correspond to the values of the sine and cosine of $+90°$ and $-90°$, respectively, and indeed that is their origin. We may generalize the 2×2 transformation matrix for any degree of rotation by using the matrices:

$$\begin{pmatrix} \cos \varphi & \sin \varphi \\ -\sin \varphi & \cos \varphi \end{pmatrix} \text{ and } \begin{pmatrix} \cos \varphi & -\sin \varphi \\ \sin \varphi & \cos \varphi \end{pmatrix}$$

and in either case $2 \cos \varphi$ corresponds to the character. Thus, in any doubly degenerate species, E, the character for the appropriate rotation of $\varphi = 2\pi/p$ either clockwise or counterclockwise around the C_p axis is $2 \cos (2\pi/p)$. In these examples we have rotated the orbitals (or, more generally, the vectors), but frequently the coordinate system is rotated instead. The choice affects the sign of the off-diagonal elements but is immaterial, since the diagonal elements which determine the character remain the same.

In the nondegenerate point groups discussed earlier, the transformation matrices are all 1×1 matrices, and so we could immediately assign $+1$ to symmetric and -1 to antisymmetric behavior.

5.23   Symmetry Properties of Translational and Rotational Motion

It will be noted that character tables such as Tables 5.1–5.4 all include in a last column the notations x, y, z and R_x, R_y, R_z. These symbols are assigned to particular symmetry species in each point group. They inform us to which symmetry species the translations (of a molecule, e.g.) along the x-, y-, z-axes belong, and to which symmetry species the rotations R_x, R_y, R_z around the x-, y-, and z-axes, respectively, belong.

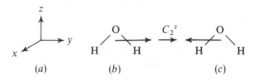

Fig. 5.30   Behavior of translational motion in the y direction on C_2^z.

Let us examine the motion of the water molecule (point group C_{2v}) in the direction of each of the Cartesian coordinate axes. First we will ascertain how a translation along the positive y-axis, Fig. 5.30a, transforms under the symmetry operations of C_{2v}. To simplify the analysis, we draw an arrow from the center of the molecule along the y-axis. Under I the molecule is unchanged, so the motion is $+1$ with respect to I. Under C_2^z we transform b, Fig. 5.30, to c, and we see that now the arrow is pointing in a direction opposite to the original, that is, the vector has changed sign and the motion in the y direction is thus antisymmetric with respect to C_2^z and hence has the character -1. On reflection in the xz-plane, the arrow would again be pointing in the direction opposite to the original, as in C_2^z, and again the character would be -1. Finally, reflection in the yz-plane leaves the arrow unchanged, and hence under this operation the character is $+1$. The behavior under the four operations is thus $+1$, -1, -1, $+1$, which belongs to symmetry species B_2. In the character table for C_{2v}, we see that in the last column of the row of characters in the representation B_2 we have the symbol y, which tells us that motion in the y direction of the water molecule transforms as B_2. Motions in the x and z directions can be similarly analyzed, and we see that such motions belong to the representations B_1 and A_1, respectively. Since the p_x, p_y, and p_z orbitals behave like translations in these directions, this notation also tells us how these orbitals transform.

Similarly, transformations of rotational motion around the three Cartesian coordinates can be assigned to symmetry species. Let us analyze how rotation around the z-axis transforms, Fig. 5.31. To help us analyze the situation, we employ arrows and show them going in and out of the plane of the paper at each hydrogen. The $\oplus$ sign indicates motion upward out of the plane; the $\ominus$ sign, motion downward away from the plane. The molecule is now rotating

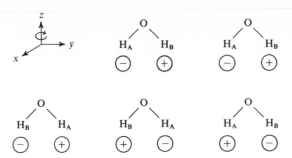

Fig. 5.31 Behavior of rotational motion around the z-axis under the symmetry operations of C_{2v}.

around the z-axis in a clockwise direction so that H_B is moving toward, and H_A away from, the observer. In performing the operations I, C_2^z, σ_{xz}, and σ_{yz}, we can focus on the behavior of the arrows; if they change direction after the operation, the character is -1; if they remain pointing in the same direction as the original, the rotation is $+1$ with respect to the operation.

The behavior of the rotational motions after each of the operations is shown in Fig. 5.31. On the C_2^z operation, we exchange H_A and H_B, but the atom on the left is still going away and the atom on the right is still coming toward the observer; thus, although we exchanged atoms, they are indistinguishable and the direction of motions of the atoms on the left and right is identical after the operation to the original direction. Hence, the C_2^z operation is $+1$ with respect to rotational motion around the z-axis. The reflections in the two planes of symmetry are shown in Fig. 5.31, and the characters belong to symmetry species A_2. If we look at the C_{2v} character table (Table 5.1) we see R_z in the last column of the A_2 species. The species assignments for R_x and R_y are also indicated in the table.

5.24 Some Features of Character Tables and Irreducible Representations

We have mentioned explicitly some properties of character tables and symmetry species and implied others, and it is perhaps useful to summarize in one place some characteristics of these tables. As a typical example we can choose the character table of point group T_d, Table 5.5. The symmetry operations spanning this point group can be appreciated by reference to a T_d molecule such as methane, Fig. 5.15c.

The following generalizations can be verified by reference to the T_d character table.

(a) In each point group, there exist as many irreducible representations (symmetry species) as there are classes of symmetry operations (five for T_d).

TABLE 5.5
Character Table for Point Group T_d

T_d	I	$8C_3$	$6\sigma_d$	$6S_4$	$3S_4{}^2 = 3C_2$		
A_1	$+1$	$+1$	$+1$	$+1$	$+1$		$x^2 + y^2 + z^2$
A_2	$+1$	$+1$	-1	-1	$+1$		
E	$+2$	-1	0	0	$+2$		$2z^2 - x^2 - y^2$; $x^2 - y^2$
T_1	$+3$	0	-1	$+1$	-1	R_x, R_y, R_z	
T_2	$+3$	0	$+1$	-1	-1	x, y, z;	xy, xz, yz

(b) For each irreducible representation, the square of the character in each class, multiplied by the order of the class and summed over all classes, is equal to the order of the group. Thus for T_2 in T_d: $3^2 \times 1 + 1^2 \times 6 + (-1)^2 \times 6 + (-1)^2 \times 3 = 24$. This is in effect a normalization condition.

(c) The sum of the squares of the characters in any class, multiplied by the order of the class, is equal to the total number of symmetry operations. For operation I, this may be stated alternatively as follows: the sum of the squares of the degeneracies of the irreducible representations is equal to the order of the group. For T_d: $1^2(A_1) + 1^2(A_2) + 2^2(E) + 3^2(T_1) + 3^2(T_2) = 24$.[g]

(d) The product obtained by multiplying the character of each class by the order of the class, summed over all classes, is zero for all irreducible representations except the totally symmetric one. Thus for species E in T_d: $2 \times 1 + (-1) \times 8 + 2 \times 3 = 0$.

(e) The sum of the products of the character of any two symmetry species multiplied by the order of the class is always zero. Thus, in T_d, $T_2 \times E = (3 \times 2 \times 1) + [0 \times (-1) \times 8] + (-1 \times 0 \times 6) + (-1 \times 2 \times 3) = 0$. In other words, any two symmetry species are orthogonal.

(f) The orbitals p_x, p_y, p_z transform as the identical symmetry species as the translations along the x-, y-, and z-axes, found in the last column of a character table. In T_d the three p orbitals transform jointly as species t_2.

(g) In the last column of most character tables appears a set of quantities expressed as α with two subscripts which are chosen from among the coordinates x, y, and z. These are the polarizabilities, which are required in applications of selection rules to Raman spectroscopy (cf. Chapter 8). These α_{ij}, where i and j are chosen from among x, y, and z, transform as the direct products of the irreducible representation to which i and j belong. Thus, for example, in point group C_{2v} (Table 5.1), x transforms as b_1, y as b_2, and consequently α_{xy} as a direct product $b_1 \times b_2 = a_2$, α_{xx} as $b_1 \times b_1 = a_1$, etc. In the degenerate point group the problem is more complicated because the direct product of two degenerate species is a reducible representation, that is,

[g] This follows from (b) and the Hermetian nature of the transformation matrices.

the sum of two or more irreducible representations (cf. Chapter 6). The proper α's or linear combinations thereof are contained in the irreducible representations of these direct products.

(h) The symmetry species of the d orbitals follow directly from those of the polarizabilities. The orbitals d_{xy}, d_{xz}, d_{yz}, and $d_{x^2-y^2}$ transform as α_{xy}, α_{xz}, α_{yz}, and $\alpha_{xx} - \alpha_{yy}$, respectively. The orbital d_{z^2} is totally symmetric in all point groups except types T and O, where it should be considered as a linear combination of $d_{z^2-x^2}$ and $d_{z^2-y^2}$ and transforms like $2\alpha_{zz} - \alpha_{xx} - \alpha_{yy}$. Thus in the T_d character table, Table 5.5, the d orbitals transform as t_2 and e.

PROBLEMS

5.1   Assign the following molecules in the configuration shown to their appropriate point group:

(a) $H-C\equiv N$

(b)

(c)

(d)

(e)

(f)

(g)

(h)

(i)

(j)

(k)

(l)

(m)

(n)

(o)

5.2 The normal modes of vibration of the water molecule are shown below. Assign each of them to their appropriate symmetry species in point group C_{2v}.

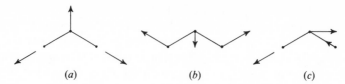

(a) (b) (c)

5.3 Place *trans*-dichloroethylene in the appropriate coordinate system and determine the symmetry species to which each of the rotations R_x, R_y, and R_z and each of the translations x, y, and z belong.

5.4 How many different proton NMR signals will be observed in each of the possible isomers of (*a*) dichlorocyclopropane, (*b*) dichloroallene, and (*c*) 1,4-dichlorocyclohexane. Assign all of the above compounds to their appropriate point group.

5.5 Determine which of the 15 compounds shown in problem 5.1 may have dipole moments.

GENERAL REFERENCES

1. F. A. Cotton, *Chemical Applications of Group Theory*, John Wiley and Sons, New York, 1963.
2. H. H. Jaffé and M. Orchin, *Symmetry in Chemistry*, John Wiley and Sons, New York, 1965.
3. M. Hamermesh, *Group Theory and Its Application to Physical Problems*, Addison-Wesley, Reading, Mass., 1962.
4. Kurt Mislow, *Introduction to Stereochemistry*, W. A. Benjamin, New York, 1965.
5. L. H. Hall, *Group Theory and Symmetry in Chemistry*, McGraw-Hill Book Company, New York, 1969.

6 Symmetry orbitals and bonding in transition-metal complexes

6.1 Group or Symmetry Orbitals

We indicated in the last chapter that the stationary properties of a molecule must remain unchanged under every symmetry operation appropriate to the molecule. Thus, the dipole moment of water is unchanged by any of the operations of point group C_{2v}. Dynamic properties, such as translations and rotations of the water molecule, must be either symmetric or antisymmetric with respect to the operations of point group C_{2v}. Since orbitals describe the wave motion of the electrons of the water molecule, the orbitals of water should also be symmetric or antisymmetric with respect to the C_{2v} operations, that is, each of the orbitals in the basis set must transform as one of the symmetry species in C_{2v}.

In order to classify the orbitals, we place water in the appropriate co-ordinate system, Fig. 6.1a, and we label the two hydrogen atoms A and B, as in Fig. 6.1b. If we now consider the $1s$ atomic orbital of H_A by itself Fig. 6.1c, we see that it does not possess the symmetry properties of the molecule, that is, on reflection in the xz-plane or on the C_2^z operation, the single orbital is neither symmetric nor antisymmetric because there is nothing opposite H_A. We simply cannot treat the individual orbital alone. In such a situation we treat the two AO's of H_A and H_B, Fig. 6.1d, together by taking linear combinations of them. The linear combinations are obtained as usual by addition and subtraction; the sum gives the wave function $(1/\sqrt{2})[1s(H_A) + 1s(H_B)]$, and the difference gives $(1/\sqrt{2})[1s(H_A) - 1s(H_B)]$. These combinations are shown in Figs. 6.1e and 6.1f. These orbitals resemble the MO's of H_2, but in H_2O the two hydrogen atoms are not within bonding distance, and consequently the energy associated with these orbitals is, in the first approximation, the same as the energy of the separate H orbitals. The two orbitals are called symmetry orbitals or group orbitals, and they both now possess the symmetry of the molecule; the orbital formed by addition belongs to species a_1 and the one resulting from subtraction to species b_2 in point group C_{2v}.

137

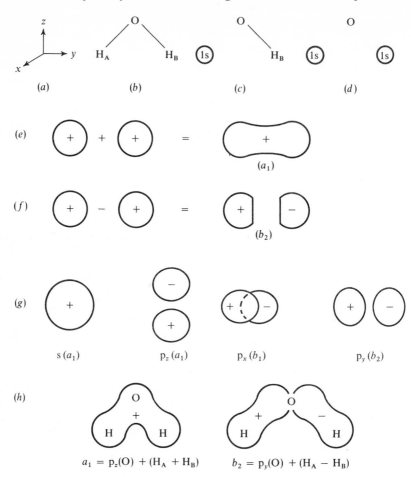

Fig. 6.1   The orbitals of the water molecule.

The oxygen atom has 2s, $2p_x$, $2p_y$, and $2p_z$ orbitals, the so-called basis set of orbitals, belonging to symmetry species a_1, b_1, b_2 and a_1, respectively, Fig. 6.1g. If we wish to hybridize the 2s and 2p orbitals, we can generate two sp hybrids, both a_1, the first, $(s + \lambda p_z)$, having the large positive lobe pointing in the $+z$ direction away from the H atoms, and the other, $(\lambda s - p_z)$, to be used for bonding, pointing in the $-z$ direction toward the H atoms. In order to generate the MO's of the system, we combine the group orbitals of H_2 with the appropriate AO's of the oxygen atom of the same symmetry. The $[1s(H_A) + 1s(H_B)]$ group orbital (a_1) combines with $(\lambda s - p_z)$ of oxygen (a_1), and the $[1s(H_A) - 1s(H_B)]$ group orbital (b_2) combines with the $2p_y$, group orbital b_2 of oxygen. These combinations give, as usual, bonding

(Fig. 6.1h) and antibonding sets of MO's. A total of eight valence electrons is available, two from the H atoms and six from oxygen. Four electrons are placed in the two bonding MO's, and the additional four electrons occupy the two nonbonding orbitals, (s + λp$_z$) and p$_x$, localized on the O atom. It is not necessary to first hybridize and then combine the hybridized AO's to obtain MO's. The MO's can be generated directly from the basis set by combining all AO's of appropriate symmetry that contribute to the particular MO. Thus, the four occupied orbitals of H$_2$O may be written (without normalization):

$$(a_1)\ \psi_1 = \left[1s(H_A) + 1s(H_B)\right] + \lambda 2s(O) - 2p_z(O)$$
$$(b_2)\ \psi_2 = \left[1s(H_A) - 1s(H_B)\right] + 2p_y(O)$$
$$(b_1)\ \psi_3 = 2p_x(O)$$
$$(a_1)\ \psi_4 = 2s(O) + \lambda 2p_z(O)$$

There are also the two antibonding orbitals corresponding to ψ_1 and ψ_2, which have the form:

$$(b_2^*)\ \psi_5^* = \left[1s(H_A) - 1s(H_B)\right] - 2p_y(O)$$
$$(a_1^*)\ \psi_6^* = \left[1s(H_A) + 1s(H_B)\right] - \lambda 2s(O) + 2p_z(O)$$

According to the above description, the two lone pairs of electrons on the O atom are nonequivalent. One lone pair is in an sp orbital concentrated on the p$_z$ axis and mostly on the side away from the H atoms. The other lone pair is in the p$_x$ orbital. The nonequivalence of these lone pairs is in accord with the experimental evidence from spectroscopic and ionization potential data. However, when these lone pairs are used to form H bonds, it appears that they are equivalent and, together with the two localized O-H bonds, describe a slightly distorted tetrahedron. This apparent contradiction with respect to the lone pairs' equivalency has no solution, since electrons cannot be distinguished. Only when the equivalence is tested by some probe can we obtain what appears to be an answer. However, the probe must perturb the system, and we may obtain different answers to the question of equivalency, depending on the nature of the perturbation. The lone-pair orbitals are each doubly occupied. It is possible, without affecting the prediction of any observable property of the molecule, to form linear combinations of these two lone-pair orbitals. The result is then two lone electron pairs that are equivalent and occupy equivalent MO's.

The concept of H$_2$ group orbitals is of importance in understanding problems of bonding and structure in certain other compounds. Consider, for example, the molecule of cyclopentadiene, Fig. 6.2. The methylene hydrogen atoms are in a plane perpendicular to the plane of the ring, and the molecule belongs to point group C_{2v}. The situation with respect to the methylene

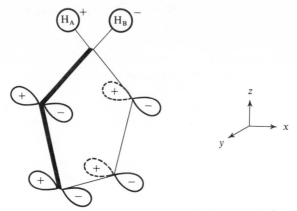

Fig. 6.2   Cyclopentadiene, showing the H_2 group orbital $[1s(H_A) - 1s(H_B)]$ and the $p\pi$ system with which it can combine.

hydrogens is analogous to that of the hydrogens of water. Accordingly, the two hydrogen 1s orbitals are combined into symmetry orbitals $[1s(H_A) + 1s(H_B)]$ and $[1s(H_A) - 1s(H_B)]$. The carbon atom of the methylene group may be considered to use sp^2 orbitals, leaving the $2p_x$ orbital of this C atom with the appropriate symmetry to overlap the π system generated by the $p\pi$ orbitals on the other four C atoms. The $[1s(H_A) + 1s(H_B)]$ orbital combines with an sp^2 on the carbon, but the $[1s(H_A) - 1s(H_B)]$ orbital has the symmetry of the $p\pi$ system of cyclopentadiene and hence can participate in bonding with the $p\pi$ system. In other words, the π system can be partially delocalized over the methylene group because of the partial π character of the CH_2 group.

The participation of the C-H bonding electrons in a $p\pi$ system attached to the carbon atom is, of course, the definition of *hyperconjugation*. That such hyperconjugation is extensive in cyclopentadiene is demonstrated by its ultraviolet spectrum, which shows λ_{max} at about 239 nm, whereas the acyclic 1,3-pentadiene absorbs at about 222 nm. Hyperconjugation is more extensive in the cyclic than in the acyclic system. All homocyclic dienes absorb at longer wavelengths than either acyclic or heteroannular dienes, and in all probability extensive hyperconjugation of the kind described above is largely responsible, although ring strain effects also play a role.

6.2   The Nature of the Bonding in Transition-Metal Complexes: Introduction

Before considering spectra, some of the basic principles regarding the structure and the nature of the bonding in transition-metal complexes will be reviewed.

The simplest model of the transition-metal complexes involves the assumption that the bonds between the metal and the surrounding ligands are completely ionic. In this model the bond energy can be considered as arising from electrostatic interactions between the positive metal ion and negative point charges of the ligands; it is called the crystal field model.

The five d orbitals which, in the isolated gaseous state of a metal or metal ion are degenerate, become split when the ion is placed in a chemical environment, that is, surrounded by ligands. The nature of the splitting caused by the electric field of the ligands was first examined by Bethe (1929) and van Vleck (1932), who found that the extent of the splitting depends on the number of ligands surrounding the metal and the geometry of the resulting complex. In both octahedral and tetrahedral surroundings the d_{xy}, d_{xz}, and d_{yz} orbitals remain equivalent but differ in energy from the $d_{x^2-y^2}$ and d_{z^2} orbitals, which also remain degenerate. In tetragonal surroundings, the d_{xz} and d_{yz} remain equivalent but the other d orbitals are not degenerate. In the electrostatic or crystal field model of the complex, the ligands are treated as point negative charges and the magnitude of the energy difference between the d orbitals can be calculated if the metal-ligand distances, the ligand charge (or dipole moment for a neutral ligand), and the radial part of the wave function for the d orbital are known; this difference in energy between sets of d orbitals is frequently referred to as $10Dq$ or Δ (*vide infra*).

Such an electrostatic model is unrealistic, however. The electrons which are supposed to be entirely in the d orbitals of the metal ion actually spend time in orbitals that belong to the ligand atoms, and vice versa, giving rise to considerable covalency in the bonding. Accordingly, a new model of the bonding in complexes which took into account the covalency of the complex was required. The theory which emerged, starting about 1950, is an amalgamation of molecular orbital and crystal field theory and is called ligand field theory (LFT). It is particularly successful in explaining the electronic spectra of complexes.

6.3   Ligand Field Theory and Octahedral Complexes

The complexing of a metal ion to six identical ligands leads to octahedral geometry and a molecule belonging to point group O_h. The octahedral arrangement minimizes the mutual repulsions of the six negative charges (or dipoles if the ligand is neutral).

In the conventional orientation, Fig. 6.3a, the ligands are placed at opposite ends of each of the three coordinates. The disposition of the five metal d orbitals in O_h is as follows. The d_{z^2} orbital, Fig. 6.3d, (which by convention is the sum of $d_{z^2-x^2}$ and $d_{z^2-y^2}$, and hence really $d_{2z^2-x^2-y^2}$) is strongly directed along the z-axis with a negative doughnut in the xy-plane. The $d_{x^2-y^2}$ orbital, Fig. 6.3b, is directed along the x- and y-axes. These two metal d orbitals are repelled more by the ligands which lie on these two axes than is the other set of orbitals, d_{xy}, d_{xz}, d_{yz} (Fig. 6.3c shows the d_{xy}), which

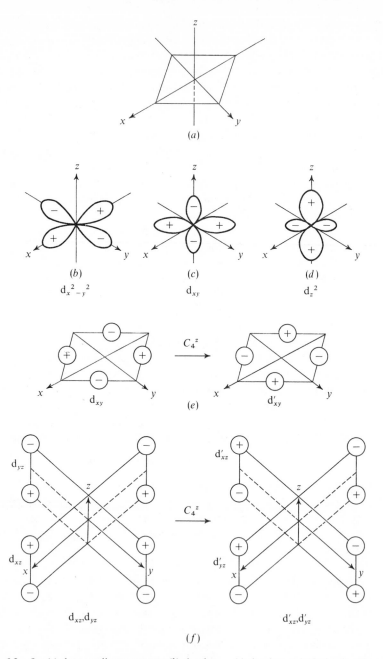

Fig. 6.3 O_h: (a) the coordinate system; (b) the $d_{x^2-y^2}$, (c) the d_{xy}, and (d) the d_{z^2}; the transformation of (e) d_{xy} and (f) d_{yz} under the $C_4{}^z$ operation.

TABLE 6.1

Character Table for Point Group O

O	I	$6C_4$	$3C_2$	$8C_3$	$6C_2'$	
A_1	1	1	1	1	1	$x^2 + y^2 + z^2$
A_2	1	-1	1	1	-1	
E	2	0	2	-1	0	$2z^2 - x^2 - y^2, x^2 - y^2$
T_1	3	1	-1	0	-1	x, y, z
T_2	3	-1	-1	0	1	xy, xz, yz

are directed halfway between the axes. The d_{z^2} and $d_{x^2-y^2}$ transform as one set, and the other three d orbitals as a second set. To determine the symmetry species to which the two sets belong, and to serve other purposes as well, it is more convenient to use point group O (Table 6.1), with exactly half the number of symmetry operations (twelve) and half the number of species (five) as point group O_h, to which the molecule belongs. In point group O there is no i operation but O_h does have i, and so for each of the species in O there is a corresponding g and u species in O_h, thus giving a total of ten symmetry species in O_h. One need only determine the species in point group O and then perform i on the orbital (or other property) to obtain the proper species (g or u) in O_h.

Let us now see how the d_{xy}, d_{xz}, d_{yz} set transforms under the various operations appropriate to point group O. Under the identity operation, all three orbitals transform into themselves and each new orbital after the operation contains all of its old self and none of the others:

$$d'_{xy} = 1d_{xy} + 0d_{xz} + 0d_{yz}$$
$$d'_{xz} = 0d_{xy} + 1d_{xz} + 0d_{yz}$$
$$d'_{yz} = 0d_{xy} + 0d_{xz} + 1d_{yz}$$

The transformation matrix for the identity operation is thus:

$$\begin{pmatrix} 1 & 0 & 0 \\ 0 & 1 & 0 \\ 0 & 0 & 1 \end{pmatrix}$$

If it is assumed that these three orbitals belong to a single irreducible representation, the character is $+3$; this assumption will be verified below. If we look at the character table for point group O, Table 6.1, under the operation I, we see that the d_{xy}, d_{xz}, d_{yz} orbitals must belong to either T_1 or T_2. These two species differ under the C_4^z and C_2' operations (the $6C_2'$ axes are the axes bisecting opposite edges), and so let us now look at the character of the transformation matrix under the C_4^z operation. Figures 6.3e and f show that, under C_4^z, $d'_{xy} = -1d_{xy}$, $d'_{xz} = -1d_{yz}$, and $d'_{yz} = +1d_{xz}$. The complete transformation is thus:

$$d'_{xy} = -1d_{xy} + 0d_{xz} + 0d_{yz}$$
$$d'_{xz} = 0d_{xy} + 0d_{xz} - 1d_{yz}$$
$$d'_{yz} = 0d_{xy} + 1d_{xz} + 0d_{yz}$$

This transformation matrix shows that d_{xz} and d_{yz} transform into one another under the C_4^z transformation, and hence must be degenerate and belong to one irreducible representation. If we also write the transformation matrix for C_4^x, we see that now d_{xz} and d_{xy} transform into one another. Hence all three orbitals are degenerate and must belong to a single irreducible representation, as assumed above. The character of the above transformation matrix is -1, which tells us (Table 6.1) that the orbitals transform as species T_2. Since all d orbitals are g, the correct species in O_h is T_{2g}.

The correct assignment of the d_{z^2} and $d_{x^2-y^2}$ set is readily made if one recognizes that the two orbitals are degenerate, but the demonstration that they belong to a single set, that is, transform together, is not so straightforward. Performing the C_4^z operation transforms d_{z^2} into itself and $d_{x^2-y^2}$ into minus itself but shows nothing about their interrelation. However, operating on d_{z^2} with C_4^x, Fig. 6.4a, gives a new orbital having the same form

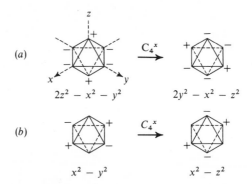

Fig. 6.4   The transformation under C_4^x of (a) the $d_{2z^2-x^2-y^2}$ and (b) the $d_{x^2-y^2}$ orbitals.

but directed in the y direction, that is, d_{y^2}. Similarly, operating on $d_{x^2-y^2}$ with C_4^x gives $d_{x^2-z^2}$. Neither of these new orbitals *appears* to be one of the original set or a linear combination of the two. We shall now demonstrate, however, that the new orbitals are indeed linear combinations of the old ones and proceed to develop the transformation matrix.

If the d_{z^2} and $d_{x^2-y^2}$ orbitals are indeed degenerate, they must be orthonormal functions. We recall that d_{z^2} is really the linear combination $(d_{z^2-x^2} + d_{z^2-y^2}) = d_{2z^2-x^2-y^2}$. However, this statement is not quite complete; the normalization constant is missing. In order to understand why the normalization is necessary and how it may be evaluated, let us start by examining the other degenerate orbital, $d_{x^2-y^2}$. This orbital, like all orbitals with an angular dependence, is a product of a radial function, $R(r)$, and an angular function, $Y_{\theta\phi}$. The angular function can also be expressed in Cartesian coordinates as $(1/\sqrt{2})(x^2 - y^2)$, and when we write $d_{x^2-y^2}$ we imply a function of the form $(x^2 - y^2)R(r)$, where the normalizing factor $1/\sqrt{2}$ is absorbed in $R(r)$. When we expand d_{z^2} as a linear combination we imply a function $(2z^2 - x^2 - y^2)R(r)$, and for consistency we must consider the $R(r)$ to contain the same normalization factor of $1/\sqrt{2}$. However, the function $(2z^2 - x^2 - y^2)R(r)$ is no longer normalized. If it is to be normalized, the sum of the squares of the coefficients of the squares of the Cartesian coordinates for the degenerate $d_{2z^2-x^2-y^2}$ and $d_{x^2-y^2}$ orbitals must be equal. This equality is achieved by multiplying $(2z^2 - x^2 - y^2)$ by $1/\sqrt{3}$, viz.:

$$1^2(x^2) + 1^2(y^2) = N^2[2^2(z^2) - 1^2(x^2) - 1^2(y^2)]$$
$$2 = 6N^2$$

whence $N = 1/\sqrt{3}$. The orthonormal degenerate set is thus $d_{x^2-y^2}$ and $1/\sqrt{3}(d_{2z^2-x^2-y^2})$.

We now return to our problem of testing whether it is possible to express the new orbitals, which are, after $C_4{}^x$, $d_{y^2} = 1/\sqrt{3}(d_{y^2-x^2+y^2-z^2})$ and $d_{x^2-z^2}$, as linear combinations of the original d_{z^2} and $d_{x^2-y^2}$ orbitals. We need to ask only if we can find constants c_{11}, c_{12}, c_{21}, and c_{22} in the following equation:

$$\frac{1}{\sqrt{3}}(2y^2 - x^2 - z^2) = \frac{c_{11}}{\sqrt{3}}(2z^2 - x^2 - y^2) + c_{12}(x^2 - y^2) \quad (6.1)$$

$$x^2 - z^2 = \frac{c_{21}}{\sqrt{3}}(2z^2 - x^2 - y^2) + c_{22}(x^2 - y^2) \quad (6.2)$$

Because the coefficients of x^2, y^2, and z^2 separately must be equal on both sides of each equation, we obtain three equations, each in two unknowns, for equation 6.1 and also for equation 6.2:

$$\left.\begin{array}{lll} \text{For } y^2 & 2/\sqrt{3} = -c_{11}/\sqrt{3} - c_{12} \\ \text{For } x^2 & -1/\sqrt{3} = -c_{11}/\sqrt{3} + c_{12} \\ \text{For } z^2 & -1/\sqrt{3} = 2c_{11}/\sqrt{3} \end{array}\right\} \text{From eq. 6.1}$$

$$\left.\begin{array}{lll} \text{For } x^2 & 1 = -c_{21}/\sqrt{3} + c_{22} \\ \text{For } y^2 & 0 = -c_{21}/\sqrt{3} - c_{22} \\ \text{For } z^2 & -1 = 2c_{21}/\sqrt{3} \end{array}\right\} \text{From eq. 6.2}$$

In both sets of equations, values for the c's can be obtained from any two of the three equations. When these are substituted into the third equation of the set, an identity results, which shows that equations 6.1 and 6.2 are valid, and that d_{y^2} and $d_{x^2-z^2}$, the rotated orbitals, are linear combinations of d_{z^2} and $d_{x^2-y^2}$, the orbitals before rotation by C_4^x. The coefficients c_{ij} found from the treatment just described are the elements of the transformation matrix which transform the old into the new orbitals:

$$\begin{pmatrix} d_{y^2} \\ \\ d_{x^2-z^2} \end{pmatrix} = \begin{pmatrix} (d_{z^2})' \\ \\ (d_{x^2-y^2})' \end{pmatrix} = \begin{pmatrix} -\frac{1}{2} & \frac{-\sqrt{3}}{2} \\ \\ -\frac{\sqrt{3}}{2} & \frac{1}{2} \end{pmatrix} \begin{pmatrix} d_{z^2} \\ \\ d_{x^2-y^2} \end{pmatrix}$$

Multiplying the matrix by the rules of matrix multiplication verifies the transformation:

$$(z^2)' = -\frac{1}{2}\left(\frac{1}{\sqrt{3}}\right)(2z^2 - x^2 - y^2) - \frac{3}{2\sqrt{3}}(x^2 - y^2)$$

$$= \frac{1}{\sqrt{3}}\left(-z^2 + \frac{x^2}{2} + \frac{y^2}{2} - \frac{3x^2}{2} + \frac{3y^2}{2}\right)$$

$$= \frac{1}{\sqrt{3}}(2y^2 - x^2 - z^2) \equiv y^2$$

$$(x^2 - y^2)' = \frac{-\sqrt{3}}{2}\left(\frac{1}{\sqrt{3}}\right)(2z^2 - x^2 - y^2) + \tfrac{1}{2}(x^2 - y^2)$$

$$= -z^2 + \frac{x^2}{2} + \frac{y^2}{2} + \frac{x^2}{2} - \frac{y^2}{2}$$

$$= x^2 - z^2$$

The transformation matrix has a character of zero, and thus we have shown that the d_{z^2} and $d_{x^2-y^2}$ are a degenerate set which belongs to species E of point group O (Table 6.1) and to E_g of O_h.

The splitting pattern for the d orbitals in the octahedral field is shown in Fig. 6.5. The energy separation between the sets of orbitals is of great interest; for example, such energy differences are related to the d → d electronic transition in the transition-metal ions. This energy difference depends on the magnitude of the charge on the ligands, q, and the polarizability of the central metal ion, D. It is generally not possible to separate these two quantities, and hence the product Dq is treated as a single, adjustable,

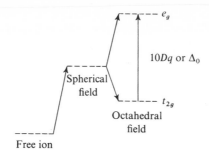

Fig. 6.5   d Orbital splitting in O_h.

empirical parameter. As mentioned in Section 6.2, the energy difference is thus indicated as $10Dq$; it is also sometimes referred to as Δ_0, where the subscript refers to the energy splitting in the octahedral field. We will treat this subject in greater detail when we discuss the electronic spectra of complexes in Chapter 7.

6.4   Symmetry Orbitals of the Ligands in Octahedral Complexes

Let us use as a model complex the ion $[\text{Co(NH}_3)_6]^{3+}$, of O_h symmetry. Our first problem is to classify the 3d, 4s, and 4p orbitals of the metal into their appropriate symmetry species. We have already done this for the d orbitals. The three p orbitals belong to a set in point group O and hence must belong to either species T_1 or T_2. These differ in the character under C_4 and C_2'. Under $C_4{}^z$, p_x and p_y are transformed into each other and hence contribute zero to the character of the matrix, but p_z transforms into itself; hence the character of the 3 × 3 transformation matrix is 1 and the symmetry species is T_1. Since all p orbitals are ungerade, the correct species in O_h is T_{1u}. The s orbital is symmetric with respect to all operations and hence is A_1 in O, and A_{1g} in O_h. The symmetry species of the metal orbitals are listed in Table 6.2.

Our next problem is to generate out of the six p orbitals (or sp^3 hybrids), one each on each nitrogen atom, six symmetry orbitals that can combine with the metal orbitals. Since each of the N atoms is at the corner of the octahedron, Fig. 6.6a, the six sp^3 ligand orbitals lie on the bond axes, Fig. 6.6b. These will generate σ-type molecular orbitals, and we can call the atomic orbitals of the ligands σ orbitals and subscript the σ with a number corresponding to the number of the ligand.

The symmetry or group orbitals that combine with metal orbitals can be determined by inspection—the preferable method—or by means of a recipe that is useful for this purpose and other ones. The coordinate system which is placed on each ligand is such that each z-axis is coincident with one

TABLE 6.2
Classification of Metal Valence Orbitals and Ligand Group Orbitals in O and O_h

Orbitals		Symmetry	Species
Ligand	Metal	O	O_h
$\dfrac{1}{\sqrt{6}}(\sigma_1 + \sigma_2 + \sigma_3 + \sigma_4 + \sigma_5 + \sigma_6)$	4s	a_1	a_{1g}
$\dfrac{1}{\sqrt{2}}(\sigma_1 - \sigma_3)$ $\dfrac{1}{\sqrt{2}}(\sigma_2 - \sigma_4)$ $\dfrac{1}{\sqrt{2}}(\sigma_5 - \sigma_6)$	$4p_x, 4p_y, 4p_z$	t_1	t_{1u}
$\dfrac{1}{\sqrt{12}}(-\sigma_1 - \sigma_2 - \sigma_3 - \sigma_4 + 2\sigma_5 + 2\sigma_6)$ $\frac{1}{2}(\sigma_1 - \sigma_2 + \sigma_3 - \sigma_4)$	$3d_{z^2}, 3d_{x^2-y^2}$	e	e_g
	$3d_{xy}, 3d_{xz}, 3d_{yz}$	t_2	t_{2g}

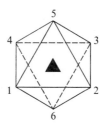

Fig. 6.6a   The O_h complex and location and numbering of ligands.

of the coordinate axes of the complex, the positive z on each ligand pointing toward the metal, Fig. 6.6b.

The inspection method is a straightforward and common-sense method. If we ask what combination of ligand orbitals will combine with, say, the $d_{x^2-y^2}$ of the metal, the answer is clearly the combination $\sigma_1 - \sigma_2 + \sigma_3 - \sigma_4$, Fig. 6.6c, all orbitals contributing equally, and so the normalized group orbital (GO) is:

$$(e_g)\,\text{GO}\,(1) = \tfrac{1}{2}(\sigma_1 - \sigma_2 + \sigma_3 - \sigma_4)$$

The $d_{x^2-y^2}$ is one of the e_g pair; the other is the d_{z^2} orbital, which correctly is $d_{2z^2-x^2-y^2}$, as pointed out earlier. By inspection, the combination of ligand

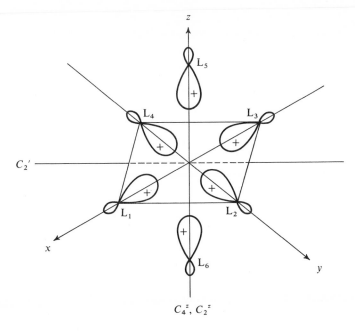

Fig. 6.6b The coordinate axes and the rotational axes of the O_h complex and the ligand σ orbitals.

Fig. 6.6c The combination $(\sigma_1 - \sigma_2 + \sigma_3 - \sigma_4)$ of ligand orbitals to match the metal $d_{x^2-y^2}$ orbitals.

orbitals which will combine with this metal orbital is $-\sigma_1 - \sigma_2 - \sigma_3 - \sigma_4 + 2\sigma_5 + 2\sigma_6$, and normalized this becomes:

$$(e_g)\ GO\ (2) = \frac{1}{\sqrt{12}}(-\sigma_1 - \sigma_2 - \sigma_3 - \sigma_4 + 2\sigma_5 + 2\sigma_6)$$

The coefficient of 2 for σ_5 and σ_6 follows from the matching with the proper form of the d_{z^2} orbital and stems from the orthogonality requirement that in a degenerate set of orbitals the contribution that each AO makes to the total electron density must be the same. (For a systematic orthogonalization procedure, *vide infra*.) Stated more generally, the sum of the squares of the coefficients of all AO's over the group orbitals of a degenerate set must be equal:

$$\sum_i c_{ir}^{2} = \text{constant independent of } r$$

where c_{ir} is the coefficient on atom r in group orbital i and the summation is over all i group orbitals of the degenerate set. The orbitals shown fulfill this requirement. Thus, for example, $c_{11}^{2} + c_{21}^{2} = \frac{1}{4} + \frac{1}{12} = \frac{1}{3}$ and $c_{15}^{2} + c_{25}^{2} = 0 + \frac{4}{12} = \frac{1}{3}$. Our example thus far has shown the two combinations of group orbitals to match the e_g set of $d_{x^2-y^2}$ and d_{z^2}. All other ligand combinations to match metal orbitals can be obtained by inspection as well.

The group orbitals can also be generated from the σ orbitals of the ligands by the following systematic procedure involving the use of a projection operator:

Step 1. Determine the reducible representation. Take each of the six ligand σ orbitals in turn and perform the operations of point group O on it. If the orbital is transformed into itself, call the result $+1$; if it is transformed into minus itself, -1; and if it is transformed into another orbital, 0. In doing these operations under a class, for example, $6C_4$, it is necessary to do the operation under only one C_4, but the same C_4 must be used for all ligand transformations. In Fig. 6.6b are shown the C_4, C_2^z, and C_2' axes that are used, and in Fig. 6.6a the C_3 axis used is indicated by a solid triangle passing through the center of faces 1, 2, 5 and 3, 4, 6. In Table 6.3a, which shows the results of these operations, the number in the first column refers to the orbital on the ligand of the same number. The sums of the vertical columns are shown on the last line; this is a reducible representation because it represents the sum of a particular combination of irreducible representations.

Step 2. Determine the irreducible representations contained in the reducible representations. If the sum of the irreducible representations corresponding to the reducible one is not obvious, the number of times, $n(\Gamma)$, that each particular irreducible representation, Γ, appears in the reducible representation can be calculated from the formula:

$$n(\Gamma) = \frac{1}{g} \sum_{(SO)} g_{(SO)}\chi_\Gamma^{(SO)}\chi_{(RR)}^{(SO)} \tag{6.3}$$

TABLE 6.3

The Reducible Representation (a), A Trial Set of Irreducible Representations (b), and the Correct Set of Irreducible Representations (c).

Orbital	I	C_4^z	C_2^z	C_3	C_2'	
1	1	0	0	0	0	
2	1	0	0	0	0	
3	1	0	0	0	0	(a)
4	1	0	0	0	0	
5	1	1	1	0	0	
6	1	1	1	0	0	
Σ	6	2	2	0	0	

0	I	C_4^z	C_2^z	C_3	C_2'	
T_1	3	1	-1	0	-1	(b)
T_2	3	-1	-1	0	0	
Σ	6	0	-2	0	-1	

0	I	C_4^z	C_2^z	C_3	C_2'	
A_1	1	1	1	1	1	
E	2	0	2	-1	0	(c)
T_1	3	1	-1	0	-1	
Σ	6	2	2	0	0	

where g is the order of the group (the total number of symmetry operations of the group), $g_{(SO)}$ is the order of the class (the number in front of each operation in the character table) to which the symmetry operation (SO) belongs, $\chi_\Gamma^{(SO)}$ is its character in the species Γ, $\chi_{(RR)}^{(SO)}$ is its character in the reducible representation, and the summation extends over all classes of symmetry operations (SO). In the present case, let us calculate the number of E species contained in the reducible representation, Table 6.3c last line. In point group $O, g = 24$ and

$$n(E) = \tfrac{1}{24}[(1 \cdot 2 \cdot 6) + (6 \cdot 0 \cdot 2) + (3 \cdot 2 \cdot 2) + (8 \cdot -1 \cdot 0) + (6 \cdot 0 \cdot 0)] = 1$$

Step 3. Determine into which new orbital one of the ligand σ orbitals transforms under all the operations of the point group. Let us take σ_1 and see

TABLE 6.4
A Step in Generating Group Orbitals, Using σ_1

Operation on σ_1		New Orbital after Operation	Character of			Product: New Orbital × Character		
			A_1	E	T_1	$\cdot A_1$	$\cdot E$	$\cdot T_1$
I		σ_1	$+1$	2	3	σ_1	$2\sigma_1$	$3\sigma_1$
$6C_4$	$C_4^{5,6}$	σ_4				σ_4	0	σ_4
	$C_4'^{(5,6)}$	σ_2				σ_2		σ_2
	$C_4^{1,3}$	σ_1	$+1$	0	1	σ_1		σ_1
	$C_4'^{(1,3)}$	σ_1				σ_1		σ_1
	$C_4^{2,4}$	σ_5				σ_5		σ_5
	$C_4'^{(2,4)}$	σ_6				σ_6		σ_6
$3C_2$	$C_2^{5,6}$	σ_3				σ_3	$2\sigma_3$	$-\sigma_3$
	$C_2^{1,3}$	σ_1	$+1$	2	-1	σ_1	$2\sigma_1$	$-\sigma_1$
	$C_2^{2,4}$	σ_3				σ_3	$2\sigma_3$	$-\sigma_3$
$8C_3$	$C_3^{1,2,5}$	σ_5				σ_5	$-\sigma_5$	0
	$C_3'^{(1,2,5)}$	σ_2				σ_2	$-\sigma_2$	
	$C_3^{1,4,5}$	σ_4				σ_4	$-\sigma_4$	
	$C_3'^{(1,4,5)}$	σ_5	$+1$	-1	0	σ_5	$-\sigma_5$	
	$C_3^{2,3,5}$	σ_4				σ_4	$-\sigma_4$	
	$C_3'^{(2,3,5)}$	σ_6				σ_6	σ_6	
	$C_3^{3,4,5}$	σ_2				σ_2	$-\sigma_2$	
	$C_3'^{(3,4,5)}$	σ_6				σ_6	$-\sigma_6$	
$6C_2'$	$C_2'^{(1,2)}$	σ_2				σ_2	0	$-\sigma_2$
	$C_2'^{(1,4)}$	σ_4				σ_4		$-\sigma_4$
	$C_2'^{(2,5)}$	σ_3	$+1$	0	-1	σ_3		$-\sigma_3$
	$C_2'^{(4,5)}$	σ_3				σ_3		$-\sigma_3$
	$C_2'^{(1,6)}$	σ_6				σ_6		$-\sigma_6$
	$C_2'^{(1,5)}$	σ_5				σ_5		$-\sigma_5$

to which orbitals it transforms, using Fig. 6.6a as a guide. The results of the transformation are shown in the second column of Table 6.4.

Step 4. Multiply the orbitals obtained in Step 3 by the characters of the symmetry species, and sum. In Table 6.4 are shown the results (last three columns) of multiplying the transformed orbital, column 2, by the characters of species A_1, E, and T_1, columns 3, 4, and 5. If we sum the results of the multiplication under A_1, we obtain:

$$GO\ (A_1) = \sigma_1 + \sigma_4 + \sigma_2 + \sigma_1 + \sigma_1 + \sigma_5 + \sigma_6 + \sigma_3 + \sigma_1 + \sigma_3 + \sigma_5$$
$$+\ \sigma_2 + \sigma_4 + \sigma_5 + \sigma_4 + \sigma_6 + \sigma_2 + \sigma_6 + \sigma_2 + \sigma_4 + \sigma_3$$
$$+\ \sigma_3 + \sigma_6 + \sigma_5$$
$$=\ 4\sigma_1 + 4\sigma_2 + 4\sigma_3 + 4\sigma_4 + 4\sigma_5 + 4\sigma_6$$

Step 5. Normalize the group orbital.[a] In the example above we normalize in the usual fashion to obtain the appropriate group orbital:

$$GO\ (A_1) = \frac{1}{\sqrt{6}}(\sigma_1 + \sigma_2 + \sigma_3 + \sigma_4 + \sigma_5 + \sigma_6).$$

We can now repeat steps 4 and 5 to obtain the GO combination of the E species. For one of the e orbitals we can go directly to Table 6.4 and sum the column under $\cdot E$, which gives us after normalization:

$$GO\ (E) = \frac{1}{\sqrt{12}}(2\sigma_1 - \sigma_2 + 2\sigma_3 - \sigma_4 - \sigma_5 - \sigma_6) = \psi_1$$

In order to obtain the other GO of species e we must repeat the entire process, using another ligand orbital. The results analogous to those in Table 6.4 are shown in Table 6.5, where σ_2 is used.

The result of adding the column of the product under E for σ_2 is:

$$\psi_2 = -2\sigma_1 + 4\sigma_2 - 2\sigma_3 + 4\sigma_4 - 2\sigma_5 - 2\sigma_6$$

On normalization this becomes

$$\psi_2 = \frac{1}{\sqrt{12}}(-\sigma_1 + 2\sigma_2 - \sigma_3 + 2\sigma_4 - \sigma_5 - \sigma_6)$$

This second wave function, however, is not orthogonal to ψ_1, and we must now find some way to orthogonalize.

There is a standard mathematical procedure for assuring orthogonality, called the *Gram-Schmidt method* of orthogonalization. Provided that ψ_1 and ψ_2 are independent, that is, are not multiples of each other, a pair of orthogonal functions can be obtained directly by forming linear combinations subject to the orthogonality condition. Our problem is to find a linear com-

---

[a] The procedure of steps 4 and 5 is known as the use of a projection operator for each species of each point group; it consists of successively operating on the operand with each symmetry operation of the point group, multiplying by the character, and summing. Thus,

$$P(a_1, 0) = 1 \cdot I + 1 \cdot \ldots$$

In the present case we operate with this operator on the orbital σ, and obtain the result shown in step 4.

TABLE 6.5
Generation of Orbitals Group/Using σ_2

Operation on σ_2	New Orbital after Operation	Character of E	Character of T_1	Product: New Orbital × Character $\cdot E$	Product: New Orbital × Character $\cdot T_1$
I	σ_2	2	3	$2\sigma_2$	$3\sigma_2$
$6C_4$ $\begin{cases} C_4^{5,6} \\ C_4'^{(5,6)} \\ C_4^{1,3} \\ C_4'^{(1,3)} \\ C_4^{2,4} \\ C_4'^{(2,4)} \end{cases}$	$\begin{matrix} \sigma_1 \\ \sigma_3 \\ \sigma_6 \\ \sigma_5 \\ \sigma_2 \\ \sigma_2 \end{matrix}$	0	1	0	$\begin{matrix} \sigma_1 \\ \sigma_3 \\ \sigma_6 \\ \sigma_5 \\ \sigma_2 \\ \sigma_2 \end{matrix}$
$3C_2$ $\begin{cases} C_2^{5,6} \\ C_2^{1,3} \\ C_2^{2,4} \end{cases}$	$\begin{matrix} \sigma_4 \\ \sigma_4 \\ \sigma_2 \end{matrix}$	2	−1	$\begin{matrix} 2\sigma_4 \\ 2\sigma_4 \\ 2\sigma_2 \end{matrix}$	$\begin{matrix} -\sigma_4 \\ -\sigma_4 \\ -\sigma_2 \end{matrix}$
$8C_3$ $\begin{cases} C_3^{1,2,5} \\ C_3'^{(1,2,5)} \\ C_3^{1,4,5} \\ C_3'^{(1,4,5)} \\ C_3^{2,3,5} \\ C_3'^{(2,3,5)} \\ C_3^{3,4,5} \\ C_3'^{(3,4,5)} \end{cases}$	$\begin{matrix} \sigma_1 \\ \sigma_5 \\ \sigma_6 \\ \sigma_3 \\ \sigma_5 \\ \sigma_3 \\ \sigma_1 \\ \sigma_6 \end{matrix}$	−1	0	$\begin{matrix} -\sigma_1 \\ -\sigma_5 \\ -\sigma_6 \\ -\sigma_3 \\ -\sigma_5 \\ -\sigma_3 \\ -\sigma_1 \\ -\sigma_6 \end{matrix}$	0
$6C_2'$ $\begin{cases} C_2'^{(1,2)} \\ C_2'^{(1,4)} \\ C_2'^{(2,5)} \\ C_2'^{(4,5)} \\ C_2'^{(1,6)} \\ C_2'^{(1,5)} \end{cases}$	$\begin{matrix} \sigma_1 \\ \sigma_3 \\ \sigma_5 \\ \sigma_6 \\ \sigma_4 \\ \sigma_4 \end{matrix}$	0	−1	0	$\begin{matrix} -\sigma_1 \\ -\sigma_3 \\ -\sigma_5 \\ -\sigma_6 \\ -\sigma_4 \\ -\sigma_4 \end{matrix}$

bination of ψ_1 and ψ_2 that is orthogonal to ψ_1. We define the new function generated by a linear combination:

$$\psi_2' = a\psi_1 + b\psi_2$$

$$= \frac{a}{\sqrt{12}}(2\sigma_1 - \sigma_2 + 2\sigma_3 - \sigma_4 - \sigma_5 - \sigma_6) +$$

$$+ \frac{b}{\sqrt{12}}(-\sigma_1 + 2\sigma_2 - \sigma_3 + 2\sigma_4 - \sigma_5 - \sigma_6)$$

with the condition:

$$\int \psi_1 \psi_2' \, d\tau = \int \psi_1 (a\psi_1 + b\psi_2) \, d\tau = a \int \psi_1^2 \, d\tau + b \int \psi_1 \psi_2 \, d\tau = 0$$

whence

$$a \int \psi_1^2 \, d\tau = -b \int \psi_1 \psi_2 \, d\tau$$

and, since $\int \psi_1^2 \, d\tau = 1$,

$$\frac{a}{b} = -\int \psi_1 \psi_2 \, d\tau$$

$$= -\int \left[\frac{1}{\sqrt{12}} (2\sigma_1 - \sigma_2 + 2\sigma_3 - \sigma_4 - \sigma_5 - \sigma_6) \right] \times$$

$$\times \left[\frac{1}{\sqrt{12}} (-\sigma_1 + 2\sigma_2 - \sigma_3 + 2\sigma_4 - \sigma_5 - \sigma_6) \right] d\tau$$

$$= \frac{1}{12} (-2 \int \sigma_1^2 \, d\tau - 2 \int \sigma_2^2 \, d\tau - 2 \int \sigma_3^2 \, d\tau - 2 \int \sigma_4^2 \, d\tau + \int \sigma_5^2 \, d\tau +$$

$$\int \sigma_6^2 \, d\tau + \ldots)$$

The dots represent the cross terms, which are zero if the six σ orbitals are assumed orthogonal. Each of the above integrals is equal to 1, since the σ orbitals were previously normalized. Hence

$$\frac{a}{b} = -\tfrac{1}{12} (-2 - 2 - 2 - 2 + 1 + 1) = \tfrac{1}{2}$$

Substituting this result into the expression for ψ_2' gives:

$$\psi_2' = \frac{1}{2} b\psi_1 + b\psi_2 = \frac{b}{2} (\psi_1 + 2\psi_2)$$

$$= \frac{b}{2} \left[\frac{1}{\sqrt{12}} (2\sigma_1 - \sigma_2 + 2\sigma_3 - \sigma_4 - \sigma_5 - \sigma_6) + \right.$$

$$\left. + \frac{1}{\sqrt{12}} (-2\sigma_1 + 4\sigma - 2\sigma_3 + 4\sigma_4 - 2\sigma_5 - 2\sigma_6) \right]$$

$$= \frac{b}{2\sqrt{12}} (3\sigma_2 + 3\sigma_4 - 3\sigma_5 - 3\sigma_6)$$

$$= \frac{3b}{2\sqrt{12}} (\sigma_2 + \sigma_4 - \sigma_5 - \sigma_6)$$

The $3b/2\sqrt{12}$ is a normalizing factor which can now be evaluated and is found to be $\frac{1}{2}$, since $\sum c_i^2 = 1$; $4 \times (3b/2\sqrt{12})^2 = 1$ and then $3b/2\sqrt{12} = \frac{1}{2}$. Thus, the pair of orthogonal functions of species E is:

$$\psi_1 = \frac{1}{\sqrt{12}}(2\sigma_1 - \sigma_2 + 2\sigma_3 - \sigma_4 - \sigma_5 - \sigma_6)$$

$$\psi_2' = \tfrac{1}{2}(\sigma_2 + \sigma_4 - \sigma_5 - \sigma_6)$$

This procedure is quite general. If we have three-fold degenerate functions, we require three independent nonorthogonal functions to start with and then we form linear combinations of all three.

We could have employed another method for finding a second group orbital of species E orthogonal to the first. In this method we transform two additional orbitals to obtain group orbitals and take combinations of them. In order to obtain ψ_1 above we used the σ_1 orbital. Let us now use σ_2 and σ_5 (a choice made by intuition and experience) to obtain GO's. We have already obtained ψ_2, the group orbital, by using σ_2, and it remains to obtain the GO by using σ_5. This is done in Table 6.6. The resulting of adding the column under $\cdot E$ for σ_5 (Table 6.6) is:

$$-2\sigma_1 - 2\sigma_2 - 2\sigma_3 - 2\sigma_4 + 4\sigma_5 + 4\sigma_6 = \psi_3$$

Now neither of the group orbitals ψ_2 and ψ_3 is suitable for the second GO of species E, since neither orbital is orthogonal to the one generated by the transformation of σ_1. However, if we subtract the result obtained by transforming σ_5 from that obtained by transforming σ_2, we obtain:

$$
\begin{array}{rrrrrr}
-2\sigma_1 & + 4\sigma_2 & - 2\sigma_3 & + 4\sigma_4 & - 2\sigma_5 & - 2\sigma_6 \\
-2\sigma_1 & - 2\sigma_2 & - 2\sigma_3 & - 2\sigma_4 & + 4\sigma_5 & + 4\sigma_6 \\
\hline
 & 6\sigma_2 & & + 6\sigma_4 & - 6\sigma_5 & - 6\sigma_6
\end{array}
$$

which after normalization gives

$$\psi_2' (E) = \tfrac{1}{2}(\sigma_2 + \sigma_4 - \sigma_5 - \sigma_6) = \psi_2 - \psi_3$$

This GO is identical to the one found by the Gram-Schmidt method and is orthonormal to $\psi_1(E) = (1\sqrt{12})(2\sigma_1 - \sigma_2 + 2\sigma_3 - \sigma_4 - \sigma_5 - \sigma_6)$. The choice of subtraction to give the right result is again a matter of experience and intuition.

We might profitably digress for a moment to discuss the orthogonality concept again. If we regard the condition of orthogonality of two functions as analogous to the orthogonality of two vectors, the two orthogonal vectors must be at right angles to each other. If we consider the two vectors to be in the same plane of the paper, we can select an infinite number of pairs of vectors in the plane such that the two are at right angles to each other. However, once we fix the direction of one of the vectors, say in a

TABLE 6.6
Generation of Orbitals Group/Using σ_5

Operation on σ_5	New Orbital after Operation	Character of E	Character of T_1	Product: New Orbital × Character $\cdot E$	Product: New Orbital × Character T_2
I	σ_5	2	3	$2\sigma_5$	$3\sigma_5$
$6C_4$ $C_4^{5,6}$	σ_5	0	1	0	σ_5
$C_4'^{(5,6)}$	σ_5				σ_5
$C_4^{1,3}$	σ_2				σ_2
$C_4'^{(1,3)}$	σ_4				σ_4
$C_4^{2,4}$	σ_1				σ_1
$C_4'^{(2,4)}$	σ_3				σ_3
$3C_2$ $C_2^{5,6}$	σ_5	2	-1	$2\sigma_5$	$-\sigma_5$
$C_2^{1,3}$	σ_6			$2\sigma_6$	$-\sigma_6$
$C_2^{2,4}$	σ_6			$2\sigma_6$	$-\sigma_6$
$8C_3$ $C_3^{1,2,5}$	σ_2	-1	0	$-\sigma_2$	0
$C_3'^{(1,2,5)}$	σ_1			$-\sigma_1$	
$C_3^{1,4,5}$	σ_1			$-\sigma_1$	
$C_3'^{(1,4,5)}$	σ_4			$-\sigma_4$	
$C_3^{2,3,5}$	σ_3			$-\sigma_3$	
$C_3'^{(2,3,5)}$	σ_2			$-\sigma_2$	
$C_3^{3,4,5}$	σ_4			$-\sigma_4$	
$C_3'^{(3,4,5)}$	σ_3			$-\sigma_3$	
$6C_2'$ $C_2'^{(1,2)}$	σ_6	0	-1	0	$-\sigma_6$
$C_2'^{(1,4)}$	σ_6				$-\sigma_6$
$C_2'^{(2,5)}$	σ_2				$-\sigma_2$
$C_2'^{(4,5)}$	σ_4				$-\sigma_4$
$C_2'^{(1,6)}$	σ_3				$-\sigma_3$
$C_2'^{(1,5)}$	σ_1				$-\sigma_1$

perfectly vertical direction, then the direction of the second vector must necessarily be in the horizontal direction in order to be perpendicular. In the case of the e group orbitals just discussed, we generated first a group orbital of e symmetry by transforming σ_1. Our second e orbital then has to be orthogonal to the first one, and neither of the orbitals generated by transforming σ_2 or σ_5 met this condition; however, a combination of the two did meet the orthogonality requirement and hence is an appropriate second e orbital.

The three group orbitals that were generated under e were as follows:

$$(\sigma_1)\,\psi_1 = \frac{1}{\sqrt{12}}(2\sigma_1 - \sigma_2 + 2\sigma_3 - \sigma_4 - \sigma_5 - \sigma_6)$$

$$(\sigma_2)\,\psi_2 = \frac{1}{\sqrt{12}}(-\sigma_1 + 2\sigma_2 - \sigma_3 + 2\sigma_4 - \sigma_5 - \sigma_6)$$

$$(\sigma_5)\,\psi_3 = \frac{1}{\sqrt{12}}(-\sigma_1 - \sigma_2 - \sigma_3 - \sigma_4 + 2\sigma_5 + 2\sigma_6)$$

and we selected as the orthonormal set ψ_1 and $(\psi_2 - \psi_3)$. We could have selected ψ_3 and $(\psi_1 - \psi_2)$ as the orthonormal set. Then for $(\psi_1 - \psi_2)$ we would obtain, after normalization:

$$\psi_4 = \tfrac{1}{2}(\sigma_1 - \sigma_2 + \sigma_3 - \sigma_4)$$

As a matter of fact, the orthonormal set of ψ_3 and ψ_4 is a more convenient set, since it matches the d_{z^2} and $d_{x^2-y^2}$ set in the coordinate system we have chosen. This second procedure that we have shown is correct and, even though laborious, leads to the right results, but care must be exercised to avoid inconsistencies.

Finally we turn to the determination of the T_1 group orbitals. If we sum the product under T_1 for σ_1 (Table 6.4), we get after normalization:

$$GO\,(t_1) = \frac{1}{\sqrt{12}}(\sigma_1 - \sigma_3)$$

Similarly, for the transformations of σ_2 and σ_5 we obtain from Tables 6.5 and 6.6:

$$GO\,(t_1) = \frac{1}{\sqrt{2}}(\sigma_2 - \sigma_4)$$

$$GO\,(t_1) = \frac{1}{\sqrt{2}}(\sigma_5 - \sigma_6)$$

In this case we arrive, accidentally, at an orthonormal set.

The information given above is summarized in Table 6.2, and the data in this table can now be used to generate our molecular orbital energy diagram (MOED).

6.5   Molecular Orbital Energy Diagrams for O_h Complexes with Sigma Bonding Orbitals Only

We can now proceed to construct the MOED for $[Co(NH_3)_6]^{3+}$, shown in Fig. 6.7. There are six σ bonding molecular orbitals, generated from the d^2sp^3 orbitals of the metal (4s, $4p_x$, $4p_y$, $4p_z$, $3d_{z^2}$, and $3d_{x^2-y^2}$), the six

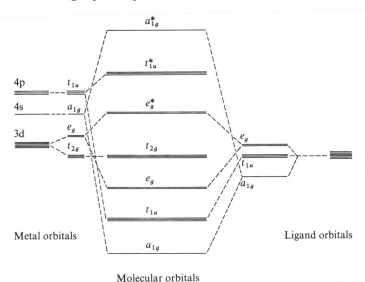

Molecular orbitals

Fig. 6.7   The molecular orbital energy diagram for the octahedral complex $[Co(NH_3)_6]^{3+}$. (All levels up to and including t_{2g} are completely filled.)

ligand group orbitals of similar symmetry, and the six corresponding σ^* molecular orbitals. The three t_{2g} orbitals of the metal are nonbonding orbitals. We can now fill the levels with the appropriate number of electrons, which in the case of $[Co(NH_3)_6]^{3+}$ is six from the d^6 of Co^{3+} and twelve from the ligands, or a total of eighteen. The eighteen fill exactly all the low-lying orbitals, including the t_{2g} nonbonding orbitals. This fact accounts for the great stability of Co^{3+} as a hexacoordinated ion, since additional electrons would necessarily have to occupy e_g^* orbitals. Furthermore, $[Co(NH_3)_6]^{3+}$ is diamagnetic. The lowest-energy transition in this ion is the $t_{2g} \rightarrow e_g^*$ transition, a $g \rightarrow g$ transition, and hence of very low intensity (cf. Section 6.8). Since the e_g^* orbital is still mostly on the metal, this transition may be considered as a d $\rightarrow$ d transition.

The energy separation $10Dq$ between the t_{2g} and e_g^* orbitals in this MO picture corresponds to the crystal field splitting of the five d orbitals into the t_{2g} and e_g sets discussed earlier, Fig. 6.5.

6.6   Low- and High-Spin Complexes

The difference in energy between the t_{2g} and e_g^* levels is of great importance. In the $[Co(NH_3)_6]^{3+}$ case just discussed, the ammonia nitrogen has a lone pair of electrons in an sp^3 orbital; because of the directional character, this orbital makes for good overlap with the d_{z^2} and $d_{x^2-y^2}$ orbitals (e_g) of the metal, and hence the splitting between t_{2g} and e_g^* is rather large. It is estimated that Δ_0 (or $10Dq$) is about 66 kcal/mole for the $[Co(NH_3)_6]^{3+}$ complex. Small ligands with a highly directional charge or dipole cause large splitting. Ligands with more spherical charge, as well as large ligands, do not cause as large a splitting. Thus, iodide has a weak ligand field effect and causes small splitting. The order of splitting for the halogens, H_2O, and NH_3 ligands is:

$$NH_3 > H_2O > F^- > Cl^- > Br^- > I^-$$

In the case of $[CoF_6]^{3-}$, there is relatively little splitting, and $10Dq$ is estimated to be about 37 kcal/mole. If, of all the fluoride electrons, we consider only those in the σ orbital directed along the bond axes, we again have eighteen electrons to place in our level diagram. We immediately dispose of twelve of them in the σ bonding a_{1g}, t_{1u}, and e_g molecular orbitals, and then place three additional ones, one each, in the t_{2g} nonbonding d orbitals. This leaves us with three electrons to dispose of.

In the $[Co(NH_3)_6]^{3+}$ case these also went into t_{2g}, and although double occupancy (pairing) of an orbital costs energy because of electron-electron repulsions, this energy is less than would be required to occupy the high-energy e_g^* orbital with parallel spin. The magnitude of the pairing energy, P, is thus of great importance. For the d^6 configuration of Co^{3+}, spectral data from the gaseous ion lead to a value of P of about 50 kcal/mole.

In $[CoF_6]^{3-}$, Δ_0 is 37 kcal/mole and the pairing energy exceeds the orbital energy difference by 13 kcal/mole. Consequently (see Fig. 6.10) the fourth and fifth electrons occupy, separately, the e_g^* orbitals, all with parallel spin, and the sixth and final electron now pairs with one of the electrons already placed in one of the t_{2g} orbitals. This means that $[Co(F_6)]^{3-}$ should be paramagnetic with a total of four unpaired electrons, and this is precisely what is found experimentally.

Accordingly, $[Co(NH_3)_6]^{3+}$ and $[CoF_6]^{3-}$ are called *low-spin* and *high-spin* complexes, respectively. Furthermore, occupancy of the e_g* orbitals in $[Co(F_6)]^{3-}$ weakens the bonding between the metal and the ligands, and the fluoride is much more ionic. In the past, the two kinds of complexes have also been called "covalent" and "ionic" complexes (Pauling) or "inner" and "outer" orbital complexes (Nyholm, Taube). The latter names arose because, in $[Co(F_6)]^{3-}$, the occupation of e_g^* molecular orbitals was (incorrectly) associated with the occupation of 4d atomic orbitals in the ion in which $4p^3 4s 4d^2$ orbitals rather than $3d^2 4s 4p^3$ orbitals were used.

6.7 Molecular Orbital Energy Diagrams for O_h Complexes with Ligands Having Filled p Orbitals

In discussing the $[CoF_6]^{3-}$ ion we were concerned exclusively with the p_z orbital of fluorine pointing toward the metal, which we called the σ orbital. On each fluoride there are also a p_x and a p_y orbital. Such orbitals have π

symmetry with respect to the metal-fluorine bond, and it is possible to generate group or symmetry orbitals from these twelve pπ's on the fluorides. Before we attempt to make the appropriate combinations of atomic orbitals we need a coordinate system. One such system that has been found to be convenient consists of a right-hand coordinate system for the cobalt and a left-hand coordinate system for each fluoride ligand, Fig. 6.8. The z-axis on each ligand coincides with one of the axes of the metal atom, with the

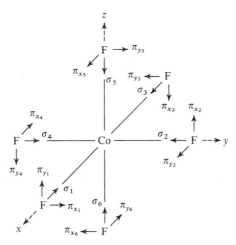

Fig. 6.8   Coordinate system for $[Co(F_6)]^{3-}$ and the symmetry species of the metal and the ligand group orbitals.

O_h	Cobalt Orbitals	Ligand Group Orbitals
A_{1g}	4s	$(1/\sqrt{6})(\sigma_1 + \sigma_2 + \sigma_3 + \sigma_4 + \sigma_5 + \sigma_6)$
E_g	$3d_{z^2}$	$(1/\sqrt{12})(2\sigma_5 + 2\sigma_6 - \sigma_1 - \sigma_2 - \sigma_3 - \sigma_4)$
	$3d_{x^2-y^2}$	$\frac{1}{2}(\sigma_1 - \sigma_2 + \sigma_3 - \sigma_4)$
T_{2g}	$3d_{xz}$	$\frac{1}{2}(\pi_{y_1} + \pi_{x_5} + \pi_{x_3} + \pi_{y_6})$
	$3d_{yz}$	$\frac{1}{2}(\pi_{x_2} + \pi_{y_5} + \pi_{y_4} + \pi_{x_6})$
	$3d_{xy}$	$\frac{1}{2}(\pi_{x_1} + \pi_{y_2} + \pi_{y_3} + \pi_{x_4})$
T_{1u}	$4p_x$	$(1/\sqrt{2})(\sigma_1 - \sigma_3)$
		$\frac{1}{2}(\pi_{y_2} + \pi_{x_5} - \pi_{x_4} - \pi_{y_6})$
	$4p_y$	$(1/\sqrt{2})(\sigma_2 - \sigma_4)$
		$\frac{1}{2}(\pi_{x_1} + \pi_{y_5} - \pi_{y_3} - \pi_{x_6})$
	$4p_z$	$(1/\sqrt{2})(\sigma_5 - \sigma_6)$
		$\frac{1}{2}(\pi_{y_1} + \pi_{x_2} - \pi_{x_3} - \pi_{y_4})$
T_{1g}	$\cdots$	$\frac{1}{2}(\pi_{y_1} - \pi_{x_5} + \pi_{x_3} - \pi_{y_6})$
		$\frac{1}{2}(\pi_{x_2} - \pi_{y_5} + \pi_{y_4} - \pi_{x_6})$
		$\frac{1}{2}(\pi_{x_1} - \pi_{y_2} + \pi_{y_3} - \pi_{x_4})$
T_{2u}	$\cdots$	$\frac{1}{2}(\pi_{y_2} - \pi_{x_5} - \pi_{x_4} + \pi_{y_6})$
		$\frac{1}{2}(\pi_{x_1} - \pi_{y_5} - \pi_{y_3} + \pi_{x_6})$
		$\frac{1}{2}(\pi_{y_1} - \pi_{x_2} - \pi_{x_3} + \pi_{y_4})$

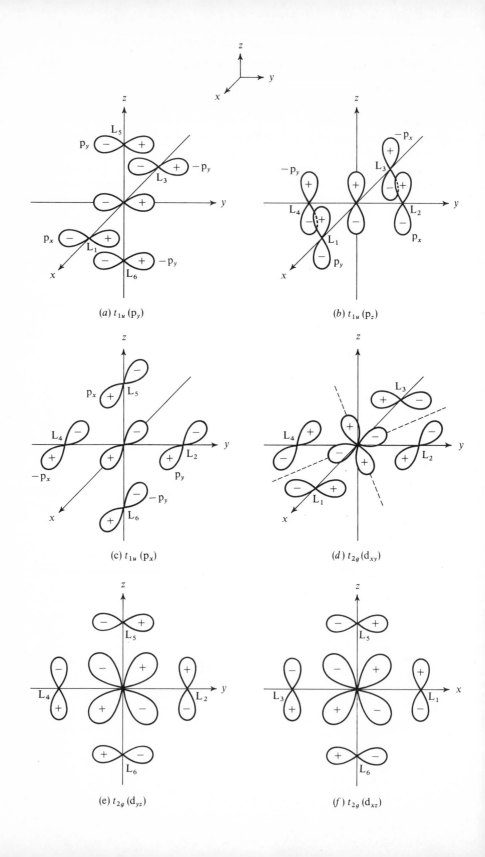

(a) t_{1u} (p_y)

(b) t_{1u} (p_z)

(c) t_{1u} (p_x)

(d) t_{2g} (d_{xy})

(e) t_{2g} (d_{yz})

(f) t_{2g} (d_{xz})

positive z on each ligand pointing toward the metal. The positive directions of the x- and y-axes are fixed by the left-hand coordinate system on each ligand, but the assignment of x and y is arbitrary.

The proper combination of ligand $p\pi$ orbitals to overlap the metal orbitals can be seen from inspection. In Figs. 6.9a, b, and c are shown the proper combinations for the members of the t_{1u} set which overlap the 3p set of the cobalt, and Figs. 6.9d, e, and f show the combinations of ligand orbitals of the t_{2g} set that overlap the d_{xy}, d_{yz}, and d_{xz} metal orbitals. (One reason for the selection of the left-hand coordinate system is the convenience of having the important t_{2g} set as a combination of AO's all of which have plus signs. This is perhaps more clearly illustrated by isolating the xy-plane of Fig. 6.9d, as in Fig. 6.9d'.) The twelve ligand $p\pi$ orbitals generate twelve ligand

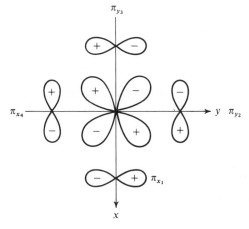

Fig. 6.9d' The combination of ligand orbitals $(\pi_{x_1} + \pi_{y_2} + \pi_{y_3} + \pi_{x_4})$ which matches the metal d_{xy} (t_{2g}).

group or symmetry orbitals; we see that six of these, Fig. 6.8, can overlap metal orbitals, whereas the other set of six, t_{2u}, t_{1g}, does not and hence is non-bonding.

We now redraw Fig. 6.7 to construct a new MOED, Fig. 6.10. To simplify the picture, we show the six σ molecular orbitals in Fig. 6.7 as degenerate in Fig. 6.10. The most important difference between the MOED diagrams is that the t_{2g} set of d_{xy}, d_{xz}, and d_{yz}, which is nonbonding in the $[Co(NH_3)]^{3+}$ case, Fig. 6.7, becomes involved in bonding-antibonding combinations in the $[CoF_6]^{3-}$ case, Fig. 6.9. Another but less important difference is that there is now a $p\pi$ set of t_{1u} orbitals of the ligands in addition to the σ set

---

Fig. 6.9 The combination of ligand p orbitals (a, b, c) matching the triply degenerates t_{1u} and (d, e, f) the t_{2g} metal orbitals.

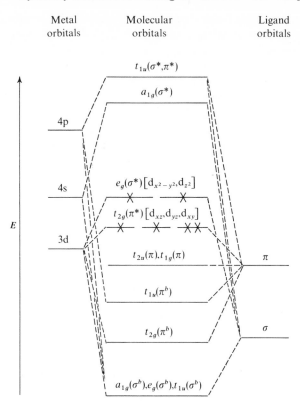

| Metal orbitals | Molecular orbitals | Ligand orbitals |

Fig. 6.10 The molecular orbital energy diagram for the high-spin O_h complex $[Co(F_6)]^{3-}$. (The levels up to and including t_{2u} and t_{1g} are completely filled.)

of t_{1u} ligand orbitals. Both of these sets of t_{1u} orbitals combine with the metal t_{1u} set to form three sets of t_{1u}, all of which have both σ and π character. Since the σ overlap is much greater than the π overlap, the bonding is probably mostly σ, and in a crude approximation the t_{1u} set from the $p\pi$ ligand orbitals can be considered to be largely nonbonding.

The utilization of the twelve orbitals of the six fluoride ions means that 24 additional electrons are brought into the system, and so we must accommodate a total of $24 + 18 = 42$ electrons. Now the energy difference between the t_{2g}^* and e_g^* levels is much smaller than would be the case if we used only σ orbitals. Hence, all the electrons are not paired; each e_g^* is singly occupied, and there are two singly occupied t_{2g}^* orbitals. Thus, $[CoF_6]^{3-}$ is a high-spin complex. The effect of the filled $p\pi$ orbitals on the ligands is to destabilize the complex and to make it more ionic.

6.8   Molecular Orbital Energy Diagrams for O_h Complexes with Ligands Having π Systems of Their Own

Finally we must consider the situation in which the ligand may have a π system of its own, as in the case where the ligand is carbon monoxide, an isocyanide, a cyanide ion, etc. Since we have already discussed the MO structure of carbon monoxide, we will use this ligand as an example. If we wish to examine an octahedral complex involving CO, it is better to use, not they hypothetical $[Co(CO)_6]^{3+}$, but a known uncharged hexacarbonyl, such as $Cr(CO)_6$. Since CO is a neutral or uncharged ligand, the valence or oxidation state of the chromium is zero and the effective atomic number (E.A.N.) around the Cr is 36, thus accounting for its unusual stability.

The new factor arising when ligands with their own π and π^* systems are considered is the involvement of the π^* orbitals of the ligands. The π system has the symmetry of the p orbitals, which we have already discussed, and these π orbitals are occupied just as the p orbitals on F^-. Accordingly, we concentrate on the new factor, the π^* system.

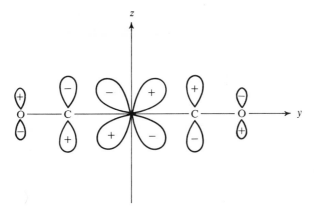

Fig. 6.11   The combination of d_{yz} with the antibonding orbitals of carbon monoxide.

We recall that in CO we have two sets of π^* orbitals orthogonal to each other. As Fig. 6.11 shows, the π^* orbitals have appropriate symmetry to combine with the d_{yz} orbital, and of course there will be combinations of π^* that can combine with the d_{xz} and d_{xy} orbitals as well. Such overlap has a profound effect on the complex. The t_{2g} orbitals in the complex of Fig. 6.10 now combine with symmetry-matching combinations of π^* on the ligands, and the t_{2g}^* is lowered to form a set of bonding orbitals with the insertion of a corresponding set of antibonding orbitals. Thus, with carbon monoxide as ligands, we will have two sets of t_{2g} orbitals: one set generated out of $p\pi$'s similar to the p's on the fluoride, and a second set generated from π^*'s on

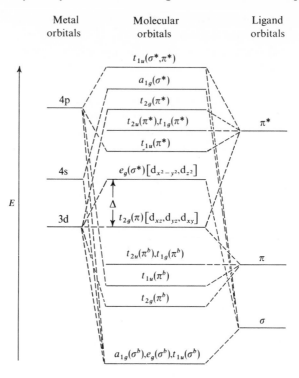

Fig. 6.12 Molecular orbital energy diagram for $Cr(CO)_6$. (All orbitals up to and including t_{2g} are filled.)

the ligand. Both of these sets will combine with the metal set (d_{xy}, d_{xz}, d_{yz}) of t_{2g} to form three sets of MO's of t_{2g} symmetry. One set of t_{2g} will be bonding, one antibonding, and one slightly bonding.

The net result is to lower the t_{2g}^* set of Fig. 6.10 (where no π^* system was involved) and to increase the energy split between t_{2g} and e_g. Since the introduction of the π^* system does not add any electrons, the highest occupied t_{2g} orbital is much lower in energy than would otherwise be the case. Accordingly, stable complexes arise when ligands with π and π^* systems are introduced. The complete MOED of $Cr(CO)_6$ is shown in Fig. 6.12. The π^* orbitals on the six CO's bring in twelve π^* orbitals; t_{2g}, t_{1u}, t_{2u}, and t_{1g}. The t_{2g} interacts strongly with the metal $t_{2g}(d)$ set, the t_{1u} interacts rather weakly with the metal $t_{1u}(p)$ set, and the t_{2u} and t_{1g} are essentially nonbonding. The total number of valence electrons in the system is 42; 6 contributed by the chromium, 12 from the lone pairs on the C atoms of the six carbon monoxides, and 24 from the π systems of the six ligands. These 42 electrons completely occupy all levels up to and including the t_{2g} level.

The low-lying π^* orbitals on the carbon monoxide and their interaction with filled metal d orbitals have a very pronounced effect on the C-O bonding in the coordinated carbon monoxide. Electron density in the π^* orbitals decreases the C-O bond order. This effect is dramatically illustrated by the infrared stretching frequencies of free CO and of CO complexed with various metals. The valence bond structure of CO is best expressed as having a triple bond between the C and O atoms. Free CO in solution has a stretching frequency of 2143 cm^{-1}. However, the stretching frequencies in the tetra-hedral series Ni(CO)$_4$, $[Co(CO)_4]^-$, $[Fe(CO)_4]^{2-}$ are, respectively, 2057, 1886, and 1786 cm^{-1}, showing that, as the carbon monoxide is coordinated to increasingly strong electron-releasing metal atoms, the bond order (as well as the bond length) between the C and O atoms progressively decreases as the π^* orbitals become increasingly populated. The infrared spectrum of a carbonyl thus provides a useful tool for evaluating the tendency of the metal atom to participate in d $\rightarrow \pi^*$ donation. We shall discuss this subject more fully in Chapter 9.

6.9   Direct Products of Representations

The material we have covered thus far in this chapter not only is essential background for understanding the modern view of bonding in transition-metal complexes but also provides the information necessary for a discussion of the spectra of such complexes. One additional concept required for the discussion of spectra is that of the direct product of representations.

It is frequently desirable to obtain the symmetry characteristics of the product of two functions. In dealing with nondegenerate point groups this is quite simple because in such point groups the only characters we must consider are ± 1. It is intuitively obvious that, just as the product of $(-1) \times (-1)$ is $+1$, the product of two antisymmetric functions is symmetric. Similarly we expect that the product of two symmetric functions is symmetric ($1 \times 1 = 1$), and the product of an antisymmetric and a symmetric function is antisymmetric ($1 \times -1 = -1$).

One of the applications of the direct product of representations is in the determination of the symmetry state of electronic configurations. To illustrate let us determine the symmetry of one of the excited states of the water molecule, point group C_{2v}. The character table for this point group is shown in Table 5.1, and the symmetry species to which the various orbitals of the molecule belong are given in Fig. 6.1. In symmetry notation we would write the electronic configuration corresponding to the promotion of an electron from the nonbonding p$_x$ orbital of species b_1 to an antibonding σ^* orbital of species b_2 as follows: $a_1{}^2b_2{}^2b_1b_2^*$, where the superscripts refer to the number of electrons in the orbitals of the indicated symmetry species. In order to obtain the direct product we first determine the species of $a_1{}^2$ by squaring the characters of a_1 under all operations of C_{2v}. Since we are dealing only with characters of ± 1, the square of a_1, as well as the square of all species in all nondegenerate groups, is totally symmetric or a_1 in C_{2v}. Thus $a_1{}^2 \times b_2{}^2 = a_1$. Next, we consider the direct product of $b_1 b_2^*$. If we consult Table 5.1, we see that the product of these two representations under the operations indicated is $I(+1)$, $C_2{}^z(+1)$, $\sigma_v{}^{xz}(-1)$. These characters correspond to species a_2, and hence $a_1{}^2 b_2{}^2 b_1 b_2^* = a_1 a_2$. But, since all characters under a_1 are $+1$, multiplying any set of characters by a_1 produces no change, and $a_1 a_2 = A_2$. Accordingly the excited state under consideration has A_2 symmetry.

The direct product of representations in point groups with degenerate species poses some difficulties. Nothing new arises if we take the product of one nondegenerate and one degenerate species. Thus, for example, the product of $E_u \times B_{2u}$, in point group D_{4h}, Table 5.4, is E_g. However, the product of two degenerate species gives a set of characters which do not correspond to any species of the point group. Thus, for example, the product of $E \times E$ in point group T_d, Table 5.5, is $I(4)$, $8C_3(+1)$, $6\sigma_d(0)$, $6S_4(0)$, $3C_2(4)$. This set of characters is a reducible representation; it does not correspond to any irreducible representation. Rather, such a reducible representation generally corresponds to the sum (called the direct sum) of several irreducible representations. We have in fact already encountered this problem in this chapter, and equation 6.3 tells us how to determine the irreducible representations. Application of this equation to our problem shows us that $E^2 = E + A_1 + A_2$. In this case the result is also obvious from inspection. The importance of the direct product will become apparent in the next two chapters.

PROBLEMS

6.1 Construct the symmetry orbitals for the sigma orbitals of the methyl hydrogens in toluene in both C_{3v} symmetry and C_s symmetry. To what symmetry species do each of these group orbitals belong in the two different point groups?

6.2 Show that d_{z^2} and $d_{x^2-y^2}$ orbitals in point group O transform as species e.

6.3 Determine the symmetry species of the 3d, 4s, and 4p orbitals of a central metal atom in a coordination complex which belongs to point group T_d.

6.4 Generate the four sigma ligand symmetry orbitals of $NH_4{}^+$ in T_d.

6.5 It was shown in this chapter that under $C_4{}^x$ in point group O_h, the d_{z^2} and the $d_{x^2-y^2}$ orbitals transform into d_{y^2} and $d_{x^2-z^2}$, respectively, and that these new orbitals are linear combinations of the old ones. Show that under $C_3{}'$ the two original orbitals transform into $d_{2x^2-z^2-y^2}(d_{x^2})$ and $d_{y^2-z^2}$, respectively.

6.6 Rationalize the relative energies of the molecular orbitals shown for the O_h complex in Fig. 6.6, i.e., $a_{1g} < t_{1u} < e_g$.

6.7 In point group D_{4h} the p_x and p_y orbitals transform as species e_u and under the C_4^z operation the transformation matrix was shown in Chapter 5 to be

$$\begin{pmatrix} 0 & 1 \\ -1 & 0 \end{pmatrix}$$

with a character of zero. (a) Show by means of the appropriate transformation matrix that the character of e_u^2 under C_4 is 0, and (b) under $S_2 = i$, it is 4.

6.8 $Ni(CO)_4$, $[Co(CO)_4]^-$, and $[Fe(CO)_4]^{2-}$ are isoelectronic tetrahedral complexes in which the metal has a d^{10} configuration. Explain why such complexes are stable using (a) a ligand field and (b) a molecular orbital explanation.

6.9 In coordination chemistry the number of d electrons in the valence shell of the metal is given as d^n. Generally the higher the value of n in d^n, the smaller the coordination number. Explain why transition metals with a d^7-d^{10} configuration do not form octahedral complexes.

6.10 Give a possible explanation for the remarkable catalytic activity of the complex $[Co(CN)_5]^{3-}$, which exhibits reactions resembling those of a free radical (e.g., its aqueous solution readily cleaves molecular hydrogen at room temperature).

6.11 Explain why metals with the d^8 configurations, e.g., Ni^{2+}, Pd^{2+}, and Au^{3+}, are particularly prone to form square planar complexes even though in such (sigma bonded) complexes the metal ion has an effective atomic number of two less than that of the next rare gas.

6.12 Construct the MOED of $[PtCl_4]^{2-}$. [It will first be necessary (Part A) to obtain the four σ and eight π group orbitals, then to determine the symmetry of the metal orbitals (Part B), and then to construct (Part C) the MOED.]

GENERAL REFERENCES

1. L. Lewis and R. B. Wilkins, *Modern Coordination Chemistry*, Interscience Publishers, New York, 1960.
2. J. S. Griffith, *The Theory of Transition-Metal Ions*, Cambridge University Press, 1961.
3. C. J. Ballhausen and H. B. Gray, *Molecular Orbital Theory*, W. A. Benjamin, New York, 1965.
4. C. K. Jørgensen, *Orbitals in Atoms and Molecules*, Academic Press, New York, 1962.
5. B. N. Figgis, *Introduction to Ligand Fields*, Wiley-Interscience, New York, 1966.

7 The electronic spectra of inorganic complexes and ions

7.1 Introduction

Interpretation of the spectra of inorganic complexes is greatly simplified and successfully integrated by the use of ligand field theory (LFT). In the octahedral complexes, the lowest-energy transition is associated with the energy difference between the t_{2g} and e_g^* orbitals. In the molecular orbital treatment we see that e_g^* is always antibonding but t_{2g} can be weakly bonding, nonbonding, or weakly antibonding, depending on the nature of the ligands. In either LFT or the complete MO treatment, the split between t_{2g} and e_g (LFT) or t_{2g} and e_g^* (MO) in O_h complexes is critical; and although the nomenclature is different in the two approaches, the two theories are closely related.

7.2 Ligand Field Stabilization Energies; Electrons and Holes

In complexes where the metal ion has the d^0, d^5, or d^{10} electronic configuration, all five d orbitals are equally occupied (for d^5 this holds only in the high-spin case) with zero, one, and two electrons in each, respectively. In all other configurations, the orbitals are not uniformly occupied and, because t_{2g} is lower in energy than e_g in the octahedral complexes, there will always be some stabilization of the complex relative to what might be expected were all the orbitals degenerate.

Let us evaluate this stabilization energy for the d^1 configuration. We ask first what fraction of a single electron could be in each of the five d orbitals if the one electron were uniformly distributed. In the doubly degenerate e_g set there would be $\frac{2}{5}(1) = \frac{2}{5}$ of the electron, and in the triply degenerate lower-energy t_{2g} set of orbitals, $\frac{3}{5}(1) = \frac{3}{5}$ of the electron. However, the electron is in fact completely in the lower-energy, t_{2g}, set, and hence the ion or complex with the metal d^1 configuration is more stable than it would be were the electron evenly distributed over the five d orbitals. If the energy difference between the t_{2g} and e_g orbitals is Δ_0 ($10Dq$), then the stabilization energy is $\frac{2}{5}\Delta_0$ ($4Dq$). The $\frac{2}{5}\Delta_0$ is called the *ligand field stabilization energy* (LFSE).

We can also arrive at this answer in another way. The five-fold degeneracy of the d orbitals is split by the ligand field into a set of three more stable and two less stable orbitals. If the energy difference between the two sets is Δ_0 or $10Dq$, then each of the three more stable orbitals is $\frac{2}{5}\Delta_0$ ($4Dq$) more stable and each of the two less stable orbitals is $\frac{3}{5}\Delta_0$ ($6Dq$) less stable than the five degenerate orbitals before the splitting, since $3 \times \frac{2}{5}\Delta_0 - 2 \times \frac{3}{5}\Delta_0 = 0$, and the difference between the two sets is $\frac{2}{5}\Delta_0 + \frac{3}{5}\Delta_0 = \Delta_0$. Single-electron occupancy of one of the t_{2g} sets then results in a net stabilization of $\frac{2}{5}\Delta_0$ ($4Dq$).

In the d^2 and d^3 configurations, the LFSE is $\frac{4}{5}$ and $\frac{6}{5}\Delta_0$, respectively. In the d^4 configuration, if there were uniform distribution of the four electrons there should be $\frac{2}{5}(4) = \frac{8}{5}$ electrons in the e_g orbitals. In the high-spin (weak-field) case (cf. Section 6.6) there is actually only one electron in this set, and accordingly $\frac{3}{5}$ of an electron has been transferred to the t_{2g} set and the LFSE is $\frac{3}{5}\Delta_0$.

In the d^9 case, we have one unpaired electron to consider, and in order to simplify the situation we take advantage of what has been called the *hole formalism*, in which d^9 and d^1 have a special relationship. In the d^1 case, the single electron could go into an e_g orbital as well as the t_{2g} orbital. Obviously, these two different electronic configurations correspond to the ground and excited states, t_{2g}^1 and $e_g{}^1$, respectively. With d^9 we again have only two possible configurations. The lower-energy ground-state electronic configuration is $t_{2g}^6 e_g{}^3$, while the other possible configuration is $t_{2g}^5 e_g{}^4$, corresponding to the excited state. In $t_{2g}^6 e_g{}^3$ all orbitals are filled except for a hole in the e_g orbital, and in $t_{2g}^5 e_g{}^4$ all are filled except for a hole in the t_{2g} orbital. In the d^1 case, all orbitals are empty except for the single electron in t_{2g} (the lower-energy configuration) and in the higher-energy configuration all are empty except for the electron in the e_g orbital. There is thus an inverse relationship between the one-electron d^1 and the one-empty-orbital or hole d^9 configuration (the relationship between d^{10-n} and d^n) which results in an exactly inverted energy relationship between the ground and excited states in the presence of the ligand field. The nine holes in the d shell of the d^1 configuration correspond to the nine electrons in the d shell of the d^9 configuration. Accordingly, when we are interested in the d^9 case, we can carry out the calculations for d^1 and then completely invert the order of the states, making the highest excited state of d^1 the ground state of d^9 and the ground state of d^1 the highest excited state of d^9. Similarly, when we wish the states for d^8, d^7, and d^6 high-spin cases, we make the calculations for d^2, d^3, and d^4, respectively, and invert the ordering of the states.

7.3   Factors Affecting Ligand Field Stabilization Energies
7.3a   The Nature of the Metal

The critical importance of the value of Δ_0 or $10Dq$ is obvious from the preceding discussion, since, depending on the electron configuration, the LFSE in the high-spin cases can vary from $\frac{2}{5}\Delta_0$ to $\frac{6}{5}\Delta_0$. In Chapter 6 we touched briefly on the factors affecting the value of Δ_0 or $10Dq$. The D term is affected mostly by the nature of the metal and is related to its polarizability. The larger the ratio of charge to radius, the more polarizable is the metal ion, and, other things being equal, the larger $10Dq$, Table 7.1. In other words, since radii are not too variable, the greater the charge on the metal, that is, the higher the oxidation state, the larger is Δ_0. Thus, $[Co(NH_3)_6]^{3+}$, with larger charge and smaller radius, has Δ_0 about twice that of the corresponding $[Co(NH_3)_6]^{2+}$ complex. The ligands can approach more closely to the smaller, more highly charged Co^{3+} than they can to the Co^{2+}, with a resultant larger splitting. The quantum number n of the d orbitals of the metal also influences Δ_0. Thus, Δ_0 is greater for $[Rh(L_6)]^{3+}$, 4d, than for $[Co(L_6)]^{3+}$, 3d, Table 7.1, because the higher d orbitals extend farther into space and thus interact more strongly with the ligands.

TABLE 7.1
Octahedral Splitting

Ion	Δ_0, cm^{-1}			Radius, A	Charge/Radius
	Cl$^-$	H$_2$O	NH$_3$		
Ni^{2+}	7,200	8,500	10,800	0.78	2.56
Cr^{3+}	13,800	17,400	21,600	0.64	4.69
Co^{3+}	...	18,200	22,900	0.65	4.62
Co^{2+}	...	9,300	11,000	0.82	2.44
Rh^{3+}	20,300	27,000	34,100	0.69	4.35

7.3b   The Effect of Geometry

Most of our discussion thus far has dealt with the octahedral complexes and Δ_0. Let us now examine how the d orbital splitting behaves in other geometries; in particular, we will examine tetrahedral and square planar geometry and focus on the d orbital splitting.

The splitting in square planar geometry can be derived by starting from a consideration of octahedral symmetry and performing the necessary perturbations to achieve the other geometries. If the ligands on the z-axis of the octahedron are pulled away from the metal so that they are at a greater distance from the metal than the ligands in the xy-plane, the resulting complex has tetragonal geometry. The resulting orbital splitting is shown in

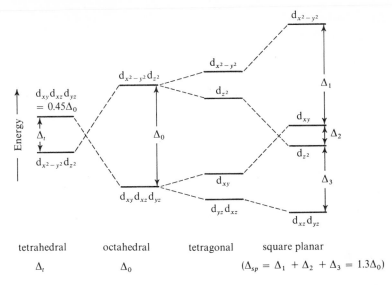

tetrahedral   octahedral   tetragonal   square planar

Δ_t Δ_0 $(\Delta_{sp} = \Delta_1 + \Delta_2 + \Delta_3 = 1.3\Delta_0)$

Fig. 7.1 Crystal field splittings of the d orbital of a central ion in complexes having different Geometries. (The subscripts to Δ identify the applicable geometry.)

Fig. 7.1. The two sets of degenerate orbitals that characterize O_h splitting are further split; the d_{z^2} now becomes lower in energy than $d_{x^2-y^2}$, and the d_{yz}, d_{xz} pair becomes lower than d_{xy} because the repulsions along the z-axis are reduced. If the ligands on the z-axis are now removed to infinity, we have a special case of a tetragonal complex, namely, the square planar complex with the splitting shown in Fig. 7.1.

In tetrahedral complexes, the splitting of the d orbitals is the reverse of the splitting in the O_h case. Perhaps this can be more readily understood by inscribing the tetrahedron in a cube, Fig. 7.2. The four ligands of the tetrahedral arrangement appear at alternating corners of the cube. The $d_{x^2-y^2}$ and d_{z^2} orbitals are degenerate and point at the centers of faces, and the d_{xy}, d_{xz}, d_{yz} orbitals are degenerate and point at the centers of edges. If we examine one face, say that on which ligands L_1 and L_2 are located, we see that $d_{x^2-y^2}$ falls at the center of the face, d_{xy} at the center of the edge, and L_1 and L_2 at opposite corners. If half the length of the edge is called a, then L_1 will be at distance a from one lobe of d_{xz} but at distance $\sqrt{2a}$ from one lobe of $d_{x^2-y^2}$. Hence, the repulsion will be greater at the orbitals closer to the ligand, the d_{xy}, d_{xz}, d_{yz} set, than at the orbitals with centers at the faces, the $d_{x^2-y^2}, d_{z^2}$ set. Accordingly, in T_d the triply degenerate t_2 set is higher in energy than the e set, just the reverse of the situation for the two sets in O_h. Thus for the same configuration, d^n, the energy relationships between O_h and

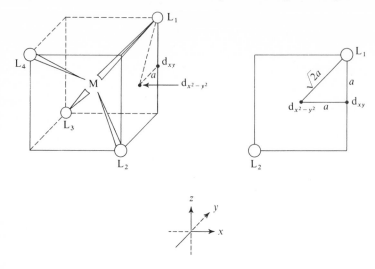

Fig. 7.2 The tetrahedral complex showing that the ligands repel the $d_{x^2-y^2}$ (and d_{z^2}) orbital less than the d_{xy} (and d_{xz}, d_{yz}) orbital.

T_d are exactly reversed. We can add this information to the electron-hole formalism and recognize that the d orbital splitting patterns are:

d^n of 0_h the same as d^{10-n} of T_d and inverted with respect to d^{10-n} for 0_h.
d^n of T_d the same as d^{10-n} of 0_h and inverted with respect to d^{10-n} for T_d.

These relationships greatly facilitate the problems of interpreting spectra.

The energy of the splitting of the d orbitals in tetrahedral geometry is less than in octahedral complexes because the d orbitals do not point directly at the ligands and hence cannot be as strongly involved in σ bonding. In square planar complexes the d orbitals are involved in a strong σ bonding. The total square planar splitting from the lowest $(d_{xz} d_{yz})$ to the highest orbital $(d_{x^2-y^2})$ is the sum:

$$[(d_{x^2-y^2}) - d_{xy}] + [d_{xy} - d_{z^2}] + [d_{z^2} - d_{xy}, d_{yz}]$$

and is larger than the octahedral splitting, since the d_{xz} and d_{yz} orbitals interact with only two ligands in the square planar complex as compared to an interaction with four ligands in the octahedral case. Analysis of a great number of experimental data indicates that the d orbital splitting in three geometries decreases as follows:

square planar > octahedral > tetrahedral

$1.3\Delta_0$ Δ_0 $0.45\Delta_0$

7.3c   The Effect of the Ligand

From our discussion of ligand field theory in the octahedral case, it is clear that a strong ligand-to-metal (L → M) σ interaction raises the σ^* (e_g) level, increasing the value of Δ_0. The L → M π interaction raises t_{2g} and therefore decreases Δ_0. Finally, M → L π^* interaction lowers t_{2g} again, increasing Δ_0. This information is summarized in Fig. 7.3 and represents the general situation for the d orbital splitting for common geometries, which, as we saw above, can be expressed as fractions of Δ_0.

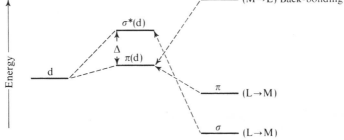

Fig. 7.3   The effect of interaction of the ligand, σ, π, and π^* orbitals on the value of Δ.

The order in which various kinds of ligands affect Δ_0 and hence the LFSE is called the *spectrochemical series* because a great deal of the information about the effect of ligands on d orbital splitting is obtained from an analysis of electronic spectral data. The good π acceptor ligands, such as CN^- and CO, have empty low-lying antibonding orbitals which facilitate strong M → L interaction. Consequently these ligands cause large splittings, in the range of 30,000 cm^{-1}. On the other hand, π donor ligands with lone pairs, such as the halogens Br^- and I^-, cause small splittings of about 10,000 cm^{-1}. Ligands such as NH_3 and H_2O with no π bonding capabilities have intermediate Δ_0 values. It may seem strange at first glance that the halogenic anions with their negative charge should cause smaller splitting than a neutral ligand such as NH_3. This result is unexpected on the basis of simple electrostatic theory, but it should be remembered that covalency is very important. The lone pair on NH_3 is in a well-directed sp^3 orbital and interacts well with the empty hybrid orbital of the metal. On the other hand, the filled p orbitals on the halogen interact quite strongly with the empty metal d orbitals. This interaction has the effect of transferring charge to the metal. The transfer of charge lowers the positive charge on the metal and reduces the D term in $10Dq$. The nature of the halogen interaction points out the difficulty of separating the D and q terms and justifies the practice of treating them together.

The spectrochemical series, which is an empirical measure of Dq, is:

$$CO \cong CN^- > NO_2 > \text{ethylenediamine (en)} >$$
$$NH_3 > H_2O > OH^- > F^- > Cl^- > Br^- > I^-$$

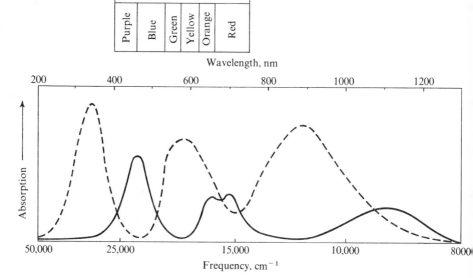

Fig. 7.4 The electronic spectra of $[Ni(H_2O)_6]^{2+}$ (———) and $[Ni\ en_3]^{2+}$ (-----). A "color map" of the visible region is also given. [From F. A. Cotton, *J. Chem. Educ.* **41**, 466 (1964) by permission.]

A simple demonstration of an application of the spectrochemical series is seen in the spectra of $[Ni(H_2O)_6]^{2+}$ and $[Ni(en)_3]^{2+}$, Fig. 7.4. An aqueous solution of $NiSO_4$ is pale green, but when ethylenediamine is added the color turns deep blue. The $[Ni(H_2O)_6]^{2+}$ has a minimum in its absorption (or a maximum in its transmittance) where green absorbs, but in all other regions of the visible spectrum it has some absorption; hence the green color. On the other hand, $[Ni(en)_3]^{2+}$ absorbs green light strongly and transmits only blue-purple and a little red: hence its deep blue color. The change in color occurs because the three ethylenediamine molecules interact more strongly with the metal than the H_2O's, pushing the σ^* (d) level (which is e_g^*), Fig. 7.3, up relative to H_2O, thereby increasing the orbital splitting and causing the observed (hypsochromic) shift to shorter wavelength.

7.4   The Spectrum of $[Ti(H_2O)_6]^{3+}$

The $[Ti(H_2O)_6]^{3+}$ ion is an example of a metal ion with a d^1 configuration. In the ligand field model the promotion energy $t_{2g} \to e_g$ should correspond to Δ_0, and one low-intensity ($g \to g$) band should be present.[a] The spectrum of this complex, Fig. 7.5, shows one band at about 20,400 cm^{-1} (500 nm). The pale red-purple color is due to the minimum of the curve in the wavelength region corresponding to these two colors, thus permitting them to be transmitted.

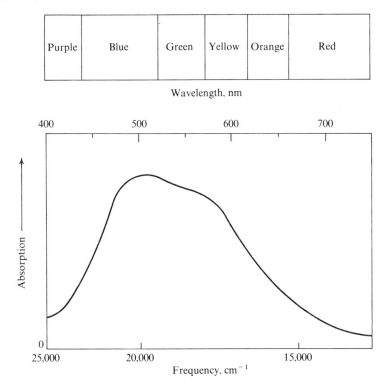

Fig. 7.5   The visible absorption spectrum of $[Ti(H_2O)_6]^{3+}$. [From F. A. Cotton, *J. Chem. Educ.* **41**, 466 (1964) by permission.]

[a] This is a typical example of a d→d transition in which an electron is excited from one d orbital to another. As long as the complex has a center of symmetry, all d orbitals are gerade and all d→d transitions are g→g. As will be shown in the next chapter, g→g transitions are quite strongly forbidden and consequently the intensities of the ligand field transition in octahedral and square planar complexes are very low. In tetrahedral complexes, no center of symmetry exists; consequently the g, u classification does not apply and the d→d transitions are substantially more intense.

In the free metal ion, all five d orbitals are, of course, degenerate. Although we can place a first electron in any of the five orbitals, any such placement leads only to one state, the ^2D state (*vide infra*), because all the d orbitals are degenerate and there are no electron-electron interactions. However, when two or more electrons are present, we can have several states which correspond to a single electron configuration. These states arise out of the many ways in which we can couple spin and orbital angular momenta. We shall now explore this subject.

7.5   Electronic Configurations and States

In Chapter 2 we discussed the four quantum numbers, n, l, m (or m_l), and s (or really m_s), of a single electron. Unfortunately, the quantum numbers l and s, and their projections m_l and m_s, are good quantum numbers only in the first approximation. Spin and orbital angular momenta (separately quantized according to s and l, respectively) are not separately observable quantities; only the total angular momentum is observable. Consequently s and l are only approximately quantized, but the total angular momentum is truly quantized, according to a further quantum number, j. This quantum number, like all angular momentum quantum numbers, gives rise to projections $m_j = j, j - 1, ..., -j$, and is related to l and s by $m_j = m_l + m_s$. Thus, if $l = 1$ (a p electron), $s = \frac{1}{2}$ for an electron, we have $m_l = 1, 0, -1$; $m_s = +\frac{1}{2}$, $-\frac{1}{2}$; and hence $m_j = \frac{3}{2}, \frac{1}{2}, \frac{1}{2}, -\frac{1}{2}, -\frac{1}{2}, -\frac{3}{2}$, or two states, one with $j = \frac{3}{2}$, the other with $j = \frac{1}{2}$. These two states may be thought of as resulting from the two possible alignments of orbital and spin angular momenta, Fig. 7.6.

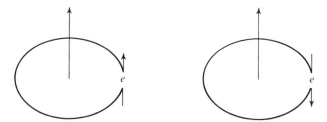

Fig. 7.6   Possible alignments of orbitals and spin angular momenta for a one-electron atom. [From B. E. Douglas and D. H. McDaniel, *Concepts and Models in Inorganic Chemistry*, Blaisdell, 1965, by permission.]

If we proceed to a similar consideration of many-electron atoms, we run into a situation in which two extreme theories have been developed, while the actual situation lies somewhere in between. One theory, called *Russel-Saunders* or *LS* coupling, is chiefly applicable to light atoms and will concern us primarily; the other, called *j-j coupling*, is most closely approximated by the heaviest atoms and will be considered only briefly.

In LS coupling, the m_l values of individual electrons combine to produce an M_L for the entire atom according to $M_L = \sum m_l$, and similarly the m_s combine to give M_S. The values of M_L and M_S are the projections of total atom orbital and spin angular momentum quantum numbers L and S, respectively. Throughout, capital letters will be applied to many-electron, lower case to one-electron, quantities. As usual, M_L and M_S take the values $L, L - 1, ..., -L$ and $S, S - 1, ..., -S$, respectively. L and S, that is, the orbital and spin angular momenta, combine to form a new, total angular momentum quantum number, J, through $M_J = M_L + M_S$, and $M_J = J$, $J - 1, ..., -J$. As a consequence J can have all integer or all half-integer values between $|L + S|$ and $|L - S|$.

The application of LS coupling implies that the interaction between spin and orbital angular momenta is negligible, that is, that spin-orbit coupling is weak. This is actually the case in light atoms.

Just as one-electron states are classified according to the value of l; s($l = 0$), p(1), d(2), f(3), etc., so the many-electron states are classified according to L; S($L = 0$), P(1), D(2), F(3), etc., continuing alphabetically. S determines the multiplicity (since the state is made up of several components of different M_S), given by $2S + 1$; the multiplicity is indicated by a preceding superscript, for example, ^{3}P means $L = 1, S = 1$.

These concepts are best illustrated by examples. Suppose that we have two electrons which we wish to place in 2p orbitals as, for example, in beryllium. The ways in which we can place the electrons are numerous; as we shall see, there are fifteen different ways in which we can do this, or fifteen different orbital wave functions (or microstates, as they are called), for 2p^2. For $n = 2$, we have $l = 1$ and $m_l = \pm 1, 0$, and we construct a table in which all possible p^2 configurations are considered. It is helpful, in constructing such a table, to group configurations having the same number of singly occupied orbitals. The orbitals with a single electron in them are indicated in our tables by a single x, whereas double occupancy of an orbital is denoted by two x's. Each x represents a spin of $\pm\frac{1}{2}$, and of course the configurations that have both p^2 electrons in the same orbital must have $M_S = 0$. The possible M_S values of the two electrons in separate orbitals are $1, 0, 0, -1$, corresponding to spin orientations in the same direction when $M_S = \pm 1$, or to orientations in opposite directions when both possibilities lead to $M_S = 0$. The results are shown in Table 7.2.

The data of Table 7.2 are summarized in Table 7.3, which shows the number of electronic configurations of p^2 which correspond to a given possible value of M_L and M_S. We see from Tables 7.2 and 7.3 that there is only one wave function with $M_L = 2$, and corresponding to $M_L = 2$, $S = 0$. Accordingly, there must be one D state with M_L values of $+2, 1, 0, -1$, and -2, and in this state the electrons are paired, leading to ^{1}D. The number

TABLE 7.2
The Wave Functions or Microstates for $2p^2$

	m_l			
1	0	-1	$M_L = \Sigma\, m_l$	$M_S = \Sigma\, m_s$
× ×			2	0
	× ×		0	0
		× ×	-2	0
×	×		1	$1, 0, 0, -1$
×		×	0	$1, 0, 0, -1$
	×	×	-1	$1, 0, 0, -1$

of microstates corresponding to a state is readily obtained by multiplying the spatial degeneracy of the state (orbital degeneracy) by the multiplicity of the state (spin degeneracy) to get the total orbital-spin degeneracy. This total is the total of the possible wave functions or microstates. For example, $^1D = 5 \times 1 = 5$, $^3P = 3 \times 3 = 9$, $^3F = 7 \times 3 = 21$, etc. If we take out of each element of our array, Table 7.3, the number of microstates (five) corresponding to the ^1D state, we have left the ten shown in Table 7.4. Now the highest remaining value of M_L is 1, and corresponding to this, the highest M_S value is 1. Hence, there must be a state $L = 1$, $S = 1$, or a ^3P state. Corresponding to this state, there is a total of nine microstates indicated by $M_L = +1, 0, -1$ and $M_S = +1, 0, -1$. When these are removed from the array of Table 7.4, we have left one microstate with $M_L = 0$ and $M_S = 0$,

TABLE 7.3
The Array of Wave Functions for $2p^2$

Note that the information to the right and below the dashed line is a repetition because of the inherent symmetry of the array.

	2		1	
	1	1	2	1
M_L	0	1	3	1
	-1	1	2	1
	-2		1	
		-1	0	1

$$M_S \rightarrow 0$$

TABLE 7.4
The Array of Wave Functions for $2p^2$ after Removing the Five Corresponding to ^1D

$$
\begin{array}{c}
\uparrow \\
M_L
\end{array}
\begin{array}{c|ccc}
1 & 1 & 1 & 1 \\
0 & 1 & 2 & 1 \\
-1 & 1 & 1 & 1 \\
\hline
 & -1 & 0 & 1
\end{array}
$$

$$M_S \rightarrow$$

obviously a ^1S state. For the p^2 electronic configuration with three degenerate orbitals, we have six possible ways in which we can place the first electron and five different ways of placing the second electron, but since we cannot distinguish which is the first and which is the second electron, we must divide the result by 2, the number of electrons to be distributed at each stage:

$$\text{Total number of microstates} = \frac{6 \times 5}{2} = 15$$

These 15 functions, with their M_L and M_S values shown in Table 7.3, correspond to three states: ^1D, ^3P, ^1S.

If we are to analyze the spectra of Ni^{2+}, which we displayed in Fig. 7.4, we recognize first that the d^8 configuration may be considered in terms of an inverted d^2 configuration by the hole-electron formalism that we discussed earlier. We must then consider how two electrons can be placed in the five degenerate d orbitals. Our problem is to determine the $(10 \times 9)/2 = 45$ microstates. Since we are discussing d orbitals, l is 2, and we have $m_l = \pm 2$, $\pm 1, 0$. We then construct Table 7.5 for the microstates and Table 7.6 to see the number of microstates for each particular value of M_L and M_S. In constructing Table 7.6, we reduce our work by recognizing that in such tables, for example, Table 7.3, there are two planes of symmetry, through the M_L and M_S values of zero, since the table repeats itself for negative values of M_L and of M_S. Table 7.6 shows a state with $L = 4$, hence a G state; since for $M_L = 4$, $M_S = 0$, this is a ^1G state. When the 9 microstates corresponding to the ^1G state are removed (one each from the nine elements with $M_S = 0$), we obtain Table 7.7, which (with the negative L and S values) lists 36 states. Table 7.7 shows a state with $M_L = 3$, $M_S = 1$, and hence there must be a ^3F state. If we remove the 21 microstates corresponding to the ^3F state, we are left with the array shown in Table 7.8; this is the same as the situation of Table 7.3, which leads to ^1D, ^3P, and ^1S states. Thus, the 45 microstates possible with the $3d^2$ (or $3d^8$) configuration lead to ^1G, ^3F, ^1D, ^3P, and ^1S states. We can check by determining the total spin-orbital degeneracies: $(1 \times 9) + (3 \times 7) + (1 \times 5) + (3 \times 3) + 1(1) = 45$.

TABLE 7.5
The 45 Wave Functions for d^2

m^l						
2	1	0	−1	−2	$M_L = \Sigma\, m_l$	$M_S = \Sigma\, m_s$
× ×					4	0
	× ×				2	0
		× ×			0	0
			× ×		−2	0
				× ×	−4	0
×	×				3	1, 0, 0, −1
×		×			2	1, 0, 0, −1
×			×		1	1, 0, 0, −1
×				×	0	1, 0, 0, −1
	×	×			1	1, 0, 0, −1
	×		×		0	1, 0, 0, −1
	×			×	−1	1, 0, 0, −1
		×	×		−1	1, 0, 0, −1
		×		×	−2	1, 0, 0, −1
			×	×	−3	1, 0, 0, −1

TABLE 7.6
A Summary of the 45 Wave Functions for d^2 for each M_L and M_S Value

(The values for negative M_L and M_S are identical to those for positive M_L and M_S but are omitted for brevity')

M_L		
4	1	
3	1	2
2	1	3
1	2	4
0	2	5
	1	0

$M_S \rightarrow$

This result is complicated enough; however, we often need the corresponding results for $d^3(d^7)$, $d^4(d^6)$, and d^5 configurations. These are tabulated in Table 7.9; the derivations, which are straightforward but tedious, are left to the student.

TABLE 7.7
Wave Functions for d^2 Remaining after Removing Microstates Corresponding to ^1G

(The values are repeated for negative M_L and M_S, but these are omitted.)

	3	1	1
	2	1	2
$\uparrow$ M_L	1	2	3
	0	2	4
		1	0

$M_S \rightarrow$

TABLE 7.8
Wave Functions for d^2 Remaining after Removing $9 + 21$ Microstates Corresponding to ^1G and ^3F

(The values are repeated for negative M_L and M_S, but these are omitted.)

	2		1
$\uparrow$ M_L	1	1	2
	0	1	3
		1	0

$M_S \rightarrow$

TABLE 7.9
The Terms Which Arise from the d^n Configurations

Configuration	Terms	Total No. of microstates
d^1, d^9	^2D	15
d^2, d^8	$^3F, {}^3P, {}^1G, {}^1D, {}^1S$	45
d^3, d^7	$^4F, {}^4P, {}^2H, {}^2G, {}^2F, 2x^2D, {}^2P$ (^2D occurs twice)	120
d^4, d^6	$^5D, {}^3H, {}^3G, 2x^3F, {}^3D, 2x^3P, {}^1I, 2x^1G, {}^1F, 2x^1D, 2x^1S$	210
d^5	$^6S, {}^4G, {}^4F, {}^4D, {}^4P, {}^2I, {}^2H, 2x^2G, 2x^2F, 3x^2D, {}^2P, {}^2S$	252

Whereas LS coupling is a very good approximation for light atoms, it breaks down more and more as spin-orbit coupling becomes stronger with increasing atomic weight, until, for the heaviest atoms in the periodic system, it is practically useless. In such cases, the atomic states are better described by j-j coupling, in which the quantum number J is determined by $M_J = \sum m_j$. Then the description of atomic states by L (i.e., as S, P, D, F, etc.) and by S (i.e., multiplicity) loses most of its meaning. Such situations will have little interest for us in what follows.

7.6 The Ground State and Energy Levels

The electronic configuration of d^2 or d^8 gives rise to the five states discussed above. We would like to know which of these states is the lowest in energy, that is, which is the ground state. The answer is readily obtained on the basis of Hund's rules, which tell us that the lowest state is the one with the highest possible multiplicity. If there are several states with this highest multiplicity, the one with highest L is lowest. Thus, our choice in the d^2 case is first between ^3F and ^3P, a decision which is then resolved in favor of ^3F. Of the components of each such state with different J, the ground state is the one with a minimum J value for subshells less than half-filled, and the one with maximum J for shells more than half-filled. On the basis of these rules, the ground state can readily be ascertained without going through the derivation of all the possible states. This is accomplished by writing an electronic configuration in which the electrons are placed in orbitals of highest m_l, each orbital being singly occupied before any double occupancy. The sum of the m_l values gives the L value, the S value is one-half the number of singly occupied orbitals, and the J values are determined and the assignment is based on Hund's rules. Thus for d^2 and d^8:

$M =$	2	1	0	-1	-2
d^2	↑	↑			
d^8	↑↓	↑↓	↑↓	↑	↑

For d^2, $L = 3$, $S = \frac{1}{2} \times 2 = 1$: hence ^3F. Possible J values are 4, 3, and 2; the lowest is selected, so the ground-state term is $^3F_2$. For d^8, $L = 3$, $S = 1$, and now, because the d's are more than half-filled, $J = 4$ for the lowest term: hence $^3F_4$.

In the absence of a magnetic field, the $2J + 1$ components of each state having different J are degenerate; this degeneracy is split by a magnetic field. In practice, a magnetic field is always present, because of the unpaired electrons, and the field causes a splitting, which generally increases, the heavier the atom. Since for most atoms usually considered this splitting is small, it is normally neglected, and the lowest state for both d^2 and d^8 is ^3F.

Although it was relatively simple to determine which of the five states of d^2 and d^8 is the ground state, the ordering and determination of the relative energies of the remaining four states is not a simple matter.[b] It requires a knowledge of the radial part of the wave function, and a numerical evaluation of electron repulsions. Usually, the problem is solved by reference to Moore's classic compilation of all known experimental atomic spectroscopic data.[c] In the d^2 case, the order of energies is $^3F < {}^1D < {}^3P < {}^1G < {}^1S$.

7.7   Free-Ion Terms and Their Splitting in a Weak Crystal Field

As a start, let us again consider the situation with a d^1 configuration, say the Ti^{3+} ion, which we have already briefly examined in Section 7.4. In the free ion, the d^1 configuration corresponds to the ^2D ground state (actually $^2D_{3/2}$ and $^2D_{5/2}$, but we neglect the J value). In the presence of a ligand field, the d degeneracy splits and in the case of O_h this split is into the t_{2g} and e_g sets. In the ground state the electronic configuration of the Ti^{3+} ion is t_{2g}^1, and in the excited state it is e_g^{*1}. In both cases the states are derived from the ^2D state of the free ion, and the transition from ground to excited state is $t_{2g}^1 \rightarrow e_g^{*1}$ or, in state symmetry notation, $^2E_g \leftarrow {}^2T_{2g}$. (Since we deal with a single electron, orbital and state notations are the same.) Thus we expect only one band in the spectrum of the $[Ti(L)_6]^{3+}$ complex, the energy of the transition depending on the nature of the ligand, L. The spectrum of $[Ti(H_2O)_6]^{3+}$ is shown in Fig. 7.5 (page 177); the energy at the position of maximum absorption is about 20,000 cm^{-1}. This energy corresponds to Δ_0; we would expect the spectrum to shift hypsochromically for $[Ti(NH_3)_6]^{3+}$ and bathochromically for $[TiCl_6]^{3-}$ as Δ_0 increases and decreases, respectively, as a function of the ligand.

In the d^2 case, such as V^{3+}, or the d^8 case, such as Ni^{2+}, the splitting is much more complicated because the d^2 free-ion configuration gives rise to five states, as derived in Section 7.6. Our problem is to find the energy of these states and to determine how each of them splits in the presence of a weak field. By definition we consider the field to be weak when the crystal field effect of the ligands is small compared to interelectronic repulsions, whereas a strong field is one that causes splitting in energies greater than the interelectronic repulsions. In the previous section we found that the ^3F state is the ground state, and we now proceed to try to answer the question of how ^3F splits in the presence of a weak crystal field.

Although we have so far defined an F state in terms of its quantum numbers L and M_L, we must recognize another definition: that F forms a basis for

---

[b] Hund's rule predicts only which state is the ground state and does not predict the order of the energies of the states.

[c] C. E. Moore, "Atomic Energy Levels," *Nat'l. Bur. Std. (U.S.) Circ.* 467, Vol. I, 1949; Vol. II, 1952; Vol. III, 1958; Washington, D.C.

TABLE 7.10

The Character Table for the Pure Rotation Group K

L		I	$(\infty C_p)^* (\infty C_\infty{}^\phi)\dagger$	∞C_2	∞C_3	∞C_4
0	S	$+1$	$+1$	$+1$	$+1$	$+1$
1	P	$+3$	$+1 + 2\cos\dfrac{2\pi}{p}$	-1	0	$+1$
2	D	$+5$	$+1 + 2\cos\dfrac{2\pi}{p} + 2\cos\dfrac{4\pi}{p}$	$+1$	-1	-1
3	F	$+7$	$+1 + 2\cos\dfrac{2\pi}{p} + 2\cos\dfrac{4\pi}{p} + 2\cos\dfrac{6\pi}{p}$	-1	$+1$	-1
4	G	$+9$	$+1 + 2\cos\dfrac{2\pi}{p} + 2\cos\dfrac{4\pi}{p} + 2\cos\dfrac{6\pi}{p} + 2\cos\dfrac{8\pi}{p}$	$+1$	0	$+1$
5	H	$+11$	$+1 + 2\cos\dfrac{2\pi}{p} + 2\cos\dfrac{4\pi}{p} + 2\cos\dfrac{6\pi}{p} + 2\cos\dfrac{8\pi}{p} + 2\cos\dfrac{10\pi}{p}$	-1	-1	$+1$
l		$2l+1$	$\left(+1 + \displaystyle\sum_{n=1}^{l} 2\cos\dfrac{2n\pi}{p}\right)\ddagger$ or $\displaystyle\sum_{n=-l}^{l} \exp(-2in\pi/p)$ or $\dfrac{\sin\left[2(l+\frac{1}{2})\pi/p\right]\S}{\sin(\pi/p)}$	$(-1)^l$		

* For all integral values of p.

† A rotation about the infinite-fold axis by an angle ϕ. Replace $2\pi/p$ by ϕ.

‡ The sum is zero for $l = 1$.

§ This formula is derived as follows:

$$\chi = \sum_{n=-l}^{+l} e^{-2in\pi/p} = e^{-2il\pi/p} \sum_{n=0}^{2l} (e^{2i\pi/p})^n$$

The sum on the right-hand side represents a geometric series, in which the ratio r of successive terms $e^{2i\pi/p}$ and the first term is 1. The sum of n terms of such a series is given by

$$s = \frac{r^n - 1}{r - 1}$$

Hence

$$\chi = \sum_{n=-l}^{+l} \exp(-2in\pi/p) = \exp(-2il\pi/p)\frac{\exp[2i\pi(2l+1)/p] - 1}{\exp(2i\pi/p) - 1}$$

$$= \frac{\exp[2i\pi(l+1)/p - \exp(-2il/p)}{\exp(2i\pi/p) - 1}$$

and, when both numerator and denominator are multiplied by $e^{-i\pi/p}/2i$,

$$= \frac{(e^{2i\pi(l+\frac{1}{2})/p} - e^{-2i\pi(l+\frac{1}{2})/p})/2i}{(e^{i\pi/p} - e^{-i\pi/p})/2i}$$

Now, since $\sin\theta = (e^{i\theta} - e^{-i\theta})/2i$, this becomes

$$\chi = \frac{\sin\left[2\pi(l+\frac{1}{2})/p\right]}{\sin(\pi/p)}$$

Frequently this formula is written as:

$$\chi = \frac{\sin(l+\frac{1}{2})\alpha}{\sin\alpha/2}$$

where $\alpha = \pi/p$.

one of the irreducible representations of the spherically symmetrical point group K_h (for *Kugel Gruppe*). In this group all symmetry operations are applicable; there are one I and one i, but an infinite number of planes (i.e., any plane is a σ), and an infinite number of axes of every order p from 2 to infinity. As usual, we can do the necessary work with the subgroup K of K_h, also called the pure rotational group R, which contains all the rotational axes and I, but not i or σ. Although the character table for this group is not usually given, it may be constructed, at least as far as we need it, by application of the various symmetry operations to the sets of atomic orbitals, s, p, d, f, etc.; cf. Table 7.10.

If we apply a weak octahedral field (say, a weak ligand field) to a set of functions which are a basis of an irreducible representation of K, the characters of the appropriate symmetry operations in K (where they are the characters of an irreducible representation) form the basis of a reducible representation in O. The irreducible representation in K which interests us is the one corresponding to the F state, made up of seven components $(M_L = \pm 3, \pm 2, \pm 1, 0)$. In point group K this set of components forms the basis of a single irreducible representation (F or $L = 3$), with the character of I as 7, that is, a seven-fold degenerate set. The characters of this irreducible representation in K correspond to the characters of a reducible representation in O. These characters (except under I, which, as indicated above, is 7) are obtained by determining the character under the class ∞C_p in Table 7.10 for $p = 2$, 3, and 4, the rotational axes which characterize point group O. Thus, for C_3, for example,

$$1 + 2\cos\frac{2\pi}{3} + 2\cos\frac{4\pi}{3} + 2\cos\frac{6\pi}{3} = 1 - 1 - 1 + 2 = 1$$

The other characters for the reducible representation are obtained similarly and are given in the first row of Table 7.11.

The resolution of the reducible representation (first row, Table 7.11) into its irreducible components can be readily achieved by the methods of Section 6.3 and equation 6.3 or by inspection. We will illustrate the inspection pro-

TABLE 7.11

The Reducible Representation of an F State in O, and Irreducible Representations of the Various Species of O

		I	C_2		$8C_3$	$6C_4$
K	F	$+7$	-1		$+1$	-1
			$6C_2'$	$3C_2''$		
O	A_1	$+1$	$+1$	$+1$	$+1$	$+1$
	A_2	$+1$	-1	$+1$	$+1$	-1
	E	$+2$	0	$+2$	-1	0
	T_1	$+3$	-1	-1	0	$+1$
	T_2	$+3$	$+1$	-1	0	-1

cedure because it is instructive with respect to the relationships between point groups K and O.

In Table 7.11 are shown the characters of the various irreducible representations of O, into which the reducible representation must be decomposed. Thus the seven degenerate functions of the F state of the free ion must split in the octahedral field into a set of functions which have a total of seven for the sum of their degeneracies. There are just eight types of possibilities for this splitting:

1. $2T + A$ 5. $3E + A$
2. $T + E + 2A$ 6. $2E + 3A$
3. $T + 4A$ 7. $E + 5A$
4. $T + 2E$ 8. $7A$

where T stands for either T_1 or T_2, and A for A_1 or A_2. We shall now see whether we can eliminate some of these possibilities.

The first and most powerful bit of information that we can use relates to the fact that all C_2 belong to a single class in K, but in O there are two classes of C_2: the three coincident with the C_4 forming one class, $3C_2''$, and the other six C_2 forming a second class, denoted as $6C_2'$. In the reducible representation, which results as the sum of the irreducible representations (of total weight seven) of the split functions in O, the sum of the characters for each of the two classes of C_2 must naturally be the same, namely -1, as shown in the first line in Table 7.11. Inspection of Table 7.11 shows that the characters under each of the two classes of C_2's are the same in T_1 and A_1, so that they may stand alone, but differ in T_2, E, and A_2. For the latter three, the difference is 2, but the sign of the difference in T_2 is opposite from that in E and A_2. Consequently, if the sum of the characters of the split functions is to be the same for each class of C_2's, each T_2 in the sum must be accompanied by either an E or an A_2, and each E and each A_2 must be accompanied by a T_2. Thus, a T_2 makes a contribution of $+1$ to the class $6C_2'$, and of -1 to the class $3C_2''$ (cf. Table 7.11). Similarly, an E makes contributions of 0 and $+2$, respectively. In sum, $T_2 + E$, they make contributions of $+1 + 0 = +1$ and $-1 + 2 = +1$, that is, equal contributions to the characters of the two classes of C_2's.

This required pairing of E and of A_2 with T_2 immediately eliminates possibilities 4, 5, 6, and 7 of our list from consideration, since they must contain a larger number of E than of T_2. Further eliminations are readily made by calculating the characters which result. Thus possibility 8 with $7A$ (all of which would have to be A_1, since there are no T) would give a character for C_2 of $+7$, which is incorrect, and hence can be eliminated. Possibility 3, which could be $T_1 + 4A_1$ or $T_2 + A_2 + 3A_1$, and possibility 2, which must be $T_2 + E + 2A_1$, both give $+3$ for the character of C_2 and are consequently eliminated. Thus we are left with only possibility 1, which could be $2T_1 + A_1$ or $T_1 + T_2 + A_2$; both combinations give -1 for C_2 and $+1$ for C_3, both of which are correct. But $2T_1 + A_1$ gives $+3$ for C_4, and only $T_1 + T_2 + A_2$ gives the correct answers for all the characters of the reducible representation. The same result could have been obtained by application of equation 6.3, using the character of -1 in the reducible representation for both classes of C_2 in the irreducible representations of point group O, Table 6.1. Thus for example:

$$n\Gamma_{T_2} = 1/24[(1 \cdot 3 \cdot 7) + (6 \cdot 1 \cdot -1) + (3 \cdot -1 \cdot -1) + 0 + (6 \cdot -1 \cdot -1)] = 1.$$

Thus, in a weak octahedral field, the F state splits up into three states, T_2, T_1, and A_2. Since the ground state of V^{3+} was a ^3F state, the three components are $^3T_2$, $^3T_1$, and $^3A_2$. Since the only unpaired electrons are d electrons and since all d orbitals are g, these components in O_h must be $^3T_{2g}$, $^3T_{1g}$, and $^3A_{2g}$.

A similar analysis can be performed on states of all L values. The results are shown in Table 7.12.

The energy differences between the various components of the state of the free ion depend entirely on the ligand field parameter $10Dq$. These are given, for the ground states of all d^n configurations, in Table 7.13.

We saw in a previous section that the configuration d^2 gives rise to, besides the ground state ^3F, a series of other states, ^1D, ^3P, ^1G, and ^1S. Their relative energies are obtained by an explicit calculation of the electron repulsions, expressed in terms of either the Slater-Condon parameters, F_2 and F_4, or the Racah parameters, $B = F_2 - 5F_4$ and $C = 35F_4$. However, the splitting pattern can be obtained in either formalism by methods exactly analogous to the one used above.

TABLE 7.12
Splitting of Russel-Saunders States in Octahedral and Tetrahedral Electrostatic Fields

States of Free Ion	States in Crystal Field
S	A_1
P	T_1
D	$E + T_2$
F	$A_2 + T_1 + T_2$
G	$A_1 + E + T_1 + T_2$
H	$E + 2T_1 + T_2$

TABLE 7.13
Diagrams for a Given Term and Configuration

The numbers under a term indicate (in parentheses) the orbital degeneracy and to its right the total, spin times orbital, degeneracy. [From B. N. Figgis, *Introduction to Ligand Fields*, Wiley, 1966, by permission.]

	Octahedral	Tetrahedral		Octahedral	Tetrahedral

$d^1, {}^2D$

Octahedral:
^2E_g
(2) 4 ↑
  10Dq
$^2T_{2g}$ ↓
(3) 6

Tetrahedral:
$^2T_2$
(3) 6 ↑
  10Dq
^2E ↓
(2) 4

$d^6, {}^5D$

Octahedral:
^5E_g
(2) 10 ↑
  10Dq
$^5T_{2g}$ ↓
(3) 15

Tetrahedral:
$^5T_2$
(3) 15 ↑
  10Dq
^2E ↓
(2) 10

$d^2, {}^3F$

Octahedral:
$^3A_{2g}$
(1) 3 ↑
  10Dq
$^3T_{2g}$ ↓
(3) 9
  8Dq
$^3T_{1g}$ ↓
(3) 9

Tetrahedral:
$^3T_1$
(3) 9 ↑
  8Dq
$^3T_2$ ↑
(3) 9
  10Dq
$^3A_2$ ↓
(1) 3

$d^7, {}^4F$

Octahedral:
$^4A_{2g}$
(1) 4 ↑
  10Dq
$^4T_{2g}$ ↓
(3) 12
  8Dq
$^4T_{1g}$ ↓
(3) 12

Tetrahedral:
$^4T_1$
(3) 12 ↑
  8Dq
$^4T_2$ ↑
(3) 12
  10Dq
$^4A_2$ ↓
(1) 4

$d^3, {}^4F$

Octahedral:
$^4T_{1g}$
(3) 12 ↑
  8Dq
$^4T_{2g}$ ↓
(3) 12
  10Dq
$^4A_{2g}$ ↓
(1) 4

Tetrahedral:
$^4A_2$
(1) 4 ↑
  10Dq
$^4T_2$ ↓
(3) 12
  8Dq
$^4T_1$ ↓
(3) 12

$d^8, {}^3F$

Octahedral:
$^3T_{1g}$
(3) 9 ↑
  8Dq
$^3T_{2g}$ ↓
(3) 9
  10Dq
$^3A_{2g}$ ↓
(1) 3

Tetrahedral:
$^3A_2$
(1) 3 ↑
  10Dq
$^3T_2$ ↓
(3) 9
  8Dq
$^3T_1$ ↓
(3) 9

$d^4, {}^5D$

Octahedral:
$^5T_{2g}$
(3) 15 ↑
  10Dq
^5E_g ↓
(2) 10

Tetrahedral:
^5E
(2) 10 ↑
  10Dq
$^5T_2$ ↓
(3) 15

$d^9, {}^2D$

Octahedral:
$^2T_{2g}$
(3) 6 ↑
  10Dq
^2E_g ↓
(2) 4

Tetrahedral:
^2E
(2) 4 ↑
  10Dq
$^2T_2$ ↓
(3) 6

7.8   Strong-Field Configurations and the Corresponding States

We now turn our attention to the strong-field case, where interelectronic repulsions are small compared to the crystal field. The strong-field configuration is obtained by assigning each d electron to either the t_{2g} or the e_g oribtal in O_h. In the d^1 (or d^9) case, there are no interelectronic repulsions and any crystal field always separates the t_{2g} and e_g orbitals sufficiently so that the lowest-energy configuration is invariably t_{2g}^1 and the higher-energy configuration $e_g{}^1$. With d^2, both electrons go into t_{2g}, giving the configuration t_{2g}^2; the next two higher-energy configurations are $t_{2g}^1 e_g{}^1$ and $e_g{}^2$. In order to find what spectroscopic states correspond to these configurations we must first recognize that we are dealing with two different cases. In $t_{2g}^1 e_g{}^1$ the two electrons are not equivalent;[d] in the case of *non-equivalent* electrons, we only need to form the direct product $t_{2g} \times e_g = T_{1g} + T_{2g}$; since the electrons are occupying different orbital sets, no restrictions on spin (or on multiplicity) occur and $t_{2g}^1 e_g{}^1$ give rise to the four states: $^3T_{1g}$, $^1T_{1g}$, $^3T_{2g}$, $^1T_{2g}$.[e] The method outlined works for any number of *non-equivalent* electrons.

In the t_{2g}^2 and $e_g{}^2$ configurations, however, the electrons are *equivalent*.[f] In this case, special problems arise since the Pauli principle places restrictions on the manner in which two electrons may be assigned to a single set of degenerate orbitals (see Section 7.5). For two equivalent electrons the problem is not too complicated because it can be shown that one state exists belonging to each species in the direct product of the species of the electrons; the only difficulty that arises is to determine which states have which multiplicity (see next Section). The state symmetry terms corresponding to t_{2g}^2, $t_{2g}^1 e_g{}^1$, and $e_g{}^2$ are found by taking the direct product and then finding the irreducible representations in point group O_h contained in such a direct product; we may again use the subgroup O, knowing that the configurations are g.

The procedure is shown for $t_2 \times t_2$ in Table 7.14, and the direct product and the corresponding species for the complete point group O are given in

TABLE 7.14
The Direct Product of $t_2 \times t_2$

O	I	$6C_4$	$3C_2''$	$6C_2'$	$8C_3$	
t_2	3	1	-1	-1	0	
$t_2 \times t_2$	9	1	1	1	0	$= T_{1g} + T_{2g} + E_g + A_{1g}$ (O_h)

[d] This may be verified by noting that $t_{2g}(1)e_g(2) \neq t_{2g}(2)e_g(1)$.

[e] The same method would be applicable to one electron each in each of two different sets of t orbitals, e.g. $1t_{2g}2t_{2g}$.

[f] Verified by $t_{2g}(1)t_{2g}(2) = t_{2g}(2)t_{2g}(1)$. Details of the formation of the direct product for the case $t_2 \times t_2$ (in O) are shown in Table 7.14, and a complete direct product table for O is given in Table 7.15.

TABLE 7.15

The Group Multiplication Table of (Results of Multiplying Symmetry) Species in Point Group O

O	A_1	A_2	E	T_1	T_2
A_1	A_1	A_2	E	T_1	T_2
A_2	A_2	A_1	E	T_2	T_1
E	E	E	$A_1 + A_2 + E$	$T_1 + T_2$	$T_1 + T_2$
T_1	T_1	T_2	$T_1 + T_2$	$A_1 + E + T_1 + T_2$	$A_2 + E + T_1 + T_2$
T_2	T_2	T_1	$T_1 + T_2$	$A_2 + E + T_1 + T_2$	$A_1 + E + T_1 + T_2$

Table 7.15. This table is readily applied to O_h by remembering that $g \times g = g = u \times u$, and $g \times u = u \times g = u$. Thus $t_{2g} \times t_{2g}$ from either Table 7.14 or Table 7.15 corresponds to $T_{1g} + T_{2g} + E_g + A_{1g}$. The other products, $t_{2g}e_g$ and $e_g{}^2$, lead to (Table 7.15) $T_{1g} + T_{2g}$ and $A_{1g} + A_{2g} + E_g$, respectively.

7.9   Determining Multiplicities; the Method of Descending Symmetries

7.9a   The Symmetric and Antisymmetric Product[g]

So far we have not written down the wave functions arising from any given configuration in detail. We have only specified that we have a configuration, for example, t_{2g}^2 in O_h, and that a number of states correspond to this configuration. Although we will consider complete wave functions in more detail in Chapters 11 and 12, we must anticipate here a few of the matters to be elaborated there; the most important of these is a more fundamental statement of the Pauli exclusion principle. According to this principle every wave function must change sign when the coordinates of any two electrons are exchanged. The complete wave functions of a two electron system, as in H_2 or in the π electron system of ethylene, in which the electrons occupy orbitals ψ_1 and ψ_2 are:

$$\frac{1}{2}[\psi_1(1)\psi_2(2) + \psi_2(1)\psi_1(2)][\alpha(1)\beta(2) - \alpha(2)\beta(1)] \qquad \text{(a)}$$

$$\frac{1}{\sqrt{2}}[\psi_1(1)\psi_2(2) - \psi_2(1)\psi_1(2)] \begin{cases} \alpha(1)\alpha(2) & \text{(b)} \\ \dfrac{1}{\sqrt{2}}[\alpha(1)\beta(2) - \alpha(2)\beta(1)] & \text{(c)} \\ \beta(1)\beta(2) & \text{(d)} \end{cases}$$

In these expressions, α and β are the two possible spin functions, and the numbers in parentheses after each function specifies the coordinates of the electron on which the function depends. It is easy to verify that wave func-

---

[g] The method outlined in this Section is given in more detail, and with proper derivations, in R. L. Ellis and H. H. Jaffé, *J. Chem. Ed.* **48**, 92 (1970).

tions (a) to (d) obey the Pauli principle since exchanging 1's and 2's in the parentheses changes the sign of each of the functions (a) to (d). However, the change occurs in different ways. In (a), the first bracket, which represents the space function, is symmetric, while the second bracket, which represents the spin function, is antisymmetric. On the other hand in (b) to (d) the first bracket, which is the space part of the function, is antisymmetric, while the spin part is symmetric. A symmetric spin function is characteristic of a triplet, an antisymmetric one of a singlet.

We now concentrate on the symmetry behavior of the space part of the singlet and triplet states derived from two electron configurations of equivalent electrons, as in our example of t_{2g}^2. The space part of the triplet is an antisymmetric (in electron coordinates) product of orbitals, that of the singlet a symmetric one (again in the electron coordinates). Our problem is to obtain the two sets of characters of each of these two products, $\{\chi^2\}$ and $[\chi^2]$, respectively. These characters are readily obtainable since they are given respectively by

$$\{\chi^2\}(G) = \tfrac{1}{2}\{[\chi(G)]^2 - \chi(G^2)\}$$
$$[\chi^2](G) = \tfrac{1}{2}\{[\chi(G)]^2 + \chi(G^2)\}$$

where G is any operation appropriate to the point groups and G^2 is the operation applied twice in succession. In other words, to find the characters of the symmetric or antisymmetric product of orbitals for any symmetry operation, all we need to know are the characters of the orbitals in our configuration under the operations, applied once, $\chi(G)$, and twice, $\chi(G^2)$.

Let us examine our example of the t_{2g}^2 configuration. The characters of t_{2g} is 3 for I, hence $\chi(G) = 3$. If the operation is applied twice, $\chi(G^2)$, the character is still 3 for I^2; hence from the formula $\{\chi^2\}(I) = \tfrac{1}{2}(9 - 3) = 3$ and $[\chi^2](I) = \tfrac{1}{2}(9 + 3) = 6$. Similarly, we obtain for the various operations of O:

G	I	C_4	C_3	C_2'	C_2''	
$[\chi(G)]$	3	-1	0	1	-1	
$[\chi(G)]^2$	9	1	0	1	1	
$[\chi(G^2)]$	3	-1	0	3	3	
$\{\chi^2\}(G)$	3	1	0	-1	$-1 = {}^3T_1$	
$[\chi^2](G)$	6	0	0	2	$2 = {}^1T_2 + {}^1E + {}^1A_1$	

The characters $[\chi(G^2)]$ are readily obtained since $I^2 = C_2'^2 = C_2''^2 = I$, $C_4^2 = C_2''$, $C_3^2 = C_3^{-1}$ (which has the same character as C_3).

Thus we have readily found that the antisymmetric product, the triplet, is just ${}^3T_1$, while the symmetric product, the singlet, has the reducible representation shown, which is readily reduced to ${}^1T_2 + {}^1E + {}^1A_1$. In O_h, then, the configuration t_{2g}^2 gives rise to $^3T_{1g} + {}^1T_{2g} + {}^1E_g + {}^1A_{1g}$ (15 micro-states).

In the strong field, the configuration d^2 splits up, as we saw above, into the three configurations t_{2g}^2, the lowest, $t_{2g}e_g$, and e_g^2. The configuration $t_{2g}e_g$, consisting of non-equivalent electrons, we saw gave rise to four states, $^3T_{2g} + {}^1T_{2g} + {}^3T_{1g} + {}^1T_{1g}$. The highest energy configuration, e_g^2, like t_{2g}^2, can be broken up into symmetric and antisymmetric products and resolved accordingly:

G	I	C_4	C_3	C_2	C_2''	
$[\chi(G)]$	2	0	-1	0	2	
$[\chi(G)]^2$	4	0	1	0	4	
$[\chi(G^2)]$	2	2	-1	2	2	
$\{\chi^2\}(G)$	1	-1	1	-1	1	$^3A_{2g}$
$[\chi^2](G)$	3	1	0	1	3	$^1A_{1g} + {}^1E_g$

Thus, e_g^2 gives rise to $^3A_{2g} + {}^1A_{1g} + {}^1E_g$.

7.9b The Method of Descending Symmetries

The method outlined in the preceding Section unfortunately is not applicable to more than two equivalent electrons, because in such cases it is not possible to write down separated space and spin functions. Accordingly, in the present Section we shall present an alternate method to assign the multiplicities to the states arising from configurations of equivalent electrons.

Let us illustrate this method first by applying it to the configuration e_g^2, which we have already solved in the preceding Section. There are 6 ($= 4 \times 3/2$) ways of distributing two electrons in two e_g orbitals, hence 6 microstates. As shown in Section 7.8, these microstates must correspond to an A_{1g}, an A_{2g} and an E_g state. Because we have two electrons and two orbitals, each state can have either singlet or triplet character:

$$e_g^2 \rightarrow {}^aE_g + {}^bA_{1g} + {}^cA_{2g}$$

where a, b, and c each are either 1 or 3, and with

$$2a + b + c = 6 \tag{7.1}$$

First, $a = 3$ is not possible, since b or c cannot be 0. Consequently, $a = 1$, and either $b = 1$, $c = 3$, or $b = 3$, $c = 1$. How do we decide between these two possibilities?

This decision can be made on the basis of the *method of descending symmetries*. In this method we correlate symmetry behavior (that is symmetry species or irreducible representations) in a highly symmetrical point group with behavior in groups of much lower symmetry. We imagine some elements of symmetry removed and determine into which symmetry species of the new, lower point group the various species of the original, higher point group go over when the symmetry is lowered. Usually, a reduction of symmetry is made from a point group having degenerate irreducible representations, for example O_h, to one in which all species are non-degenerate, for example C_{2v}. Unfortunately, reduction of symmetry usually can be made in more than one way (see Appendix I), and one particular way of making the reduction has been defined as the *principal correlation*. Table 7.16 and 7.17 give the principal correlations of all symmetry species in the more common degenerate point groups with those in C_{2v} and C_{2h}, respectively.

TABLE 7.16
Correlation Table of Various Point Groups with C_{2v}

C_{2v}	O_h	T_d^*	D_{6h}	C_{6v}	D_{4h}	C_{4v}	D_{2d}^*	D_{3h}^*
A_1	A_{1g}, A_{2g}	A_1	A_{1g}, A_{2u}	A_1	$A_{1g}, A_{2u}, B_{1g}, B_{2u}$	A_1, B_1	A_1, B_2	A_1'
A_2	A_{1u}, A_{2u}	A_2	A_{1u}, A_{2g}	A_2	$A_{1u}, A_{2g}, B_{1u}, B_{2g}$	A_2, B_2	A_2, B_1	A_1''
B_1			B_{1g}, B_{2u}	B_1				A_2'
B_2			B_{1u}, B_{2g}	B_2				A_2''
$2A_1$	E_g							
$2A_2$	E_u							
$A_1 + A_2$		E	E_{2g}, E_{2u}					
$B_1 + B_2$			E_{1g}, E_{1u}	E_1, E_2	E_g, E_u	E	E	
$A_1 + B_1 + B_2$	T_{1u}, T_{2u}	T_2						
$A_2 + B_1 + B_2$	T_{1g}, T_{2g}	T_1						
$A_1 + B_1$								E'
$A_2 + B_2$								E''

* Not a principal correlation.

TABLE 7.17
Correlation Table of Various Point Groups with C_{2h}

C_{2h}	O_h	D_{6h}^*	C_{6h}^*	D_{4h}^*	C_{4h}^*
A_g	A_{1g}, A_{2g}	A_{1g}, A_{2g}	A_g	$A_{1g}, A_{2g}, B_{1g}, B_{2g}$	A_g, B_g
A_u	A_{1u}, A_{2u}	A_{1u}, A_{2u}	A_u	$A_{1u}, A_{2u}, B_{1u}, B_{2u}$	A_u, B_u
B_g		B_{1g}, B_{2g}	B_g		
B_u		B_{1u}, B_{2u}	B_u		
$2A_g$	E_g	E_{2g}	E_{2g}		
$2A_u$	E_u	E_{2u}	E_{2u}		
$A_g + 2B_g$	T_{1g}, T_{2g}				
$A_u + 2B_u$	T_{1u}, T_{2u}				
$2B_g$		E_{1g}	E_{1g}	E_g	E_g
$2B_u$		E_{1u}	E_{1u}	E_u	E_u

* Not a principal correlation.

Unfortunately, it turns out that, in the *principal* correlation, E_g, A_{1g} and A_{2g} of O_h, the three components of the configuration e_g^2, all correlate with A_1 in C_{2v}, and hence this correlation provides no help in our problem. Thus we must search for another correlation which will allow us to make the desired decision.

Any of the three other correlations given in Table A1.2 (p. 378) will resolve the present problem. Let us take the third one in that table, based on C_2, σ_h, and σ_d. In this correlation, $E_g \rightarrow A_1 + B_1$, $A_{1g} \rightarrow A_1$, $A_{2g} \rightarrow B_1$. Now, the configuration e_g^2, in reduced symmetry, gives rise to the three configurations:

$$a_1^{\,2} \quad\text{------}\quad {}^1A_1$$
$$a_1 b_1 \quad\text{------}\quad {}^3B_1,\,{}^1B_1$$
$$b_1^{\,2} \quad\text{------}\quad {}^1A_1$$

which in turn give rise to the states indicated. We see that the only triplet is ${}^3B_1$, which correlates with ${}^3A_{2g}$; hence $c = 3$, $a = b = 1$, and e_g^2 give rise to ${}^3A_{2g} + {}^1A_{1g} + {}^1E_g$. This is the same result as obtained in Section 7.9a.

Let us now apply the method of descending symmetry to the configuration t_{2g}^2. This gives rise to states of T_{2g}, T_{1g}, E_g and A_{1g} symmetry (cf. Section 7.8), and to a total of $(6 \times 5/2) = 15$ microstates. Hence,

$$t_{2g}^2 \rightarrow {}^aT_{2g} + {}^bT_{1g} + {}^cE_g + {}^dA_{1g}$$
$$3a + 3b + 2c + d = 15 \tag{7.2}$$

Trial and error shows three possibilities:

$$(1)\ {}^1T_{2g} + {}^1T_{1g} + {}^3E_g + {}^3A_{1g}$$
$$(2)\ {}^1T_{2g} + {}^3T_{1g} + {}^1E_g + {}^1A_{1g}$$
$$(3)\ {}^3T_{2g} + {}^1T_{1g} + {}^1E_g + {}^1A_{1g}$$

Again, the principal correlation does not help, and we use the same one used above for e_g^2. We need two additional correlations from Table A1.2, $T_{2g} \rightarrow A_1 + A_2 + B_2$, $T_{1g} = A_2 + B_1 + B_2$. Hence, $t_{2g}^2 \rightarrow (a_1 + a_2 + b_2)$ $(a_1 + a_2 + b_2)$:

$$a_1 b_2 \quad\text{------}\quad {}^1A_2,\,{}^3A_2$$
$$a_1 b_2 \quad\text{------}\quad {}^1B_2,\,{}^3B_2$$
$$a_2 b_2 \quad\text{------}\quad {}^1B_1,\,{}^3B_1$$
$$a_1^{\,2} \quad\text{------}\quad {}^1A_1$$
$$a_2^{\,2} \quad\text{------}\quad {}^1A_1$$
$$b_1^{\,2} \quad\text{------}\quad {}^1A_1$$

Thus we find three triplet components, $^3A_2$, $^3B_1$, and $^3B_2$. Alternative (d) above would require $2^3A_1 + {}^3B_1$, alternative (3), $^3A_1 + {}^3A_2 + {}^3B_2$, and alternative (2) is the only one that requires $^3A_2 + {}^3B_1 + {}^3B_2$, as found. Thus, the answer is $a = c = d = 1$, $b = 3$, and t_{2g}^2 gives rise to $^3T_{1g} + {}^1T_{2g} + {}^1E_g + {}^1A_{1g}$, as found in Section 7.9a.

So far the method of descent of symmetry has not provided us with any new information which we could not obtain easier by the method of Section 7.9a. However, we shall now examine the configuration t_{2g}^3. This configuration could not be handled by the method of symmetric and antisymmetric products because, for three electrons, the wave function cannot be factored into spin and space functions. The descent of symmetry method is also complicated since some symmetry species in the direct product will no longer correspond to any state. This arises because the direct product includes such configurational components as three electrons in a single component of the triply degenerate orbital, which is forbidden by the Pauli principle. As a consequence we cannot write the equations, such as 7.1 and 7.2, in which each possible state has either of the possible multiplicities, and the degeneracies obtained in this fashion must add up to the total number of microstates.

TABLE 7.18
Correlation Table of O_h with D_{4h}

O_h	D_{4h}
A_{1g}	A_{1g}
A_{1g}	A_{1u}
A_{2g}	B_{1g}
A_{2u}	B_{1u}
E_g	$A_{1g} + B_{1g}$
E_u	$A_{1u} + B_{1u}$
T_{1g}	$A_{2g} + E_g$
T_{1u}	$A_{2u} + E_u$
T_{2g}	$B_{2g} + E_g$
T_{2u}	$B_{2u} + E_u$

The configuration t_{2g}^3 has $6 \times 5 \times 4/6 = 20$ microstates. The triple direct product $t_{2g} \times t_{2g} \times t_{2g}$ has $4T_{2g} + 3T_{1g} + 2E_g + A_{2g} + A_{1g}$ as components, and this list represents the *maximum* of states of any given symmetry possible, although we know that not all are going to occur. We shall now write down all the configurations (and the corresponding states) into which t_{2g}^3 breaks up upon reduction of symmetry by the principal correlation to D_{4h}, see Table 7.18; the choice is made because D_{4h} produces the cleanest separation in this particular case. The number of microstates represented by any of these subconfigurations is given in parentheses:

$$(4)\ b_{2g}^2 e_g \qquad -{}^2E_g$$
$$(12)\ b_{2g} e_g^2 \qquad -{}^4B_{1g} + {}^2A_{1g} + {}^2A_{2g} + {}^2B_{1g} + {}^2B_{2g}$$
$$(4)\ e_g^3 \qquad -{}^2E_g$$

The states corresponding to $b_{2g} e_g^2$ require a little elaboration. The configuration of e_g^2 of D_{4h} gives rise to four states, ${}^1A_{1g} + {}^3A_{2g} + {}^1B_{1g} + {}^1B_{2g}$, as can be shown by the methods of Section 7.9a or by reduction of symmetry (see Problem 7.11). Multiplication of these states (that is the configuration) by a b_{2g} function (remembering that this function brings with it a spin of $m_s = \pm 1/2$) gives $b_{2g} \times {}^1A_{1g} = {}^2B_{2g}$; $b_{2g} \times {}^3A_{2g} = {}^4B_{1g} + {}^2B_{1g}$; $b_{2g} \times {}^1B_{1g} = {}^2A_{2g}$; and $b_{2g} \times {}^1B_{2g} = {}^2A_{1g}$. The configuration e_g^3 is readily treated by the hole formalism; since e_g can accommodate at most four electrons, e_g^3 is equivalent to one e_g hole, and hence to one e_g electron, with ${}^2E_g$ as the only state.

Let us now look back to Table 7.18 and see if we can correlate the various states of the D_{4h} reduced configurations to O_h states. First, we note that the only quartet is ${}^4B_{1g}$ (in D_{4h}), which obviously must correlate with a ${}^4A_{2g}$ of O_h. Further, the ${}^2E_g$ of D_{4h} do not correlate by themselves, but only in combination with either ${}^2A_{2g}$ or ${}^2B_{2g}$; ${}^2E_g + {}^2A_{2g}$ with ${}^2T_{1g}$ and ${}^2E_g + {}^2B_{2g}$ with ${}^2T_{2g}$. This then accounts for all the D_{4h} states except a ${}^2A_{1g}$ and a ${}^2B_{1g}$. These could separately correlate with ${}^2A_{1g}$ and ${}^2A_{2g}$ of O_h, or jointly with ${}^2E_g$. Fortunately, the triple direct product of t_{2g}^3 only contained one A_{2g} representation, which was already used up in the quartet. Thus the only possibility is the ${}^2E_g$. Thus

$$t_{2g}^3 \rightarrow {}^4A_{2g} + {}^2E_g + {}^2T_{1g} + {}^2T_{2g}$$

To verify, $4 \times 1 + 2 \times 2 + 2 \times 3 + 2 \times 3 = 20$ microstates as predicted.

7.10   Correlation Diagrams

Let us summarize the last few pages. A free ion with a d^2 configuration possesses $10 \times 9/2 = 45$ possible orbital wave functions (or microstates), which correspond to ${}^3F$, ${}^1D$, ${}^3P$, ${}^1G$, and ${}^1S$ states. The total, orbital times spin, degeneracy of each of these states is, in the same order, $7 \times 3 = 21$, $5 \times 1 = 5$, $3 \times 3 = 9$, $9 \times 1 = 9$, and $1 \times 1 = 1$, for the total of 45. In the weak octahedral field each of these states splits into symmetry states which belong to point group O_h (Table 7.9). In an infinitely strong field where the t_{2g} and e_g orbitals are very far apart in energy, there are three possible electronic configurations for d^2; these are, in order of increasing energy, t_{2g}^2, $t_{2g}^1 e_g^1$, and e_g^2. The reducible representation for each of these configurations is obtained by taking the direct product, and the irreducible representations contained in these direct products are ascertained. The t_{2g}^2 configuration results in $6 \times 5/2 = 15$ wave functions, and the symmetry species corresponding to these 15 wave functions must have a total, orbital plus spin degeneracy of 15. The requirement and application of the method of

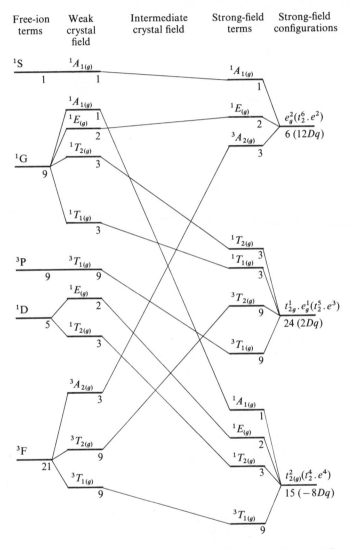

Fig. 7.7 Correlation for free-ion → strong-field configuration for: d^2 in O_h (V^{3+}, oct) and d^8 in T_d (Ni^{2+}, tet). The diagram is not to scale, except on the far right-hand side. T_d symmetry does not require the g subscripts. The numbers under a term indicate the total degeneracy. [From B. N. Figgis, *Introduction to Ligand Fields*, Wiley, 1966, by permission.]

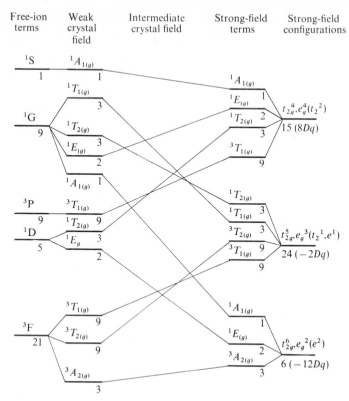

Free-ion terms	Weak crystal field	Intermediate crystal field	Strong-field terms	Strong-field configurations

Fig. 7.8 Correlation diagram for free-ion terms → strong-field configurations for d^2 in T_d (V^{3+}, tet) and d^8 in O_h (Ni^{2+}, oct). The diagram is not to scale except on the far right-hand side. The g subscript does not apply in T_d symmetry. The numbers under terms indicate their total degeneracy. [From B. N. Figgis, *Introduction to Ligand Fields*, Wiley-Interscience, 1966, by permission.]

descending symmetries yields, out of t_{2g}^2, the states $^3T_{1g} + {}^1T_{2g} + {}^1E_g + {}^1A_{1g}$. Appropriate states corresponding to the other two electronic configurations, $t_{2g}e_g$ and $e_g{}^2$, are obtained similarly.

The problem is now to connect or correlate the states in the weak field with those in the strong field, so that all intermediate fields can be spanned and the spectral properties of all ligands correlated. This is done by drawing lines from each state in the weak field to a corresponding state in the strong field. In drawing these connecting lines for the construction of these diagrams, called *Orgel diagrams*, three rules are observed. (1) Lines between states of identical designation never cross. This is the *noncrossing rule*. (2) The states in the strong field must be connected to states of the same multiplicity as the free-ion states from which they originate. (3) States which are the only ones of their type have energies that depend linearly on the crystal field strength, whereas when there are two or more states of identical designation, their lines will in general curve toward each other with respect to the ends as reference points. This is so because such states interact with one another.

With these rules in mind, we can construct the correlation diagram for the d^2 case in O_h (which is identical with d^8 in T_d); the results are shown in Fig. 7.7. Having established this correlation diagram, we can convert it to the d^8 case in O_h (or d^2 in T_d) by interchanging the most and the least stable electronic configuration in the strong-field side but keeping the order of energies of the states of each of the configurations. The result for d^8 in O_h (and d^2 in T_d) is shown in Fig. 7.8.

Now that we have developed the energy level diagram for Ni^{2+}, let us return to an examination of the spectra of $[Ni(H_2O)_6]^{2+}$ and $[Ni(en)_3]^{2+}$, shown in Fig. 7.4, p. 176. We see that each of these spectra consists of three rather well-defined bands. The energy level diagram tells us that the ground state, with configuration $t_{2g}^6 e_g^2$, is $^3A_{2g}$. Since only transitions between states of similar multiplicity are allowed (readily observable), the two lowest-energy transitions (the two long-wavelength bands) correspond to $t_{2g}^6 e_g^2 \rightarrow t_{2g}^5 e_g^3$ transitions and are $^3T_{1g} \leftarrow {}^3A_{2g}$ (lowest wavelength) and $^3T_{2g} \leftarrow {}^3A_{2g}$. The third and shortest-wavelength band has been assigned to $t_{2g}^6 e_g^2 \rightarrow t_{2g}^4 e_g^4$, another $^3T_{1g} \leftarrow {}^3A_{2g}$ transition. However, a two-electron promotion should be of very low intensity, and the third band may be a multiplicity-forbidden band observed because of spin-orbit coupling.

7.11   Charge-Transfer Bands

We have been discussing almost exclusively the promotion of an electron from an orbital in t_{2g} (in O_h) to e_g. Both of these sets of orbitals are mostly on the metal and represent the so-called $d \rightarrow d$ transitions. Because the d orbitals are all gerade, they are $g \rightarrow g$ transitions, and according to the selection rules (cf. Section 8.3) such transitions are forbidden and hence are of very low intensity. There are, of course, other transitions which can occur in the metal complexes. Reference to the O_h case involving σ bonds only, Fig. 6.6, shows that it should also be possible to promote an electron from t_{1u} to e_g^*. Such a transition is a $u \rightarrow g$ transition and is not forbidden by parity. Furthermore, it should occur at higher energies than the $d \rightarrow d$ transitions. In fact, there are strong short-wavelength bands in the spectra of most complexes.

It will be noted that the t_{1u} molecular orbitals receive their major contribution from the ligand, while the e_g^* level receives its major contribution from the metal. In the transition, therefore, the electron is promoted out of a predominantly ligand orbital into a predominantly metal orbital. Such transitions are called *charge-transfer transitions*, and the high-intensity, short-wavelength bands which frequently obscure the d → d transitions are called *charge-transfer bands*. The relative ease of electron transfer from the ligand to the metal is related to the "oxidizing" power of the metal ion, that is, its tendency to accept electrons, and to the "reducing" power of the ligand, that is, its tendency to give up electrons. In the hexahalo series, the order of the halogens in regard to their decreasing tendency to act as donors is I > Br > Cl > F. In general the greater the oxidizing power of the metal ion and the greater the reducing power of the ligand, the lower is the energy at which charge-transfer bands occur.

PROBLEMS

7.1 Explain the d orbital splitting of the central metal in a trigonal bipyramidal complex and give the order of the energies of these orbitals.

7.2 In square planar complexes a ligand *trans* to an unsaturated ligand possessing good π-accepting properties is particularly prone to nucleophilic substitution. Thus, e.g., $[C_2H_4PtCl_3]^-$ + Py → *trans*-$C_2H_4PtCl_2$(Py) + Cl^- is very, very fast. The transition state in such substitution reactions is assumed to be five coordinated trigonal bipyramidal in which the entering group, the leaving group, and the *trans* labilizing ligand occupy the trigonal plane. Such a transition state is favorable, presumably because the trigonal bipyramidal intermediate possesses four d orbitals with appropriate symmetry for some π interactions as compared to only the three d orbitals in D_{4h} symmetry that can interact with ligand p or π orbitals. In problem 7.1 the d orbital splitting in D_{3h} was determined. (*a*) Assume a stable metal complex with D_{3h} symmetry and assign the d orbitals to their appropriate symmetry species. (*b*) Show how the d_{xy} and $d_{x^2-y^2}$ set (*E'* in D_{3h}) interacts with p orbitals on ligands in the trigonal plane.

7.3 Determine the spectral terms arising from a p^3 configuration. Which of the terms is the ground state?

7.4 Determine the spectral states arising from a d^1 configuration. Which is the ground state?

7.5 Determine the spectral terms arising from an sd configuration. Which of the terms is the ground state?

7.6 Determine the spectral terms arising from the p^5d configuration. Which is the ground state?

7.7 (*a*) Verify that the T_{2g} species in O_h decomposes to $B_{2g} + E_g$ in D_{4h}, T_2 in T_d, and $A_2 + B_1 + B_2$ in C_{2v}. (*b*) Verify that T_{1u} in O_h goes over to $A_1 + B_1 + B_2$ in C_{2v}. (*c*) Verify that T_{1g} goes into $B_1 + B_2 + A_2$ in C_{2v}.

7.8 Indicate in state symmetry notation the lowest-energy, spin-allowed transition in a d^8 square planar complex.

7.9 Complexes $L_4Pt(II)$ and $L_4Au(III)$ are d^8 complexes of D_{4h} symmetry. Why does the ligand → metal charge transfer band appear at lower frequency in the Au(III) complex?

7.10 Although the spectra of $[PtBr_4]^{2-}$ and $[PtCl_4]^{2-}$ are very similar, the corresponding bands of the bromide occur at lower energies. Thus the L → M band in the former is at $36,000$ cm^{-1}, while in the chloride it is at $44,000$ cm^{-1}. What might be a resonable explanation for this difference?

7.11 Construct the Orgel diagram for the p^2 configurations of Xe in the square planar complex XeF_4.

7.12 Construct the Orgel diagram for the d^3 configuration in O_h.

7.13 Find the states corresponding to the configuration t_{1u}^2 by the method of Section 7.9a.

GENERAL REFERENCES
1. L. E. Orgel, *An Introduction to Transition-Metal Chemistry: Ligand Field Theory*, John Wiley and Sons, New York, 1966.
2. E. U. Condon and G. H. Shortley, *The Theory of Atomic Spectra*, Cambridge University Press, 1957.
3. J. C. Slater, *Quantum Theory of Atomic Electronic Structure*, Vol. I, McGraw-Hill Book Company, New York, 1960.
4. B. N. Figgis, *Introduction to Ligand Fields*, Wiley-Interscience, New York, 1966.
5. B. E. Douglas and D. H. McDaniel, *Concepts and Models of Inorganic Chemistry*, Blaisdell Publishing Co., Waltham, Mass., 1965.

8 Intensities and selection rules for electronic absorption spectra

8.1 General Discussion

It is customary in ultraviolet spectrophotometry to discuss and to describe the intensities of absorption bands in terms of the *molar absorptivities*, ϵ_{max} (frequently called the *molar extinction coefficients*). These quantities are of great convenience, since they are readily ascertained from the accurate measurement of the absorbance at the single wavelength at which maximum absorption occurs. Unfortunately, however, the value of ϵ_{max} is not directly related to any quantity obtainable from theory.

A spectral transition (i.e., an absorption line) corresponds to the energy difference between two well-defined states of the absorbing molecule. If the molecule absorbed light only at single wavelengths, the spectrum would consist of individual lines, like the emission spectra of atoms, and band widths would depend predominantly on the resolving power of the monochromator. Unfortunately, however, the situation is complicated by the fact that molecules possess, in addition to electronic energy, vibrational and rotational energy. Changes in these energies accompany the changes in electronic energy and give rise to the familiar band spectra of gases. Thus, a single electronic transition comprises many individual lines (bands), and the quantity of interest, the total energy transferred, must be the sum of the contributions from all these lines (bands).

In the case of liquids and solutions, the energy levels are broadened. Therefore the spectrum, which in the vapor state frequently consists of a multitude of rather sharp lines, is broadened into diffuse bands.

Thus the quantity of interest in theory is not ϵ_{max} but the integrated (absolute) intensity, I, which is the area under the absorption curve; its evaluation requires that the band under consideration be isolated from all other bands, and that the area under the absorption curve be measured. Such measurement can readily be achieved by the usual methods of counting squares (when the spectrum is plotted on graph paper), cutting out and weighing, or using a planimeter. Segregating and isolating an absorption band is much more difficult than measuring the area under it. Ultraviolet spectra rarely contain isolated bands, and we must generally extrapolate the band to the abscissa in a somewhat arbitrary manner. Since the experimental integrated intensity is mathematically defined by the integral

$$I = \int \epsilon \, dv$$

it is obvious that the spectrum must be plotted with an abscissa linear in frequency (or wavenumbers) and an ordinate linear in ϵ.

When comparing experimental and theoretically estimated intensities, it is often practical to use ϵ_{max} as a measure of integrated intensities. Particularly when comparing the "same" band (i.e., bands of substantially the same electronic origin) in the spectra of a series of closely related compounds, we frequently assume that the shapes or widths of the bands in the spectra under comparison are so similar that ϵ_{max} and I are proportional. A good approximation to the integrated intensity in these cases is achieved by multiplying ϵ_{max} by $v_{1/2}$, where $v_{1/2}$ is the *half-width*, that is, the width of the band in reciprocal centimeters (cm^{-1}) at $\epsilon = \frac{1}{2}\epsilon_{max}$.

In the ultraviolet region, strongly overlapping bands are frequently encountered and the molar absorptivities of such bands are frequently taken to be measured values. The intensities of such bands should not be taken for theoretical purposes, since they represent a superposition of ϵ_{max} for the main band, plus a contribution from one or more overlapping bands. The problem is particularly serious in connection with intensities of partly submerged bands appearing in the spectrum only as shoulders or inflection points. Evaluation of the integrated intensity of such bands requires analysis into separate bands, which is always somewhat arbitrary. The availability of commercial band-resolving instruments has greatly simplified this procedure.

8.2   Theoretical Treatment of Absorption Intensity

In principle, the problem of the calculation of the absolute intensity of an absorption (or emission) band is handled in the following way. The probabilities of spontaneous emission (A) and absorption (B) between two electronic states, i (initial) and f (final), are given by the Einstein coefficients:

$$A_{if} = \frac{64\pi^4 v^3 e^2}{3h} G_f D_{if}$$

and

$$B_{if} = \frac{8\pi^3 e^2}{3h^2 c} G_f D_{if}$$

where e is the charge of an electron, h Planck's constant, c the velocity of light, v the frequency of the emission, G_f the statistical weight of the final state (i.e., the number of degenerate wave functions to which absorption or emission can lead), and D_{if} the dipole strength, elaborated below. Mulliken has transformed the quantity B_{if} into a measure of intensity, the oscillator strength f, which is given by:

$$f = v \frac{8\pi^2 mc}{3h} G_f D_{if} = 1.096 \times 10^{11} v G_f D_{if}$$

and f is related to the absolute intensity by

$$f = 0.102 \frac{mc^2}{N\pi e^2} \int \epsilon \, dv = 4.315 \times 10^{-9} \int \epsilon \, dv$$

where m is the mass of the electron, N is Avogadro's number, ϵ is the molar absorptivity, and the integration extends over the entire absorption band. In the spectra generally encountered, the statistical factor G_f is unity and need not be considered further.

The dipole strength, D, thus is proportional to the intensity, and is the square of a relatively simple-appearing integral:

$$\sqrt{I} \propto \sqrt{D} = \int \Xi_i \mathbf{M} \Xi_f \, d\tau \tag{8.1}$$

where Ξ_i and Ξ_f are the total wave functions of the initial and final states of the molecule, respectively, $d\tau$ represents the product of the volume elements in the coordinates of all the nuclei and electrons, and $\mathbf{M}$ is the *dipole moment* vector, which needs some elaboration. The dipole moment of a molecule is defined as the distance between the centers of gravity of the positive and negative charges multiplied by the magnitude of these charges. Since, in the approximation usually used, location of the nuclei poses no problem, the center of gravity of positive charge is readily fixed. Electrons, however, cannot be located, but are described by a probability function, and the center of gravity of the electrons is consequently an average over the probability function. The average distance between the centers of gravity of the positive and negative charge is evaluated by averaging the distance (with direction, i.e., the vector), $\mathbf{r}$, from the center of gravity of positive charge to the electron. According to quantum mechanics, the desired average is given by the integral

$$\int \Psi_i \sum e\mathbf{r}\Psi_i \, d\tau = \int \Psi_i \mathbf{M}\Psi_i \, d\tau$$

where the summation extends over all electrons. For this reason the quantity $\mathbf{M} = \sum e\mathbf{r}$ is called the *dipole moment vector*. Equation 8.1 does not represent a dipole moment exactly but may be considered roughly to represent a charge migration or displacement during the transition; it is frequently called the *transition moment*.

The symmetry properties of the dipole moment vector $\mathbf{M}$ are of considerable interest. This vector, like all vectors, can be analyzed into three components, $M^2 = M_x^2 + M_y^2 + M_z^2$, in the direction of the three Cartesian coordinate axes. We have already analyzed the symmetry behavior of translations in the x, y, and z directions in point group C_{2v} and found these to transform as B_1, B_2, and A_1, respectively (Table 5.3). The usual difficulties occur when considering symmetry and antisymmetry with respect to rotation by more than two-fold axes, and in these cases the components of $\mathbf{M}$ belong in pairs to doubly degenerate, or in groups of three to triply degenerate, species, as we have discussed earlier.

Equation 8.1 permits the evaluation of the intensity, provided the total wave functions, Ξ_i and Ξ_f, are known. This is, however, never the case, and a long series of approximations is required. First it is assumed that rotational and vibrational wave functions can be factored out, and equation 8.1 reduces to

$$\sqrt{D} = \int \Psi_i \mathbf{M} \Psi_f \, d\tau \tag{8.2}$$

where Ψ_i and Ψ_f are the total electronic wave functions of initial and final states, and $d\tau$ is the product of the volume elements in terms of the coordinates of all the electrons.

Integrals of the form of equation 8.2 can be evaluated, and the calculations have been performed for a series of simple molecules. But such numerical computations become prohibitive in complexity for all but the simplest molecules, and further approximations become necessary.

Next, we frequently assume that (1) the Ψ can be factored into a series of one-electron functions (orbitals), ψ_j; (2) only a single electron is excited; and (3) the ψ_j are the same in the ground and excited states. Assumption 2 is generally valid and is further discussed below, but assumptions 1 and 3 are quite serious approximations. They do, however, permit reduction of the problem to manageable proportions, and in particular allow far-reaching qualitative and semiquantitative conclusions. In addition, the approximation frequently made, that the geometries of ground and excited states are equal, is serious. Using these assumptions, and arbitrarily letting the electron being excited be number 1, we can reduce equation 8.2 to

$$\sqrt{D} = \int \psi_i(1) \mathbf{M}_1 \psi_f(1) \, d\tau_1 \tag{8.3}$$

Equation 8.3 can be evaluated numerically as will be elaborated in Section 8.4, but the assumptions made above, as well as the approximate nature of the available ψ_j, mean that the results are very crude. Fortunately, however, the right-hand side of equations 8.2 and 8.3 frequently vanish identically, and this fact alone yields most important information.

8.3   Selection Rules

When the integral of equation 8.2 is equal to zero, the transition is called *forbidden* and should not, according to the *approximate* theory, occur at all. Actually, forbidden transitions do occur; and, as might have been expected, calculation by more refined theories gives small but nonvanishing values for the intensities. Observed intensities of forbidden transitions are generally quite small, much smaller than those of allowed transitions (for which equations 8.2 and 8.3 give nonvanishing intensities).

Fortunately, it is usually possible to tell, from the symmetry properties of wave functions, whether or not equations 8.2 and 8.3 will vanish. Relatively simple rules can be established which predict the vanishing or nonvanishing of the intensity integral; these rules are referred to as *selection rules*. The next few paragraphs will deal with the manner in which such selection rules may be established.

It is a well-known fact of mathematics that $\int y \, dx$ represents the area under the curve if y is plotted against x, areas below the abscissa being counted negative. Let us first examine $y(x)$, in which y is symmetric in x: for example, $y(x) = y(-x)$ (i.e., for two numerical values of x which differ only in sign, y has the same value). One such function, $y = x^2$, is shown in Fig. 8.1a; such a function is called *even*. The (shaded) area under the parabola can be expressed

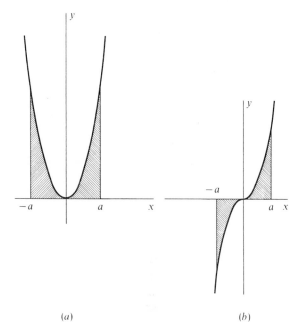

(a)   (b)

Fig. 8.1   Even (*a*) and odd (*b*) functions $y(x)$.

by the integral $\int_{-a}^{+a} y \, dx = 2\int_0^a y \, dx$, since the areas under the two halves of the curve both are above the abscissa and are equal. Next, let us examine a function antisymmetric in x. A plot of such a function, $y = x^3$, is shown in Fig. 8.1b, where it can be seen that, for every positive value of x, there is a positive y value, and for every negative value of x a corresponding negative value of y. Here $y(x) = -y(-x)$; such a function is called an *odd* function. Integration of $y = x^3$ gives $\int_{-a}^{+a} y \, dx = 0$, since the (shaded) areas under the curve for positive and negative x are equal, but above and below the abscissa, respectively, and thus cancel. The product of an even and an odd function is always odd, while the products of two even or of two odd functions are always even. The integral over all space of an odd function always vanishes identically, whereas the integral over all space of an even function may or may not vanish, but generally does not.

The integrals of equations 8.2 and 8.3 can be expanded into the sum of a series of integrals, each involving only one coordinate (i.e., one of the components of $\mathbf{M}$), and it is necessary only to demonstrate that the integrand is an odd function of the integration coordinate to demonstrate that the integral vanishes. Provided the molecule has some elements of symmetry, it is possible to demonstrate that equation 8.3 vanishes for certain transitions.

The first selection rule to be discussed is concerned only with molecules which have a center of symmetry (e.g., ethylene). All wave functions (orbitals) are either symmetric or antisymmetric with respect to the center [i.e., either gerade (g) or ungerade (u)], and all components of the vector $\mathbf{M}$ are of necessity ungerade. The product of two functions is ungerade only if one is gerade and the other ungerade. Hence the integrand of equation 8.3 can be gerade only if $\mathbf{M}$ is multiplied by an ungerade quantity, which can happen only if ψ_i, the ground-state wave function (orbital), and ψ_f, the excited-state wave function, are of unequal parity (gerade-ungerade character). Hence we have the following selection rule: $g \rightarrow u$ or $u \rightarrow g$, but $g \nrightarrow g$ and $u \nrightarrow u$. The crossed arrows indicate that the transitions are forbidden.

The second selection rule to be considered is concerned with the multiplicity of states. Thus a transition from a singlet to a triplet state, a singlet $\rightarrow$ triplet transition, will be considered. It is assumed that the initial wave function (electron configuration) is given by ψ_k^2, and the final one by $\psi_k\psi_j$. In ψ_k^2, the spins of the two electrons, since they occupy the same orbital, must be antiparallel; this is indicated by writing $\psi_k^2\alpha\beta$, where α and β are spin functions of opposite direction. In the excited triplet state, the function will be $\psi_k\psi_j\alpha^2$ (or β^2), since now the spins are parallel. [As will be shown in Chapter 11, a further spin function, $\alpha(1)\beta(2) + \alpha(2)\beta(1)$, also corresponds to parallel spins and makes up the third component of the triplet.] Equation 8.3 becomes

$$\int \psi_i\alpha\mathbf{M}\psi_f\beta \, d\tau \, d\sigma = \int \psi_i\mathbf{M}\psi_f \, d\tau \int \alpha\beta \, d\sigma$$

($d\sigma$ is the element of volume in the spin coordinates), and the integrand $\alpha\beta$ in the last integral is odd; hence the intensity vanishes, and singlet $\rightarrow$ triplet transitions are forbidden. In singlet $\rightarrow$ singlet transitions the spin functions can be safely neglected, since they reduce to a factor $\int \alpha^2 \, d\sigma = 1$.

The third selection rule deals with the symmetry of states; it says that a transition is forbidden unless the product $\psi_i \psi_f$ of equation 8.3 transforms as the same irreducible representation as at least one of the components of $\mathbf{M}$. Alternatively expressed, the third selection rule states that a transition is forbidden unless the direct product of the representation of $\psi_i \mathbf{M} \psi_f$ contains the totally symmetric representation. Before we consider the application of this selection rule, we must digress for a discussion of one-electron wave functions and symmetry states. We can do this by means of an example to develop first the molecular orbital energy diagram and then the symmetry states for the formaldehyde molecule.

TABLE 8.1

Symmetry Classification of the Basis Set of Atomic Orbitals of CH_2O

Species	O	C	H_2
a_1	s, p_z	s, p_z	$\dfrac{1}{\sqrt{2}}(s_1 + s_2) = \phi_s$
a_2	...	...	...
b_1	p_x	p_x	...
b_2	p_y	p_y	$\dfrac{1}{\sqrt{2}}(s_1 - s_2) = \phi_a$

In developing the MO's of formaldehyde, H_2CO, we use the H_2 group orbitals (GO's), just as we did in a discussion of the symmetry orbitals of the H_2 group in water, which, like formaldehyde, belongs to C_{2v}. The basis set of orbitals and their symmetry are given in Table 8.1. In an accurate MO treatment we would now form and solve a secular determinant (cf. Chapter 10) for each of the species: a 5×5 for a_1, a 3×3 for b_2, and a 2×2 for b_1. These solutions would lead to five MO's of species a_1, two of which are bonding, two are antibonding, and one is approximately nonbonding; all of them would extend over all four atoms. Similarly, in species b_2 a bonding, an antibonding, and a nonbonding MO extending over all four atoms would result. In species b_1, only a bonding and an antibonding π molecular orbital extending over the C and O atoms would be formed. However, an approximate, partially localized treatment will suffice for the present discussion.

The s and p_z orbitals of carbon and of oxygen can be combined into hybrid orbitals, one of the hybrids on each pointing up $(s + p_z)$ and one on each

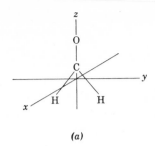

(a)

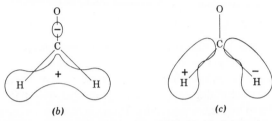

(b)   *(c)*

Fig. 8.2 Nonlocalized orbitals of CH_2 in formaldehyde. *(a)* Orientation of the molecule. *(b)* The σ bonding molecular orbital of species a_1 and *(c)* of species b_1.

pointing down $(s - p_z)$. Combination of $s - p_z$ of oxygen and $s + p_z$ of carbon forms a σ bonding a_1 molecular orbital, ψ_1, between these atoms. This orbital may be represented as:

$$\psi_1 = 2s(C) + 2p_z(C) + \lambda_1[2s(O) - 2p_z(O)]$$

(There is of course a corresponding antibonding molecular orbital, ψ_{10}, which must also be of species a_1, since it is generated from two AO's of this species.) The other hybrid of carbon, the $s - p_z$ orbital, combines with the group orbital ϕ_s of the H atoms to form the σ bonding a_1 molecular orbital, ψ_2 (and the corresponding antibonding orbital). This bonding MO is shown in Fig. 8.2*b*. It can be represented approximately by

$$\psi_2 = 2s(C) - 2p_z(C) + \lambda_2\phi_s$$

The p_y atomic orbital of carbon (species b_2) combines with the second of the GO's of the H atoms ϕ_a (species b_2) to form the bonding b_2 molecular orbital, ψ_3, shown in Fig. 8.2*c*. This MO can be represented approximately as

$$\psi_3 = 2p_y + \lambda_3\phi_a$$

A simple linear transformation, $\psi_2 \pm \psi_3$, assuming $\lambda_2 = \lambda_3$, leads to localized orbitals as follows:

$$\psi_+ = \psi_2 + \psi_3 = 2s(C) - 2p_z(C) - 2p_y(C) + \lambda(\phi_s + \phi_a)$$
$$= sp^2(C) + \lambda 1s(H_1)$$

where the last step follows by substituting the definition of $\phi_a + \phi_s$ given above. Similarly,

$$\psi_- = \psi_2 - \psi_3 = 2s(C) - 2p_z(C) - 2p_y(C) + \lambda(\phi_s - \phi_a)$$
$$= sp^2(C) + \lambda 1s(H_2)$$

Thus it is seen that ψ_+ and ψ_-, which are here written without regard to normalization factors, are localized on the carbon and one H atom each. A more realistic approach would have been to assume $\lambda_2 \neq \lambda_3$; this would have led to orbitals which were largely but not completely localized and are often called *equivalent orbitals*. They would then be of the form

$$\psi_\pm = sp^2(C) + (\lambda_2 \pm \lambda_3)1s(H_A) + (\lambda_2 \mp \lambda_3)1s(H_B)$$

In order to complete the description of formaldehyde, the two lone pairs on oxygen and the π bond of the carbonyl group must be introduced. One lone pair on oxygen is in the s + p_z orbital on the z-axis, pointing up from O and hence belonging to species a_1. The other lone pair on oxygen is in the p_y orbital and hence belongs to b_2. This description is exactly analogous to the lone pair description in H_2O. The C-O π bond, ψ_4, is antisymmetric with respect to the C_2^z axis and with respect to $\sigma(yz)$ and hence the lone pair belongs to species b_1. The set of MO's for formaldehyde is collected in Table 8.2.

TABLE 8.2
The Approximate Molecular Orbitals of Formaldehyde*

$$\psi_1 = 1a_1 = c_{11}[2s(C) + 2p_z(O)] + c_{12}[2s(O) - 2p_z(O)]$$
$$\psi_2 = 2a_1 = c_{21}[2s(C) - 2p_z(O)] + c'_{22}\phi_s$$
$$= c_{21}[2s(C) - 2p_z(O)] + c_{22}[1s(H_A) + 1s(H_B)]$$
$$\psi_3 = 1b_2 = c_{31}2p_y(C) + c'_{32}\phi_a$$
$$= c_{31}2p_y(C) + c_{32}[1s(H_A) - 1s(H_B)]$$
$$\psi_4\dagger = 1b_1 = c_{41}2p_x(O) + c_{42}2p_x(C)$$
$$\psi_5 = 3a_1 = [2s(O) + 2p_z(O)]/\sqrt{2} = n_2$$
$$\psi_6 = 2b_2 = 2p_y(O) = n_1$$
$$\psi_7\dagger = 2b_1^* = c_{71}2p_x(O) - c_{72}2p_x(C)$$
$$\psi_8 = 4a_1^* = c_{81}[2s(C) + 2p_z(O)] - c_{82}[2s(O) - 2p_z(O)]$$
$$\psi_9 = 3b_2^* = c_{91}2p_y(C) - c_{92}[1s(H_A) - 1s(H_B)]$$
$$\psi_{10} = 5a_1^* = c_{10,1}[2s(C) - 2p_z(O)] - c_{10,2}[1s(H_A) - 1s(H_B)]$$

* These MO's are cast in the usual LCAO form with coefficients c_{ij} rather than weighting factors λ_i. All c_{ij} are positive.
† In addition, there is the relation $c_{41} = c_{72}$ and $c_{42} = c_{71}$.

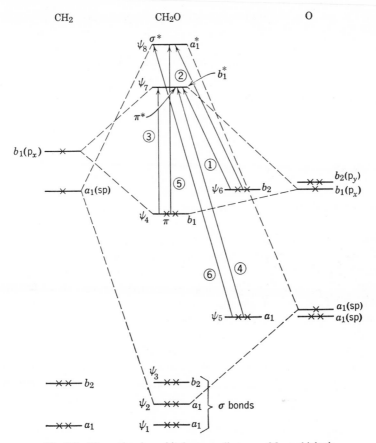

Fig. 8.3 The molecular orbital energy diagram of formaldehyde.

In addition to the occupied MO's described above, the following anti-bonding unoccupied MO's are of interest: b_1^*, the antibonding π-type orbital, ψ_7, corresponding to the b_1 π-type filled orbital; and possibly a_1^*, ψ_8, the antibonding σ-type orbital corresponding to the C-O bonding a_1-type orbital. The MOED is shown in Fig. 8.3. The ground state is then described (omitting the two a_1 and the b_2, σ bonding orbitals) as:

$$a_1{}^2 b_1{}^2 b_2{}^2 \quad \text{or} \quad 3a_1{}^2 1b_1{}^2 2b_2{}^2$$

This is the description of the electronic configuration in terms of the symmetry of the individual MO's; the direct product (cf. Section 6.9) corresponds to A_1. Since all electrons are paired, our symmetry state is a singlet and therefore is characterized as $^1A_1$. In the MOED we show six possible one-electron transitions. The electronic configurations of the six resulting

excited states and the symmetry states to which the configurations belong
are:

(1) $a_1{}^2b_1{}^2b_2b_1^*$: $^1A_2, {}^3A_2$

(2) $a_1{}^2b_1{}^2b_2a_1^*$: $^1B_2, {}^3B_2$

(3) $a_1{}^2b_1b_2{}^2b_1^*$: $^1A_1, {}^3A_1$

(4) $a_1b_1{}^2b_2{}^2b_1^*$: $^1B_1, {}^3B_1$

(5) $a_1{}^2b_1b_2{}^2a_1^*$: $^1B_1, {}^3B_1$

(6) $a_1b_1{}^2b_1{}^2a_1^*$: $^1A_1, {}^3A_1$

Now we have the six symmetry states corresponding to the six electronic
configurations, and we can redraw our MOED in terms of these symmetry
states, Fig. 8.4.

A comparison of Figs 8.3 and 8.4 brings out the distinction between
configuration (one-electron) notation and molecular-state notation. In
particular, it should be noted that the one-electron notation implies that
transitions 1, 3, and 4 of Fig. 8.3, $b_2 \rightarrow b_1^*$, $b_1 \rightarrow b_1^*$, and $a_1 \rightarrow b_1^*$, involve
excitation of electrons from different orbitals in the *same* ground state to
the same orbital in excited states, which, as states, are different, as is seen
from their configurations. This is probably clearer in Fig. 8.4, where the
ground state for all transitions has the symbol N.

It should be noted that the arrows marked 1, 2, and 3 in Fig. 8.3, the orbital
diagram, and Fig. 8.4, the term level diagram, represent the same energy

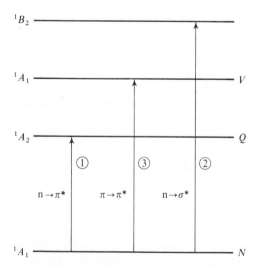

Fig. 8.4 Term level diagram for formaldehyde, showing three possible transitions.

quantities, at least in the crude approximation used here. This is true even though in the context of their respective figures they *appear* to represent different things. In the MOED, Fig. 8.3, these arrows represent the energy required to promote an electron from one orbital to another. In the term level diagram, Fig. 8.4, however, the arrows represent the excitation of the entire molecule from one state to another. Actually both processes are identical, since promoting an electron from one orbital to another excites the molecule from one state to another.

Let us now return to the application of the selection rule regarding symmetry of states and consider only the first three transitions in formaldehyde. Transitions 1, 2, and 3 shown in Figs. 8.3 and 8.4 may be designated as follows:

(1) n $\to$ π^* or $Q \leftarrow N$ or $^1A_2 \leftarrow {}^1A_1$

(2) n $\to$ σ^* or $^1B_2 \leftarrow {}^1A_1$

(3) $\pi \to \pi^*$ or $V \leftarrow N$ or $^1A_1 \leftarrow {}^1A_1$

The letters N, Q, and V refer, respectively, to the normal or ground state, a state involving a n $\to$ π^* transition, and the upper (valency) state of the high-intensity $\pi \to \pi^*$ transition. In the point group C_{2v}, to which formaldehyde belongs, the three components, M_x, M_y, and M_z, of **M** belong to the symmetry species B_1, B_2, and A_1, respectively. For the three transitions listed above, the products $\psi_i\psi_f$ belong to the species $A_1A_2 = A_2$, $A_1B_2 = B_2$, and $A_1A_1 = A_1$, respectively. Consider first the n $\to$ π^* transition, where the product $\psi_i\psi_f$ is A_2. Since none of the components of **M** belongs to species A_2, the integrand $\psi_i\mathbf{M}\psi_f$ in the intensity-determining integral, equation 8.2, is an odd function and hence the integral vanishes. The longest-wavelength n $\to$ π^* transition thus is forbidden, explaining the low intensity observed. The $\pi \to \pi^*$, the $V \leftarrow N$ transition, is very intense; the $\psi_i\psi_f$ product belonging to species A_1 makes this transition allowed, and the geometric rough coincidence of the π and π^* orbitals leads to very large values of $\psi_i\psi_f$ and hence to the usual high intensity of $V \leftarrow N$ transitions. Transition 2, designated as n $\to$ σ^* or $^1B_2 \leftarrow {}^1A_1$, is allowed, but its energy is such that it does not occur in the accessible ultraviolet, as would be true also of transitions 4, 5, and 6, Fig. 8.3. It should now be apparent that the first selection rule, the $g \to u$ rule, is just a special case, although a particularly important one, of this third selection rule.

In discussions of the spectral behavior of particular compounds we frequently encounter statements such as the following: "Because of the high symmetry, many of the transitions in this molecule are forbidden." The statement is seldom amplified and may seem puzzling because the relationship between high symmetry and forbiddenness is not immediately obvious. Accordingly, we shall now discuss this relationship explicitly.

In order for a transition to be allowed, the product $\psi_i\psi_f$ must belong to the same symmetry species as at least one of the components of $\mathbf{M}$. A molecule with no elements of symmetry belongs to point group C_1, in which there is only one symmetry species. The product $\psi_i\psi_f$ must belong to the same species as all the components of $\mathbf{M}$, and hence all transitions are allowed. In point groups C_s and C_2 at least one component of $\mathbf{M}$ belongs to each of the two symmetry species, and again the product $\psi_i\psi_f$ must belong to the same species as one of the components of $\mathbf{M}$ and all transitions are allowed. In point group C_{2v} there are four symmetry species, and the three components of $\mathbf{M}$, namely M_x, M_y, M_z, belong to (or transform as) symmetry species B_1, B_2, A_1, respectively. Since each of the three components of $\mathbf{M}$ can belong to only one symmetry species and there is a total of four such species, there must be at least one which no component of $\mathbf{M}$ belongs. Then some products of $\psi_i\psi_f$ may belong to this species and hence correspond to a forbidden transition. This is the example we have just discussed.

It follows that, the "higher" the symmetry of a molecule, the larger is the number of symmetry species. It is of course not strictly possible to assign a unique scale of increasing degree of symmetry, but, in general, the number of symmetry species increases with increasing symmetry. Products of $\psi_i\psi_f$ in no more than three symmetry species can lead to allowed transitions, since the three components of $\mathbf{M}$ can at most belong to three species (actually, frequently two components belong to one species, usually a doubly degenerate species, or all three belong to a single triply degenerate species as in point group T_d). Hence the larger the number of symmetry species, the greater is the probability of finding $\psi_i\psi_f$ products belonging to a species to which no component of $\mathbf{M}$ belongs and, accordingly, corresponding to forbidden transitions.

Summarizing the selection rules, we can say that there are at least three basic types of forbidden transitions:

1. Parity-forbidden transitions, that is, $g \nrightarrow g$ and $u \nrightarrow u$. The visible spectra of simple transition-metal complexes are examples: the molar absorptivities (extinction coefficients) usually are less than 100.

2. Spin-forbidden transitions, as the singlet-triplet transitions in ethylene. Their intensity is very low, generally $\epsilon < 1$.

3. Symmetry-forbidden transitions. Depending on the degree of forbidden-ness, they may have values of ϵ ranging from 100 to 5000.

So far the discussion has been restricted to transitions involving the excitation of a single electron. It can be shown that the integral in equation 8.2 vanishes when Ψ_i and Ψ_f, expressed as products of one-electron functions ψ, differ in more than one of the ψ's, and hence many-electron excitations are forbidden. Only the fact that the many-electron configurations cannot strictly be expressed as products of one-electron orbitals ψ_j, but are linear combinations of such configurations (configuration interaction; cf. Chapter 12), leads to a nonvanishing probability of many-electron excitations.

The proof that two-electron excitations are forbidden goes as follows. Using

$$\Psi_i = \psi_1(1)\psi_2(2) \dots \psi_{n-1}(n-1)\psi_n(n)\psi_{n+1}(n+1)$$
$$\Psi_f = \psi_1(1)\psi_2(2) \dots \psi_{n-1}(n-1)\psi_{n+2}(n)\psi_{n+3}(n+1)$$

and

$$\mathbf{M} = M(1) + M(2) + \dots + M(n-1) + M(n) + M(n+1) = \sum M(j)$$

equation 8.2 becomes the integral over two products of wave functions, multiplied by the sum of the $M(j)$. This can be separated into a sum of integrals, each of which has the form

$$\int \psi_1(1)\psi_2(2) \dots \psi_{n-1}(n-1)\psi_n(n)\psi_{n+1}(n+1)M(j)\psi_1(1)\psi_2(2) \dots$$
$$\times \psi_{n-1}(n-1)\psi_{n+2}(n)\psi_{n+3}(n+1) \, d\tau_1 \, d\tau_2 \dots d\tau_{n-1} \, d\tau_n \, d\tau_{n+1}$$

Each of this set of integrals may be factored into a product of integrals, of which the one for $j = n$ is

$$\int \psi_1(1)\psi_1(1) \, d\tau_1 \int \psi_2(2)\psi_2(2) \, d\tau_2 \dots \int \psi_{n-1}(n-1)\psi_{n-1}(n-1) \, d\tau_{n-1}$$
$$\times \int \psi_n(n)M(n)\psi_{n+2}(n) \, d\tau_n \int \psi_{n+1}(n+1)\psi_{n+3}(n+1) \, d\tau_{n+1}$$

In this product, the first $n-1$ integrals are normalization integrals, and hence unity, but the last one is an orthogonality integral, and hence zero; consequently the whole product vanishes. The same is true for $j = n + 1$; if $j < n$, there will be two vanishing factors. Consequently each term of the sum of partial integrals vanishes, the predicted intensity is zero, and any two-electron transition is forbidden.

8.4   Numerical Calculation of Intensities

In the preceding section we have used equation 8.3 to derive selection rules, that is, to determine in what cases the intensity of absorption is identically zero according to the approximate theory used. In this section we shall examine how equation 8.3 can be used to make actual calculations of intensities. For this purpose we rewrite the equation as follows:

$$\sqrt{D} = e \int \psi_i \mathbf{r} \psi_f \, d\tau \tag{8.3a}$$

where all quantities refer to the single electron being excited. Complete numerical solutions for such integrals are difficult to find and require a knowledge of the ψ involved. We shall use, as usual, LCAO functions, and substitute these for the ψ. This gives

$$\sqrt{D} = e \int \sum_r c_{ir}\phi_r \mathbf{r} \sum_s c_{fs}\phi_s \, d\tau$$

$$= e \sum_r \sum_s c_{ir}c_{fs} \int \phi_r \mathbf{r} \phi_s \, d\tau$$

The integral in the last line then produces the only problem, and such integrals are approximated as follows; the integral represents an averaging of the radius vector $\mathbf{r}$ from the center of gravity to the position of the electron over the electron density represented by $\phi_r \phi_s$. If r and s are different orbitals, $r \neq s$, then the neglect of overlap integrals, or better the neglect of differential overlap (cf. Chapter 12), suggests that the entire integral should be neglected. If r and s are the same orbital, $r = s$, the integral reduces to $\mathbf{r}_r \int \phi_r \phi_r \, d\tau = \mathbf{r}_r$, that is, the vector distance to the atom on which ϕ_r is centered. Thus,

$$\int \phi_r \mathbf{r} \phi_s \, d\tau = \mathbf{r}_r \quad \text{if } r = s$$
$$= 0 \quad \text{if } r \neq s$$

Since only terms with $s = r$ do not vanish, the double summation over r and s reduces to a single summation and

$$\sqrt{D} = e \sum_r c_{ir}c_{fr}\mathbf{r}_r \qquad (8.4)$$

Let us now apply this formula to formaldehyde, for which we wrote wave functions (Table 8.2) and derived selection rules in the preceding section. Let us focus on the following three transitions: (1) $b_2 \to b_1^*$ (or $n_1[p_y(O) \to \psi_7(\pi^*)]$), (3) $b_1 \to b_1^*$ (or $\psi_4(\pi) \to \psi_7(\pi^*)$), and (4) $a_1 \to b_1^*$ (or $n_2[s(O) + p_z(O) \to \psi_7(\pi^*))$, and make the necessary substitutions. For transition 1, the only atomic orbital for which c_{ir} is not zero is $n_1 = p_y(O)$; for this orbital c_{fr} is zero. Thus, the sum in equation 8.4 has all zero terms, and the transition is forbidden, as we saw before. For transition 4, the only two orbitals for which c_{ir} are nonzero are the orbital n_2, that is, the s and p_z orbitals of oxygen, which do not appear in ψ_7, so that the corresponding c_{fr} are zero. Consequently, the intensity is again zero, but not, this time, because of symmetry. We shall return to this transition later. Finally let us examine transition 3. Here c_{ir} and c_{fr} for the oxygen $2p_x$ orbital are both nonzero, and of equal sign, and for the carbon $2p_x$ orbital are both nonzero, and of opposite sign. Furthermore, in this special case, orthonormality requires the special relation given by

$$\psi_4 = c_{41}p_x(O) + c_{42}p_x(C)$$
$$\psi_7 = c_{71}p_x(O) - c_{72}p_x(C)$$

with $c_{41} = c_{72} = c_1, c_{71} = c_{42} = c_2.$

Substituting these values into equation 8.4 gives

$$\sqrt{D} = e(c_1 c_2 \mathbf{r}_O - c_1 c_2 \mathbf{r}_C)$$

Now, the radius vectors to oxygen and carbon are just their z coordinates,[a] and consequently

$$\sqrt{D} = e c_1 c_2 (z_O - z_C)$$

This expression shows that it is immaterial from which point the z coordinate is measured, a result which is quite general. Since $z_O - z_C$ is just the bond distance, d_{CO}, the intensity finally is

$$\sqrt{D} = e c_1 c_2 d_{CO}$$

This is a fairly large quantity, since both c_1 and c_2 are substantial. Actually, the more nearly equal they are, the less polar the bond, and the larger the product $c_1 c_2$, and consequently the intensity. As a result, the $\pi \to \pi^*$ transition in ethylene is more intense than that in formaldehyde. Thus, we have shown that the $b_1 \to b_1^*$ transition in formaldehyde is allowed and intense.

Let us note also that the intensity depends only on the z component of $\mathbf{r}$ and consequently is z-polarized (cf. the next section).

The method here outlined is quite general and gives rather reasonable intensities, particularly relative values. As long as crude approximations are used in calculating the wave functions, the method is as good as the quality (accuracy) of the wave functions justifies.

Now let us return briefly to the two forbidden transitions, 1 and 4, and also to equation 8.3a. The integrals may now be written, using c_1 and c_2 as defined above, as follows:

(1) $\sqrt{D}_{(1)} = e \int p_y(O) \mathbf{r} [c_2 p_x(O) - c_1 p_x(C)] \, d\tau$

 $= e [c_2 \int p_y(O) \mathbf{r} p_x(O) \, d\tau - c_1 \int p_y(O) \mathbf{r} p_x(C) \, d\tau]$

(4) $\sqrt{D}_{(4)} = e \int [s(O) + p_z(O)] \mathbf{r} [c_2 p_x(O) - c_1 p_x(C)] \, d\tau$

 $= e [c_2 \int s(O) \mathbf{r} p_x(O) \, d\tau + c_2 \int p_z(O) \mathbf{r} p_x(O) \, d\tau -$

 $- c_1 \int s(O) \mathbf{r} p_x(C) \, d\tau - c_1 \int p_z(O) \mathbf{r} p_x(C) \, d\tau]$

The last term in (1) and the last two in (4) involve integration over orbitals on different atoms, and consequently can be neglected with reasonable safety. We are then left with two types of integrals to evaluate:

$\int s(O) \mathbf{r} p_x(O) \, d\tau$

$\int p_y(O) \, [\text{or } p_z(O)] \, \mathbf{r} p_x(O) \, d\tau$

[a] The simplification made here is a special case. It should be noted that equation 8.4 represents a *vector sum*, which must be evaluated according to the rules of vector addition. We will come back to this problem in the next section.

Each of these involves only orbitals of a single atom, and its value consequently is an atomic property. Since the atom has a center of symmetry, even
though the molecule does not, and since both the p orbitals and the radius
vector **r** are u quantities, the second integral has a u integrand and vanishes
by *local symmetry*. The first integral, however, has an integrand which is
g, hence does not vanish by symmetry, and actually has a small but significant nonzero value. Consequently transition 4, which involved this integral,
turned out as forbidden only because we chose to neglect this small integral.
We may also note that, since p_x behaves like x, it is the x component, and only
the x component, which gives the integral a nonzero value, and hence
transition 4 is x-polarized (cf. Section 8.5).

These facts about n $\rightarrow$ π^* transitions have been given recognition in
Platt's nomenclature scheme, where states like the upper state of the $n_1 \rightarrow \pi^*$
transition (1) are called U (for **u**nallowed), and those like the upper state
of the $n_2 \rightarrow \pi^*$ transition (4) are called W (for allowed). In general, n $\rightarrow$ π^*
states are U if the lone-pair n electrons are pure p electrons, and they are
W if the lone pair occupies an sp hybrid orbital.

8.5   Polarization of Absorption Bands

The theory so far developed applies to ordinary electromagnetic radiation.
If the incident radiation is polarized (say by passage through a Nicol prism),
the dipole moment vector **M** in equation 8.1 (and in subsequent equations)
must be replaced by only its component in the direction of polarization of the
light. Since, normally, molecules are oriented randomly with respect to
the incident light, all possible relations between the axis of the light and the
axes of the molecules occur, and the intensity is unaffected. But by use of a
single crystal or of a monomolecular film of oriented molecules, it is possible
to achieve a fixed relation between the axis (or plane) of the molecule and
the direction of polarization of the incident light. Consequently it is of
interest to consider the absorption of light polarized in different directions
with respect to the molecular symmetry elements.

We now need to elaborate on equations 8.1 and 8.3. The right-hand side of
these equations is obviously a *vector*, which results from the integration over
the dipole moment operator, which is a vector quantity. Consequently, the
left-hand side, which we have denoted by $\sqrt{D}$, is also a vector quantity and
should really be written as such; let us use the symbol $\mathbf{D}_{1/2}$ for this quantity.
Then, D is not the square of the right-hand side expressions, but rather their
dot product:[b]

$$D = \mathbf{D}_{1/2} \cdot \mathbf{D}_{1/2}$$

[b] The dot product of two vectors is defined by $\mathbf{r} \cdot \mathbf{s} = rs \cos \theta$ where r and s are the lengths of
the vectors **r** and **s**, respectively, and θ is the angle between them.

The dipole moment vector, $\mathbf{M}$, may be decomposed into three Cartesian components:

$$\mathbf{M} = e\mathbf{r}$$
$$= \mathbf{M}_x + \mathbf{M}_y + \mathbf{M}_z$$
$$= e[\mathbf{i}x + \mathbf{j}y + \mathbf{k}z]$$

where $\mathbf{i}$, $\mathbf{j}$, and $\mathbf{k}$ are *unit vectors*, that is, vectors of unit length, pointing in the x, y, and z direction, respectively.

With these substitutions, we can rewrite equation 8.4 as:

$$\mathbf{D}_{1/2} = e\sum_r c_{ir}c_{fr}(\mathbf{i}x_r + \mathbf{j}y_r + \mathbf{k}z_r)$$
$$= e\Big[\mathbf{i}\sum_r c_{ir}c_{fr}x_r + \mathbf{j}\sum_r c_{ir}c_{fr}y_r + \mathbf{k}\sum_r c_{ir}c_{fr}z_r\Big]$$

and

$$D = \mathbf{D}_{1/2}\cdot\mathbf{D}_{1/2} = e^2\Big[\mathbf{i}\cdot\mathbf{i}(\sum_r c_{ir}c_{fr}x_r)^2 + \mathbf{j}\cdot\mathbf{j}(\sum_r c_{ir}c_{fr}y_r)^2 + \mathbf{k}\cdot\mathbf{k}(\sum_r c_{ir}c_{fr}z_r)^2\Big]$$

since $\mathbf{i}\cdot\mathbf{j} = \mathbf{i}\cdot\mathbf{k} = \mathbf{j}\cdot\mathbf{k} = 0$. But $\mathbf{i}\cdot\mathbf{i} = \mathbf{j}\cdot\mathbf{j} = \mathbf{k}\cdot\mathbf{k} = 1$, and consequently the intensity, which is proportional to D, may be expressed as the sum of three components, analogous to x, y, and z components. This formulation, of course, provides the easy formalism for evaluating the vector sum of equation 8.4 in the preceding section.

It now becomes of interest to examine each of the three components. As long as at least one of these components is nonzero, the transition is called allowed, and unpolarized light is absorbed. Imagine now a particular transition in a molecule for which $\int \Psi_i \mathbf{M}_z \Psi_f \, d\tau \neq 0$, but $\int \Psi_i \mathbf{M}_x \Psi_f \, d\tau$ and $\int \Psi_i \mathbf{M}_y \Psi_f \, d\tau = 0$, and further imagine that all the molecules are arranged, say in a single crystal, with their z-axes parallel. If we now shine light polarized in the z direction at this crystal, the light will be absorbed. If, however, the crystal is rotated so that its z-axis becomes perpendicular to the direction of polarization of the light, no light is absorbed. The particular absorption band would be said to be polarized in the z direction.

The argument can be illustrated by application to formaldehyde. In Section 8.3 the $^1A_2 \leftarrow {}^1A_1$, $^1A_1 \leftarrow {}^1A_1$, and $^1B_2 \leftarrow {}^1A_1$ transitions were discussed, and it was shown that $\mathbf{M}_z$ is of species A_1, and $\mathbf{M}_y$ of species B_2; consequently the $^1A_1 \leftarrow {}^1A_1$ transition is z-polarized, and the $^1B_2 \leftarrow {}^1A_1$ transition y-polarized. Since the $^1A_2 \leftarrow {}^1A_1$ transition is forbidden, all three component integrals vanish, and the polarization of this band will depend on the nature and symmetry species of the vibration which makes this transition slightly allowed. It is usually possible to choose Cartesian axes in such a way that only one of the three component integrals is nonzero, and every absorption band is polarized.

To determine which of the component integrals for an electronic transition vanish, it is necessary only to compare the symmetry species of the $\Psi_i \Psi_f$ product with those of the components of $\mathbf{M}$. Hence we can readily determine the expected polarization of any predicted absorption band, and polarization measurements thus should be of great help in interpreting spectra in terms of the underlying electronic transitions.

8.6 Vibrational Interaction

In the discussion of the intensities of absorption bands the assumption was made that the wave function of a molecule can be factored into a vibrational and an electronic component. This assumption is not strictly valid; and, since there are cases in which this interaction is of extreme importance, some qualitative discussion seems indicated.

Benzene is a highly symmetrical molecule (D_{6h} symmetry), and because of this symmetry some of its spectral transitions are forbidden. Thus, in particular, one of the transitions which may be ascribed to the excitation of one of the electrons occupying the highest MO to the lowest unoccupied orbital, the lowest $\pi \rightarrow \pi^*$ transition ($^1B_{2u} \leftarrow {}^1A_{1g}$), is forbidden by symmetry. However, the statement "the benzene molecule is of D_{6h} symmetry" is not quite correct. The molecules are constantly undergoing vibrations; and, although some of the vibrational modes are symmetric, during others the symmetry is distorted. Since electronic phenomena are vastly more rapid than the motion of nuclei, benzene behaves with respect to light absorption as if it were a mixture of many different molecules; depending on their vibrational states, some have D_{6h} symmetry, but others are vibrationally distorted. However, since, in the lower vibrationally excited states, the nuclei never move very far from their equilibrium positions, the molecules always behave approximately as if they had D_{6h} symmetry; that is, although the first $\pi \rightarrow \pi^*$ transition is not completely forbidden for the individual molecules in vibrational states distorting the D_{6h} symmetry, this transition remains of low intensity. Consequently the longest-wavelength absorption is not completely forbidden, and the relatively weak band of benzene near 256 nm is this forbidden band, which becomes allowed by vibrational interaction. The intensity of this band is approximately 230.

A quantitative treatment of vibronic interaction indicates that the weakly allowed character of this transition exists if the electronic excitation is accompanied by vibrational excitation of the particular mode that appropriately distorts the symmetry. One of the outstanding features of the spectrum of benzene in solution is the structure of the long-wavelength band, arising from the transformation of varying numbers of quanta into vibrational energy. Actually, the band corresponding to no vibrational excitation (the $0 \rightarrow 0$ band) is missing from the spectrum.

Vibrational interaction also is of considerable interest in relation to the polarization of symmetry-forbidden bands. In such bands, all three components of the intensity-determining integral vanish, and hence polarization cannot be determined directly by the methods outlined in the preceding section. Forbidden bands are observed because of vibrational interaction, and the intensity-determining integral (equation 8.1) fails to vanish completely because of this interaction. Separation of the intensity-determining integral (equation 8.1) into a product of two integrals—one over the electronic coordinates only and containing the ψ's, and the other over the nuclear coordinates and involving the vibrational wave functions, χ, assumes only equal geometry of ground and excited states and ψ's independent of the nuclear coordinates. If this assumption is not made, the ψ's are dependent on the nuclear coordinates. For an evaluation of the intensity and polarization of forbidden bands this separation thus cannot be made. Each wave function Ξ_i and Ξ_f, however, may be treated as a product of a vibrational and an electronic function, χ_i and Ψ_i, χ_f and Ψ_f, respectively. The Ξ will then belong to the symmetry species of the product of the appropriate χ and Ψ.

As an example, the forbidden $^1A_2 \leftarrow\, ^1A_1$ transition of formaldehyde will be considered. As will be explained in Chapter 9, this molecule has $3n - 6 = 6$ normal vibrations, of which three belong to species a_1, one to b_1, and two to b_2. In the nondegenerate point groups, the vibrational wave functions for even vibrational quantum numbers belong to the totally symmetric species, a_1 in the case of C_{2v}, and for odd quantum numbers to the species characteristic of the vibration which they describe. Consider first one of the vibrations belonging to species a_1. All χ belong to a_1, and the product $\Psi_i\Psi_f$ to A_2; hence $\Xi_i\Xi_f = \Psi_i\chi_i\Psi_f\chi_f$ belongs to A_2. Since neither M_x, M_y, nor M_z belongs to A_2, all three components of equation 8.1 vanish, even under this vibronic interaction. But for excitation of one of the b_1 or b_2 vibrations, the products $\Xi_i\Xi_f$ belong to B_2 and B_1, respectively, and hence one component of 8.1 is nonzero. Consequently, interaction of the $^1A_2 \leftarrow\, ^1A_1$ transition with the b_1 vibration leads to $\Psi_i\chi_i\Psi_f\chi_f = A_1 \cdot a_1 \cdot A_2 \cdot b_1 = B_2$ and to vibronic bands which are y-polarized, and interaction with the b_2 vibrations leads to x-polarized components. No z-polarized bands occur, since no a_2 vibrations exist, which would give a $\Xi_i\Xi_f$ product of a_1 symmetry.

8.7 The Franck-Condon Principle

One additional factor affecting the intensities of absorption bands has been neglected so far. Since nuclear motion is much slower than electronic phenomena (cf. Chapter 11), the geometry of a molecule after absorption must be almost identical to its geometry before absorption. Similarly, it is necessary that the momentum (or the kinetic energy of the motion) of the nuclei remain substantially unchanged during the transition. These

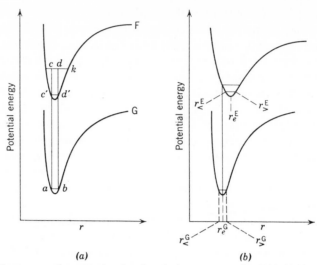

Fig. 8.5 Morse curves for ground and and excited states, assuming $U(r)$ (a) identical and (b) different in the two states. [From J. A. Pople and J. W. Sidman, *J. Chem. Phys.*, **27**, 1270 (1957), by permission.]

statements, due to J. Franck and elaborated by E. U. Condon, are known as the *Franck-Condon principle.*

The principles involved are simply illustrated by examining a diatomic molecule. The potential energy, U, of such a molecule is a function of the internuclear distance, r, and this functional relationship is approximated by the Morse curve, examples of which for the ground state are shown in the lower curves of Fig. 8.5.

The potential energy of the diatomic molecule has a minimum at the "equilibrium internuclear distance," r_e; at shorter distances the curve rises with increasing steepness because of nuclear repulsion, and goes to infinity since the nuclei cannot approach to zero distance. At distances longer than r_e, corresponding to stretching the bond, the potential energy increases. Sooner or later, when the atoms become substantially separated, however, U approaches a final value asymptotically; this value corresponds to the energy of the separated atoms and is commonly used as the zero point of the energy scale. The difference between U at r_e, $U(r_e)$, and U at $r = \infty$, $U(\infty)$, is, except for the zero-point energy, the energy required to dissociate the molecule (*dissociation energy*).

The potential energy of excited states can, of course, also be plotted against internuclear distance. If the excited state is stable to dissociation, the resulting curve also has the appearance of a Morse curve. For the special case in which this equilibrium internuclear distance is identical in the ground and excited states, the curves are given in Fig. 8.5a.

The much greater speed of electronic motion (as little as 10^{-18} sec may be required for excitation of an electron; see Chapter 11), compared with nuclear or vibrational motion (appreciable displacement of atoms does not occur in less than about 10^{-12} to 10^{-13} sec), indicates that the bond distance cannot change appreciably during excitation. At room temperature the vast majority of molecules of a diatomic compound are in the lowest (0, zero) vibrational level of the electronic ground state, and hence the atoms are at a distance r between a and b in Fig. 8.5a. Then, after excitation to the lowest (0) vibrational level of the excited state, the $0 \rightarrow 0$ transition, the distance r between the atoms will have changed only infinitesimally and hence will be between c' and d'. Since it was assumed that the potential energy curve for the excited state is the same (except for the vertical displacement) as that for the ground state, there is a point on the line $c'd'$ representing the 0 level of the excited state directly above each point between a and b of the line representing the 0 level of the ground state, and the kinetic energy is identical for such corresponding points. The kinetic energy, which is the difference between the total energy (given by the horizontal line representing the state) and the potential energy (given by the point on the appropriate Morse curve at the appropriate value of r), is the vertical distance of the line representing the state from the Morse curve at the particular value of r. Thus, a molecule may be excited from any point in its ground state to a corresponding point in its excited state, and the Franck-Condon principle produces no limitation for the $0 \rightarrow 0$ transition.

Consider next the transition from the same ground state to the kth vibrational level of the excited state, the $0 \rightarrow k$ transition (Fig. 8.5a). According to the Franck-Condon principle, the internuclear distance after excitation must be somewhere between c and d. But examination of Fig. 8.5a shows that the kinetic energy in this region is appreciably larger than that for corresponding points in the ground state. Thus, the kinetic energy of a molecule in the kth vibrational level of the excited state at point d is equal to the distance dd', whereas the kinetic energy of the molecule in the 0 level of the ground state at b is 0. Because of this difference in kinetic energy, the $0 \rightarrow k$ transition is forbidden by the Franck-Condon principle.

In the example just discussed (Fig. 8.5a) the equilibrium internuclear distances were equal in the ground and excited states, the $0 \rightarrow 0$ transition was intense, and excitations corresponding to $0 \rightarrow 1, 0 \rightarrow 2, \ldots, 0 \rightarrow k$ were forbidden. Usually, however, in going from a ground to an excited state, an electron is promoted from a bonding or nonbonding orbital to an antibonding orbital; consequently the bond of the molecule in the excited state is weakened, and the equilibrium bond distance in this state is greater than the corresponding distance in the ground state (i.e., $r_e^E > r_e^G$). The resulting pattern is shown in the upper curve of Fig. 8.5b, where the shallower and broader minimum reflects the loosening of the binding forces. In the transition to such an excited state, the Franck-Condon principle becomes of great importance both for the wavelength of absorption and for its intensity. The transition from the lowest vibrational level of the ground state to the lowest vibrational level of the excited state can, of course, occur only when the photon strikes the molecule at a moment in which the internuclear distance, r, corresponds to a classically allowed distance in the excited state, that is, if r is between $r_<^E$ and $r_<^G$, where $r_<^E$ refers to the minimum internuclear distance r in the excited state, and $r_>^G$ to the maximum r in the ground state (see Fig. 8.5b). Since for part of the time r is less than $r_<^E$, not all collisions with photons, but only that fraction occurring while the molecule is in the favorable geometric arrangement, can be effective, and in a more detailed analysis it would be important to examine the probability of finding the molecule at any given value of r.

In this case, transitions from the lowest vibrational state of the electronic ground state to higher vibrational states of the excited state become allowed, and the total absorption intensity becomes distributed between these transitions. A more detailed discussion of this redistribution will occupy us in the next section.

The Franck-Condon principle is of considerable importance in connection with intensities only if the extinction coefficient at the maximum wavelength is used as the measure of intensity. The absolute intensity is independent of the Franck-Condon restriction, and the Franck-Condon effects influence only the relative intensities of various vibrational sub-bands, but not the total integrated intensity of a band.

8.8   Franck-Condon Factors

Although we have explained and justified the Franck-Condon principle in the preceding section, on the basis of classical arguments, a full treatment, and particularly quantitative conclusions, require a quantum-mechanical discussion.

Let us return to equation 8.1, and again separate Ξ_i and Ξ_f into products $\Psi_i \chi_i$ and $\Psi_f \chi_f$, respectively. For the moment we shall consider each χ as the wave function of a single vibration; no generality is thereby lost against the real case of a polyatomic molecule, where each χ represents a product of $3n - 6$ vibrations. Furthermore, let the initial state, Ψ_i, represent the electronic ground state of the molecule, and χ_i the lowest vibrational function of the molecule *in its electronic ground state*, with $v = 0$, χ_0^G. Here the superscript indicates the vibrational function of the electronic ground state, and the subscript, $v = 0$. We are now interested in the probability of a transition to each vibronic state, with the electronic state described by Ψ_f and the vibrational function by $\chi_{v'}^E$. We are particularly interested in the relative values of this probability (i.e., the intensity) for all the various vibrational functions characterized by different v', but belonging to a single electronic transition.

The integral

$$\int \Xi_i \mathbf{M} \Xi_j \, d\tau \tag{8.1}$$

then becomes

$$\int \Psi_i \chi_0{}^G \mathbf{M} \Psi_f \chi_{v'}{}^E \, d\tau_\xi = \int \Psi_i \mathbf{M} \Psi_f \, d\tau \int \chi_0{}^G \chi_{v'}{}^E \, d\xi \tag{8.5}$$

Here ξ is the vibrational coordinate discussed further in the next chapter. It is apparent that the integral still vanishes if the transition is forbidden, that is, if the first integral on the right-hand side of equation 8.5 vanishes, and one must then have recourse to methods based on the logic of Section 8.5. But for an allowed transition, the second integrals on the right side of equation 8.5, $\int \chi_0{}^G \chi_{v'}{}^E \, d\xi$, called the *Franck-Condon factors*, provide the distribution of absorption energy over all vibrational component bands.

First, let us consider the idealized case, illustrated by Fig. 8.5a, where the potential energy curves of Ψ_i and Ψ_f have identical shapes. Since the vibrational functions are completely determined by the potential energy curve, the sets of functions $\chi_v{}^G$ and $\chi_{v'}{}^E$ are identical sets. Since the vibrational wave functions, like electronic ones, form an orthonormal set, the only Franck-Condon factor which is nonzero in this case is $\int \chi_0{}^G \chi_0{}^E \, d\xi$, that is, the one giving the intensity of the $0 \to 0$ band, and this factor is a normalization integral and hence equal to 1. Thus, in this special case, all the intensity is concentrated in the $0 \to 0$ band. At higher temperatures, where, in the ground state, some functions $\chi_v{}^G$ with $v > 0$ are populated, higher $v \to v$ $(1 \to 1, 2 \to 2,$ etc.) transitions should appear, but these will coincide with the $0 \to 0$ band and, if anything, only add something to its width.

Now let us consider Fig. 8.5b, the case in which the potential energy curve changes between ground and excited states. The sets of functions $\chi_v{}^G$ and $\chi_{v'}{}^E$ are no longer identical, and no Franck-Condon factors are necessarily zero. However, the potential energy curves of ground and excited states are always qualitatively similar, and consequently the χ^G and χ^E are similar. This leads generally to a largest Franck-Condon factor for the $0 \to 0$ and other lower transitions, with the intensities falling off rapidly. It seems intuitively obvious that the differences between the χ^G and χ^E sets of vibrational functions will be larger the larger the difference in potential energy curves, and these in turn are expected to be larger the larger the differences in geometry (bond lengths and valence angles) between the ground and excited states. This relationship accounts for the frequently encountered statement that the vibrations which involve motion in the same direction as the transformation of the ground-state geometry into that of the excited state appear most prominently in the vibrational structure of the spectrum.

At the end of Section 8.7 we made the statement that the Franck-Condon principle does not affect the absolute intensity, but influences only its distribution among vibrational components. We can now readily prove this statement. The intensity of each individual vibrational component of the absorption band is given by

$$\int \Psi_i \mathbf{M} \Psi_j \, d\tau \int \chi_0{}^G \chi_{v'}{}^E \, d\xi \tag{8.6}$$

It is always possible to expand any function in terms of the members of a complete orthonormal set of functions, and as long as the function expanded is close to one member of the expansion set, the expansion converges rapidly (i.e., not many terms are needed). We shall expand $\chi_0{}^G$ in terms of the set $\chi_{v'}{}^E$ (except that we shall call the set $\chi_u{}^E$, to avoid confusion of indices):

$$\chi_0{}^G = \sum_u c_u \chi_u{}^E$$

and substitute this in equation 8.6:

$$\int \Psi_i \mathbf{M} \Psi_j \, d\tau \int \chi_0{}^G \chi_{v'}{}^E \, d\xi$$

However, the intensity of absorption of each vibrational component is given by the square of this expression; this can be expressed as the square of the first integral, which we shall call D, multiplied by the square of the second integral. The expression thus becomes:

$$D(\int \sum_u c_u \chi_u{}^E \chi_{v'}{}^E \, d\xi)^2 = D(\sum_u c_u \int \chi_u{}^E \chi_{v'}{}^E \, d\xi)^2$$

In this equation all integrals for $v' \neq u$ vanish; consequently the summation can be dropped, and $\chi_u{}^E$ be changed to $\chi_{v'}{}^E$ and c_u to $c_{v'}$. Now, however, each integral $\int \chi_{v'}{}^E \chi_{v'}{}^E \, d\tau = 1$, so that the expression becomes

$$D c_{v'}{}^2$$

Summing over all the intensities of all the vibrational components, that is, over all values of v', gives

$$\sum_{v'} D c_{v'}{}^2 = D^2 \sum_{v'} c_{v'}{}^2 \tag{8.7}$$

But the last sum is equal to unity because of the normalization of $\chi_0{}^G$ and the $\chi_u{}^E$:

$$
\begin{aligned}
1 = \int \chi_0{}^{G2} \, d\xi &= \int [\sum_u c_u \chi_u{}^E]^2 \, d\xi \\
&= \int \sum_u c_u \chi_u{}^E \cdot \sum_w c_w \chi_w{}^E \, d\xi \\
&= \sum_u \sum_w c_u c_w \int \chi_u{}^E \chi_w{}^E \, d\xi \\
&= \sum_u c_u{}^2
\end{aligned}
$$

by reasoning exactly analogous to the above. Thus, equation 8.7 shows that the total integrated intensity of an allowed transition is not affected by the Franck-Condon principle; only the distribution of the intensity among the vibrational components is so affected.

PROBLEMS

8.1   The ground state π electron configuration of 1,3-butadiene can be represented as: $\Psi_i = \pi_1{}^2\pi_2{}^2$, and four possible excited states as: $\Psi_1 = \pi_1{}^2\pi_2\pi_3^*$; $\Psi_2 = \pi_1{}^2\pi_2\pi_4^*$; $\Psi_3 = \pi_1\pi_2{}^2\pi_3^*$; and $\Psi_4 = \pi_1\pi_2{}^2\pi_4^*$. (*a*) Determine the symmetry of the ground state and of the four indicated excited states in each of the three geometries: *s-trans*, *s-cis*, and linear. (*b*) From the symmetry selection rules determine which of the four electronic transitions, from the ground to each of the excited states, are allowed and which are forbidden in each geometry.

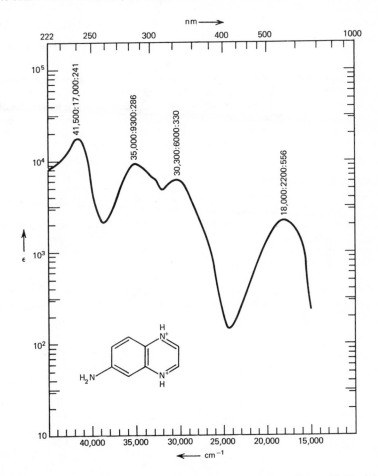

Fig. 8.6   The ultraviolet spectrum of 6-aminoquinoxaline dication. [From U. V. Atlas of Organic Compounds, Vol. IV, Butterworth, Verlag Chemie, 1968, by permission.]

8.2   Explain the effect that vibrational interaction (vibronic coupling) has on the spectral transitions of a typical centrosymmetric octahedral complex.

8.3   The n $\to$ σ^* transition in formaldehyde, shown as transition 6 in Fig. 8.3, is the $\psi_5 \to \psi_8$ transition, $3a \to 4a_1^*$ (Table 8.2), namely:

$$\frac{1}{\sqrt{2}}[2s(O) + 2p_z(O)] \to c_{81}[2s(C) + 2p_z(O)] - c_{82}[2s(O) - 2p_z(O)]$$

Set up the equation for the calculation of the intensity of this transition in terms of the coefficients c.

8.4   Given the ultraviolet spectrum of the quinoxaline ion (Fig. 8.6), calculate the oscillator strength, f, for the long wavelength absorption band using the formula

$$f = 4.32 \times 10^{-9} \, \epsilon_{max} \Delta \nu_{1/2}.$$

8.5   Assume the *s-trans* configuration for glyoxal $\begin{smallmatrix} H & & O \\ \diagdown & & \diagup \diagup \\ & C{-}C & \\ \diagup\diagup & & \diagdown \\ O & & H \end{smallmatrix}$. In order to consider the

n $\to$ π^* transitions, assume two n MO's constructed from pure in-plane p orbitals on each O atom (each of these orbitals will also contain some contributions from C and H orbitals, but these are neglected for the present purpose). Examine these two n orbitals and the corresponding n $\to$ π^* transition, and find whether they are allowed or forbidden by molecular symmetry. For any transition that is not forbidden by molecular symmetry, set up the intensity-determining integral and find any nonzero terms.

GENERAL REFERENCES

1. H. H. Jaffé and Milton Orchin, *Theory and Applications of Ultraviolet Spectroscopy*, John Wiley and Sons, New York, 1962.
2. J. N. Murrell, *The Theory of the Electronic Spectra of Organic Compounds*, John Wiley and Sons, New York, 1963.
3. G. Herzberg, *Molecular Spectra and Molecular Structure: Spectra of Diatomic Molecules*, Vol. 1, D. Van Nostrand Co., Princeton, N.J., 1966.
4. G. Herzberg, *Molecular Spectra and Molecular Structure: Electronic Spectra of Polyatomic Molecules*, Vol. 3, D. Van Nostrand Co., Princeton, N.J., 1966.
5. G. Barrow, *Introduction to Molecular Spectroscopy*, McGraw-Hill Book Company, New York 1962.

9 Infrared spectroscopy

9.1 Introduction

In order to understand and predict the infrared absorption spectrum of a particular compound we require the answers to at least two important questions: (1) how many absorption bands may we expect to observe; and (2) where in the energy scale, that is, at what frequencies, should each of the absorption bands occur? The first question may be subdivided into two parts: what transitions are predicted from theory; and, of these, which can we expect to observe with reasonable intensity? These are of course the same questions that we ask about any absorption spectrum. Since the infrared spectra we are about to discuss are due to vibrational motion of atoms in the molecule, we must first examine the physics of such motion.

9.2 The Harmonic and Anharmonic Oscillator

When a ball (or any particle) is held by springs between two fixed points (Fig. 9.1) and moved in the direction of one of the fixed points, it is constrained to move linearly. Displacement of the ball from its equilibrium position generates a restoring force which tends to return the ball to its position of equilibrium. The restoring force is proportional to the displacement, x (Hooke's law):

$$f = kx \tag{9.1}$$

where k is a proportionality constant called the *force constant*, that is, the restoring force per unit displacement. The force constant is a measure of the stiffness of the springs; strong, inflexible springs give rise to large k. If the ball

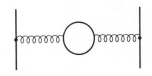

Fig. 9.1 A particle attached to fixed points by springs.

231

is displaced and then released, it undergoes vibrational motion, which is harmonic if Hooke's law is strictly applicable. The frequency of the oscillation is expressed by:

$$v = \frac{1}{2\pi}\sqrt{\frac{k}{m}} \tag{9.2}$$

where m is the mass of the ball or particle. If we wish to express frequency in wavenumbers, equation 9.2 becomes

$$\tilde{v} = \frac{1}{2\pi c}\sqrt{\frac{k}{m}} \tag{9.3}$$

where c is the speed of light.

We would like to inquire into the potential and kinetic energy of the particle undergoing oscillation. The total energy must be constant, neglecting frictional losses, and at any time will be equal to the sum of the potential (PE) and kinetic (KE) energies. The potential energy at point x_1, PE, follows from the integration of equation 9.1:

$$PE = \int_0^{x_1} f\,dx = \int_0^{x_1} kx\,dx = \tfrac{1}{2}kx_1{}^2 \tag{9.4}$$

A plot of the potential energy as a function of displacement is thus a parabola, Fig. 9.2. If the initial displacement of the ball is to the point x_0, then at this point the KE is zero since the ball is not moving any longer and all the energy is PE. As the ball is released, it travels to the point $-x_0$, where it momentarily comes to rest again, and at this point the KE again is zero. At the midway

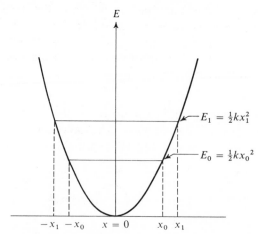

Fig. 9.2 The potential energy of a harmonic oscillator as a function of displacement from equilibrium.

point, the PE is zero and all the energy is KE. The horizontal line thus represents the sum

$$KE + PE = \text{total energy}$$

which is a constant. In this classical harmonic oscillator any value for the energy is allowed since the total energy depends only on the force constant and the magnitude of the displacement.

As is well known, the dynamic behavior of small particles such as atoms requires a quantum-mechanical description. When the quantum-mechanical restrictions are applied to our harmonic oscillator, we have as the allowed energy values:

$$E_v = (v + \tfrac{1}{2}) \frac{h}{2\pi} \sqrt{\frac{k}{m}} \qquad (9.5)$$

where v, the vibrational quantum number, is any integer $\geqslant 0$. If we substitute for the value of the frequency, v, in equation 9.2, equation 9.5 becomes:

$$E_v = (v + \tfrac{1}{2}) hv, \quad v = \frac{1}{2\pi} \sqrt{\frac{k}{m}} \qquad (9.6)$$

When frequency is expressed as wave numbers, $\tilde{v}$, equation 9.6 becomes:

$$E_v = (v + \tfrac{1}{2}) hc\tilde{v} \qquad (9.7)$$

Equation 9.7 gives us the following information: (1) the energy of the harmonic oscillator can have values only of positive half-integral multiples of hv; (2) the energy levels are evenly spaced; and (3) the lowest possible energy, that is the value when v is zero, is $\tfrac{1}{2}hv$, so that, even at $0°K$, there is present this *zero-point energy* consisting of a half quantum of vibrational energy.

For an analysis of the vibrational behavior of the simplest molecular system, that of a diatomic molecule, it is instructive to consider the molecule as two balls joined by a spring. When two balls or two particles of masses m_1 and m_2 are connected by a spring, and are then pulled apart and released, the system oscillates with motion which is nearly harmonic. The frequency of oscillation, assumed to be harmonic and expressed in wavenumbers, is given by an equation similar to equation 9.3:

$$\tilde{v} = \frac{1}{2\pi c} \sqrt{\frac{k}{\mu}} \qquad (9.3a)$$

The mass of the single particle which appears as m in equation 9.3 is replaced by the reduced mass, μ, which is defined as:

$$\mu = \frac{m_1 m_2}{m_1 + m_2}$$

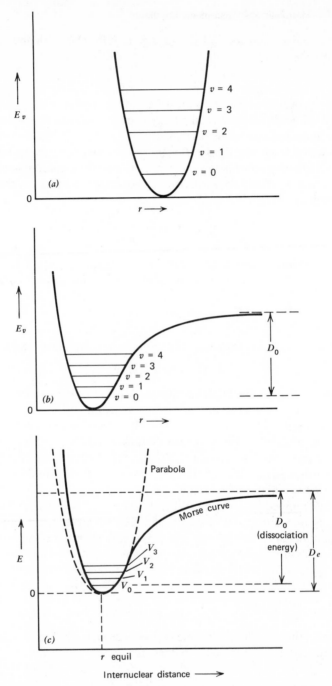

Fig. 9.3 Harmonic (*a*) and anharmonic (*b*) oscillator and a comparison of the two (*c*).

The reduced mass replaces the masses m_1 and m_2 of the two particles by an equivalent single mass, so that the motion of the two particles can be expressed as the motion of a single mass. However, in the real situation of a diatomic molecule, the PE cannot be described exactly by equation 9.4, which is the equation of a perfect parabola. The motion of the real diatomic molecule is slightly anharmonic, and the PE is not a perfect parabola even though, in the low-energy regions, there is a close correspondence between the two curves, Fig. 9.3c. The energy of the various vibrational levels in an anharmonic oscillator is generally less than that of the corresponding harmonic oscillator. The potential energy of the anharmonic oscillator can be described by a series equation:

$$E_v = (v + \tfrac{1}{2})hv - (v + \tfrac{1}{2})^2 hvx_e + (v + \tfrac{1}{2})^3 hvy_e - \ldots \qquad (9.8)$$

where the constants x_e, y_e, ... are *anharmonicity constants*, all small, usually positive numbers, and generally $|x_e| > |y_e| > |z_e| > \ldots$.

TABLE 9.1

Energies of Vibrational Levels (in units of h_v)

v	E_v Harmonic	Anharmonicity Correction (2nd term, eq. 9.8)	E_v Anharmonic
0	0.50	0.03	0.47
1	1.50	0.23	1.27
2	2.50	0.63	1.87
3	3.50	1.23	2.27
4	4.50	2.03	2.47

To illustrate the effect of anharmonicity, let us assume an arbitrary value for x_e of 0.01 and calculate E_v from equation 9.8, neglecting all terms after the second. The results for five levels of v, harmonic and anharmonic, are given in Table 9.1 and plotted in Fig. 9.3. These data and figures show that the levels in the anharmonic case are not evenly spaced but converge at the higher vibrational levels.

According to the selection rules which apply to the *harmonic* oscillator, only transitions corresponding to $\Delta v = \pm 1$ are allowed. Transitions from $v = 0$ to $v = 1$, from $v = 1$ to $v = 2$, etc., are all allowed; but because all the levels are evenly spaced, all transitions are superimposed on the $v = 0 \rightarrow 1$ transition and should occur with the same energy (frequency). Thus only one intense band should appear in the absorption spectrum, corresponding to the energy required for promotion from the vibrational level in the ground state, $v = 0$, to the vibrational level of the first excited state, $v = 1$, the *fundamental vibration*.

Even in the anharmonic cases, transitions from $v = 1$ and higher vibrational levels are not important because, at room temperature and below, the population of molecules in $v > 0$ levels is quite small. However, the anharmonicity of the real molecule allows the occurrence, besides the fundamental, of transitions from $v = 0$ to $v = 2$ and $v = 3$, etc., and these transitions are known as the first and second *overtones*, etc., respectively. These transitions occur with much lower probability than the fundamental and are thus weak bands; the first overtone is about an order of magnitude weaker than the fundamental, and the second overtone another order of magnitude weaker than the first. These intensities are related to the anharmonicity constants x_e, y_e in equation 9.8; the smaller these constants, the weaker the overtones. Usually the first overtone is found at a frequency slightly less than twice the frequency of the fundamental. This occurs because the anharmonicity constant x_e is almost always positive. If the first overtone occurs at a frequency slightly greater than twice the fundamental, the spacing between successive vibrational energy levels in the potential energy curve increases, and the anharmonicity constants are negative.

Overtones (and combination bands; cf. the end of this section) sometimes are observed at frequencies substantially different from the expected values, and often with unexpected intensity. The reason for such differences frequently is *Fermi resonance*. Fermi resonance occurs when such an overtone or combination band occurs near another frequency, usually a fundamental, belonging to the same symmetry species. In that case, interaction between the two modes permits a mixing, the two are pushed apart, and intensity is transferred from one to the other.

An example is found in the spectrum of benzene, which has fundamental frequencies at 1485 cm^{-1}, 1585 cm^{-1}, and 3070 cm^{-1}. The combination band of the first two is also expected at 3070 cm^{-1}; this combination band undergoes Fermi resonance with the fundamental at the same frequency to yield two equally intense bands at 3099 and 3045 cm^{-1}.

Equation 9.8 is only one approximation to the actual potential energy curve; it is particularly useful when discussing anharmonicity, since this is introduced as a perturbation, which is small near the equilibrium distance. A better approximation to the total potential energy curves is the *Morse curve* (Fig. 9.3c).

$$\text{PE} = D_e \exp[-2\beta(r - r_e)] - 2\beta \exp[-\beta(r - r_e)]$$

Here, D_e is the spectroscopic dissociation energy, larger than the observable chemical dissociation energy, D_0, by the zero-point energy, $\frac{1}{2}hv$; r_e is the equilibrium distance; and $\beta = 2\pi^2 v\mu D_e$; cf. Fig. 9.3c.

The anharmonicity of real molecules gives rise not only to overtones but also to *combination bands*, which arise from the sum or difference of two (or more) fundamentals. An example will be discussed in Section 9.4.

9.3   The Force Constant

In the example of the balls attached to a spring we mentioned that the force constant was related to the stiffness of the spring. In a diatomic molecule, the force constant is related to the bond strength between the two atoms. We would expect the force constant for the H_2 molecule to be considerably less than the force constant for the multiple-bonded molecule CO because of the stronger bonding in CO, and this expectation is verified by experiment.

The force constant is actually a measure of the curvature of the potential well near the equilibrium position:

$$k = \frac{d^2 V}{dq^2} \quad (q = 0) \tag{9.9}$$

where V is the potential energy and q is the displacement from the equilibrium position. The dissociation energy, D_0, Fig. 9.3c, is given by the depth of the well. A large force constant means a sharp curvature near the bottom but does not necessarily mean a deep well. Usually a large force constant is interpreted as an indication of a strong bond, and this is especially true if we compare compounds of similar structure. From equation 9.3a we see that, if the reduced mass μ were relatively constant for a series of diatomic molecules, the frequency at which absorption occurs should be related to $\sqrt{k}$. In the series of hydrogen halides, for example, μ does not vary greatly because hydrogen is light compared to all halogens:

$$\mu(HF) = \frac{1 \times 19}{20} = 0.95; \mu(HCl) = \frac{1 \times 35}{36} = 0.97;$$

$$\mu(HBr) = \frac{1 \times 80}{81} = 0.99; \mu(HI) = \frac{1 \times 127}{128} = 0.99$$

TABLE 9.2
Infrared Spectra of Hydrogen Halides and Other Diatomics

AB	μ	k, dynes/cm	$\sqrt{k}$	$\tilde{\nu}$, cm^{-1}	DE, kcal/mole
HF	0.95	8.8	2.97	3958	134
HCl	0.97	4.8	2.19	2885	102.2
HBr	0.99	3.8	1.95	2559	86.5
HI	0.99	2.9	1.70	2230	70.5
CO	6.86	18.4	4.29	2143	255.8
NO	7.53	15.2	3.87	1876	162

Accordingly in this series, Table 9.2, there is a good correlation between $\sqrt{k}$ and $\tilde{\nu}$. Table 9.2 also gives data for two other diatomics simply to show the difference in k for multiple-bonded molecules.

In polyatomic molecules, the relationship between the observed vibrational frequencies and force constants does not hold. The force constants must be calculated by a complicated coordinate analysis of the molecule. However, if only small changes are made from one compound to another, k may be considered relatively constant, and then from equation 9.3a the frequency should be inversely proportional to μ. This is the case in which one C-H bond of a particular organic molecule is replaced by C-D, that is, in deuterium substitution. In general the carbon atom is part of a fairly large radical, R, of mass M_R. Then $\mu(R-H)$ is $1 \times M_R/(1 + M_R)$, and $\mu(R-D)$ is $2 \times M_R/(2 + M_R)$. For most radicals, M_R is large compared to 1 or 2, and, in the limit for very large M_R, the ratio $\tilde{v}(C-D)/\tilde{v}(C-H)$ becomes

$$\frac{\tilde{v}(C\text{-}D)}{\tilde{v}(C\text{-}H)} = \frac{\sqrt{\mu(R\text{-}H)}}{\sqrt{\mu(R\text{-}D)}} = \sqrt{\frac{1 \times M_R/(1 + M_R)}{2 \times M_R/(2 + M_R)}} \approx = \sqrt{\frac{1 \times M_R/M_R}{2 \times M_R/M_R}} = \sqrt{\frac{1}{2}}$$

This, then, is an upper limit. When M_R is not large enough to neglect the 1 and 2 in the denominator, slightly lower values are obtained, with an absolute lower limit of 1.37 in the radical species CH.

Although force constants between specific atoms in a polyatomic molecule theoretically are influenced by neighboring atoms, empirical results indicate that vibrational frequencies are associated with specific groups of atoms, nearly independently of the molecule in which the group is present, and that the effects due to environment usually amount only to small shifts. Accordingly, the frequencies characteristically associated with these groups, the *group frequencies*, have been powerful tools in elucidating structure from infrared absorption spectra.

9.4   The Number of Vibrations in a Polyatomic Molecule

In a diatomic molecule, there is only one possible vibration, the one that results from stretching or compressing the distance between the two atoms along the internuclear axis, a symmetric stretching vibration. In order to describe the various possible vibrations (stretching, in-plane bending, out-of-plane bending) in molecules with more than two atoms it is necessary to define the position of the atoms with respect to each other.

In a molecule of n atoms, the complete location of all atoms requires $3n$ coordinates, the so-called $3n$ degrees of freedom, that is, 3 coordinates for each atom. The position of the entire molecule in space, that is, the position of its center of gravity in space, is determined by 3 coordinates, or 3 degrees of freedom. Three more degrees of freedom, or 3 more coordinates, are needed to define the orientation of the molecule; two angles to locate the principal axis and one more angle to define the rotational position about this axis (the three Euler angles). If the molecule is linear, this last angle is unnecessary since rotation about the lengthwise axis is not an observable process. This leaves $3n - 6$ degrees of freedom in a nonlinear molecule and $3n - 5$ in a linear molecule, which define the positions of the atoms relative to one another. And it is these $3n - 6$ or $3n - 5$ degrees of freedom or changes in the bond distances and bond angles, that is, the vibrations of the molecule, that interest us.

Just as a translation can be resolved into three components corresponding to the degrees of freedom that the translation represents, so a vibration can be resolved into as many components as the number of vibrational degrees of freedom. Such a resolution can be made in an infinite number of ways, just as a single vector can be resolved into three components in an infinite number of ways. [The components of a vector are normally defined in terms of projections on the axes of a normal (rectilinear) coordinate system, x, y, z. They can be equally defined, however, in terms of projections on any arbitrarily rotated coordinate system, or even on a nonorthogonal or curvilinear coordinate system.] But there is a preferred way for resolving the vector, namely, by letting the *components* (projections) be along the axis of an appropriately chosen coordinate system.

Similarly, there is a preferred way of resolving an arbitrary vibration into components, a resolution such that the periodic motions of each of the atoms in any given component occur with precisely the same frequency. These $3n$ components are called the *normal vibrations* of the molecules. The normal vibrations accordingly involve in-phase motions, that is, all atoms pass simultaneously through their equilibrium configurations and all reach their extreme positions at the same time. This is illustrated in the totally symmetric stretching vibration of water (one of its three possible vibrations) in Fig. 9.4, where the equilibrium positions of the atoms is shown in a, the maximum stretching positions of the atoms in b, and the minimum stretching positions in c. The equilibrium positions of the original atoms are shown as broken atoms in b and c in order to emphasize the new positions in relation to the original ones.

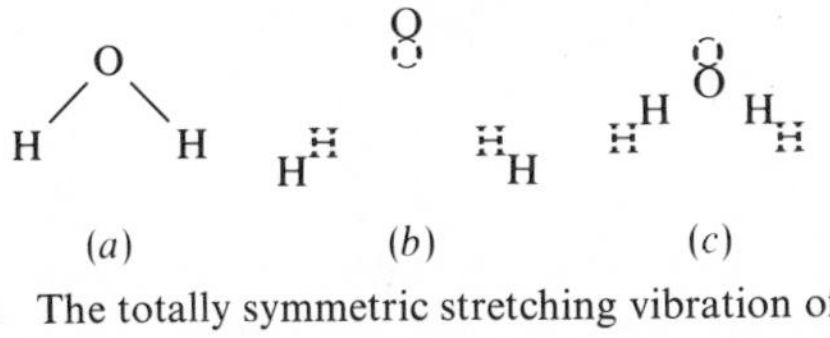

(a) (b) (c)

Fig. 9.4   The totally symmetric stretching vibration of H_2O.

Although the normal coordinates define the best set of components of the vibrations, they are not easy to determine and particularly require a knowledge of the attractive or repulsive forces between all pairs of atoms as a function of interatomic distance. A simpler set of components is available, however: the *symmetry vibrations* or *symmetry coordinates*. These are in many ways analogous to the symmetry orbitals and are determined in much

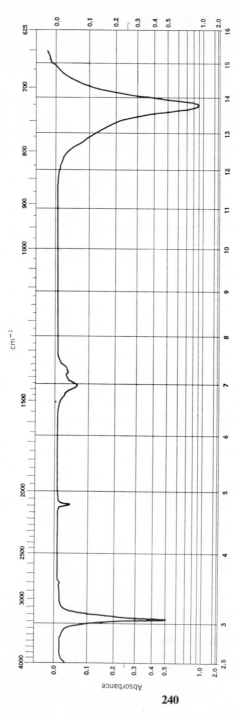

Fig. 9.5 The absorption spectrum of HCN in solution. [From W. J. Potts, Jr., Chemical Infrared Spectroscopy, John Wiley and Sons, 1963, p. 27.]

the same way by the use of projection operators, a process that requires no knowledge beyond the symmetry of the molecule. Most of what we will have to say here will relate only to symmetry coordinates; for the problem of normal coordinate analysis the reader is referred to specialized textbooks.

In a molecule of n atoms there are $3n$ degrees of freedom, the three translations, the three rotations, and the $3n - 6$ vibrations. Because all $3n$ degrees of freedom are often treated together, the three translations and the three rotations (when the molecule is nonlinear), which do not change the relative positions of the atoms, are often called *nongenuine modes*. Hence there are $3n - 6$ *genuine vibrations* for a nonlinear and $3n - 5$ *genuine* vibrations for a linear molecule.

For a diatomic molecule, $n = 2$, $3n - 5 = 1$, there is only a single vibration which is a stretching mode. In general, in nonring compounds of n atoms, there are always $n - 1$ stretching modes, since there are $n - 1$ bonds, each of which can be stretched independently. Thus, in a linear molecule with n atoms, of the $3n - 5$ normal vibrations, there are $n - 1$ stretching and $2n - 4$ bending vibrations. Thus, in HCN we expect that, of the total of four normal vibrations, there will be two stretching and two bending vibrations. In anharmonic vibrations, as mentioned earlier, we can expect bands corresponding to transitions other than the fundamental vibration, that is, *overtones*, which are near integral multiples (in cm^{-1}) of the fundamentals, and *combination bands*, which arise from the sum, or difference, of pairs of fundamentals.

The infrared spectrum of HCN is shown in Fig. 9.5. The two fundamental stretching frequencies occur at $3270\ cm^{-1}$ (H—C) and $2085\ cm^{-1}$ (C≡N). The two bending frequencies are degenerate and are jointly responsible for the band at $727\ cm^{-1}$; since the two modes must correspond to bending the $180°$ angle ∢ HCN, their degeneracy is easy to recognize, as may be seen from the following argument. The molecule, belonging to point group $C_{\infty v}$, has an infinite number of (vertical) planes of symmetry. The bending motions in any two perpendicular (orthogonal) planes are two independent vibrations but are obviously of the same energy.[a] If the molecule is placed in the plane of the paper, the bending mode can then either be in the plane of the paper (xz) or be perpendicular to the plane of the paper (yz). The weak band at $1433\ cm^{-1}$ is the first overtone of the bending vibration (harmonically expected at $1454\ cm^{-1}$). The weak band at $2800\ cm^{-1}$ is a combination band of the C≡N stretch plus the bending vibration ($2085 + 727 = 2812\ cm^{-1}$), while the weak band at $1370\ cm^{-1}$ is the difference band ($2085 - 727 = 1358\ cm^{-1}$).

[a] This statement is strictly true only in the absence of rotational motion. If the molecule is rotating, the vibrations in the plane of rotation and at right angles to the plane are slightly different. The splitting of the degenerate vibration due to this effect is known as *Coriolis coupling*.

Although we will consider later in this chapter the selection rules as applied to infrared spectroscopy, we can state some of them here. Consider the linear triatomic molecule CO_2. We again expect four normal vibrations, two stretching and two bending. The two stretching frequencies can be related to the displacements sketched in Figs. 9.6a and b. The CO_2 molecule has a center of symmetry, and we see that the symmetric stretching vibration of Fig. 9.6a retains the center of symmetry. This transition and *all* its overtones are forbidden and are not observed in the infrared. Also, the *odd* overtones of the asymmetric stretching and the bending frequencies, Fig. 9.6b, and combination bands between them are forbidden if the molecule has a center of symmetry. The two bending frequencies, Fig. 9.6c, are degenerate (just as they were with HCN), and thus we find that in the infrared spectrum of CO_2 there are only two strong bands, the asymmetric stretch at 2349 cm^{-1} and the doubly degenerate bend at 667 cm^{-1}.

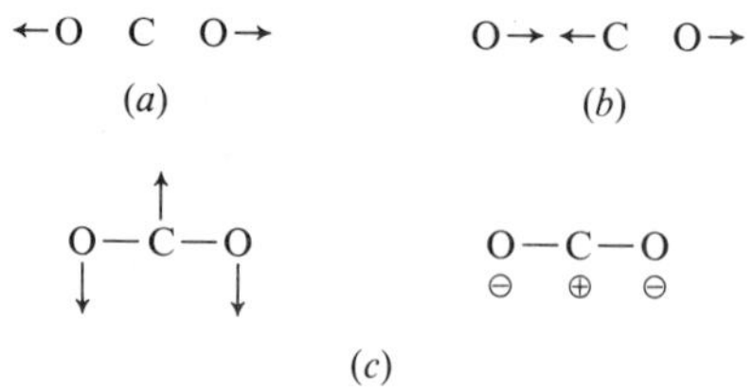

$$\leftarrow O \quad C \quad O \rightarrow \qquad\qquad O \rightarrow \leftarrow C \quad O \rightarrow$$
$$(a) \qquad\qquad\qquad\qquad (b)$$

$$O - \overset{\uparrow}{\underset{\downarrow}{C}} - O \qquad\qquad \underset{\ominus}{O} - \underset{\oplus}{C} - \underset{\ominus}{O}$$
$$(c)$$

Fig. 9.6 (a) The symmetric and (b) the antisymmetric stretching vibrations, and (c) the doubly degenerate bending vibrations of carbon dioxide.

The third fundamental in CO_2, which is forbidden in the infrared but allowed in the Raman spectrum (for definitions and reasons, cf. Section 9.7), is expected to lie near 1300 cm^{-1}, very near the first overtone of the fundamental at 667 cm^{-1}. Considerable interaction occurs by Fermi resonance (cf. Section 9.2) between these, shifting one component of the overtone to 1388 cm^{-1} and the fundamental to 1286 cm^{-1}. That this is a shift is shown by the fact that the other component of the overtone (there are two components since the fundamental is degenerate) occurs at 1338 cm^{-1}.

9.5   Symmetry Species of the Normal Vibrations

The symmetry species to which the various normal vibrations of each particular molecule belong are readily determined. Let us use H_2O as an example. Our problem is to determine the symmetry species of the $3n - 6 = 3$ genuine normal vibrations of the water molecule.

The water molecule belongs to point group C_{2v}; the character table is shown in Table 9.3. We set up parallel coordinate systems on each of the

TABLE 9.3
Character Table for C_{2v}

C_{2v}	I	C_{2z}	σ_{xz}	σ_{yz}	
A_1	$+1$	$+1$	$+1$	$+1$	z; α_{xx}, α_{yy}, α_{zz}
A_2	$+1$	$+1$	-1	-1	R_z; α_{xy}
B_1	$+1$	-1	$+1$	-1	x, R_y; α_{xz}
B_2	$+1$	-1	-1	$+1$	y, R_x; α_{yz}

three atoms as shown in Fig. 9.7. We next generate a reducible representation by transforming the molecule under the symmetry operations appropriate to C_{2v}. In performing the operations we actually transform the coordinates, but, if atoms are shifted by the operation, the coordinates on each atom shifted cannot transform into themselves or any combination of themselves. Thus the character of the transformation matrix is zero, and, as we shall see, the transformation need not be recorded for atoms shifted by the operations.

Under the identity operation I, all atoms are unshifted; and, since on each atom we have three coordinates, we generate a 9×9 matrix, shown as Table 9.4. In this table the subscripts indicate the atom number, Fig. 9.6, and the primes refer, as usual, to the coordinates after the transformation. The numbers in the table are obtained in the usual fashion. Thus under I, the new x_1, called x_1', contains 1 of the old x_1; the new y_1 and z_1, called y_1' and z_1', both contain zero of the old x_1; and thus we have the numbers 1, 0, 0 as the first three members in the first column of the matrix. Since none of the atoms and hence of their coordinates transforms to other atoms under I, but all transform only into themselves, we have the blocks of zeros shown in the table, and the entire matrix is a unit matrix of order 9. The character for the transformation is 9. We could have simplified this procedure greatly by recognizing that all three atoms transform under I in the same way; thus we need only consider the submatrix of one atom with a character of 3 and multiply by the number of submatrices to obtain the character of 9.

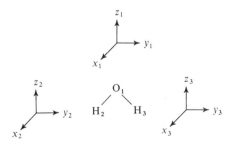

Fig. 9.7 Coordinates for water.

TABLE 9.4

Transformations for the Nine Coordinates of Water (Fig. 9.7) under the Operation I in C_{2v}

	x_1	y_1	z_1	x_2	y_2	z_2	x_3	y_3	z_3
x_1'	1	0	0	0	0	0	0	0	0
y_1'	0	1	0	0	0	0	0	0	0
z_1'	0	0	1	0	0	0	0	0	0
x_2'	0	0	0	1	0	0	0	0	0
y_2'	0	0	0	0	1	0	0	0	0
z_2'	0	0	0	0	0	1	0	0	0
x_3'	0	0	0	0	0	0	1	0	0
y_3'	0	0	0	0	0	0	0	1	0
z_3'	0	0	0	0	0	0	0	0	1

Under C_2^z, only the oxygen atom remains unshifted, and we have the 9×9 matrix of Table 9.5a with a character of -1. Examination of this matrix shows that the only nonzero diagonal elements come from the transformation of the oxygen coordinates into themselves, that is, the diagonal elements in the 3×3 submatrix corresponding to the transformation of the coordinates on the O atom. Under σ_{xz} again only the oxygen remains unshifted, and the character, Table 9.5b, of this submatrix is $+1$. Under σ_{yz} all three atoms remain unshifted, and because the coordinates on all atoms transform similarly we can develop a 3×3 submatrix and multiply

TABLE 9.5a

Transformation Matrix for Water under C_2^z

	x_1	y_1	z_1	x_2	y_2	z_2	x_3	y_3	z_3
x_1'	-1	0	0	0	0	0	0	0	0
y_1'	0	-1	0	0	0	0	0	0	0
z_1'	0	0	1	0	0	0	0	0	0
x_2'	0	0	0	0	0	0	-1	0	0
y_2'	0	0	0	0	0	0	0	-1	0
z_2'	0	0	0	0	0	0	0	0	1
x_3'	0	0	0	-1	0	0	0	0	0
y_3'	0	0	0	0	-1	0	0	0	0
z_3'	0	0	0	0	0	1	0	0	0

its character by 3. If we do this, we get a character of $+1$ for the submatrix and $+3$ for the 9×9 matrix. The reducible representation is then:

	I	C_2^z	σ_{xz}	σ_{yz}
	9	-1	$+1$	$+3$

This reducible representation corresponds to $3A_1 + A_2 + 2B_1 + 3B_2$ (equation 6.3). Now, of these nine irreducible representations, the three translational and three rotational nongenuine modes must be subtracted. Reference to the last column of the character table for point group C_{2v}, Table 9.3, shows that these correspond to $1A_1$, $1A_2$, $2B_1$, and $2B_2$, leaving, as normal modes, $2A_1$ and $1B_2$ for a total of three vibrations. Since all of these are allowed (cf. the next section) one may expect, and actually finds, three fundamental bands in the spectrum of H_2O.

TABLE 9.5b
The submatrix for the transformation of the coordinates on the Oxygen Atom in Water under σ_{xz} in C_{2v}

σ_{xz}	x_1	y_1	z_1
x_1'	1	0	0
y_1'	0	-1	0
z_1'	0	0	1

9.6 Selection Rules in Infrared and Raman Spectra

The intensity of an infrared absorption band is proportional to the square of the integral

$$e \int \chi^0 \, \mathbf{r}\chi^j \, d\tau = e[\int \chi^0 x\chi^j \, d\tau + \int \chi^0 y\chi^j \, d\tau + \int \chi^0 z\chi^j \, d\tau] \qquad (9.10)$$

where χ^0 and χ^j are the initial and the final vibrational wave functions of the molecule, between which the transition occurs. In these integrals, x, y, and z are the Cartesian coordinates. The intensity is zero, that is, the band is not observed, unless at least one of these partial integrals on the right-hand side of equation 9.10 is not zero. For any of these integrals to be nonvanishing, its integrand must be totally symmetric. Because the vibrational wave function in the ground state of a molecule, χ^0, is always totally symmetric, that is, belongs to an A or A_1 species, one of the coordinates x, y, or z and the wave function of the excited state, χ^j, must belong to the same representation in order that the representation (symmetry species) of their direct product will contain the totally symmetric representation. The wave function χ^j belongs to the same symmetry species as the normal mode, which is undergoing its fundamental transition. (Only one vibration changes in a given transition.)

This leads to the important symmetry selection rule: A fundamental will be infrared active if the normal mode involved belongs to the same symmetry species as at least one of the Cartesian coordinates. An alternative but equivalent way of stating this is to say that a vibration is active in the infrared only if the dipole moment of the molecule changes during the vibration.

It was shown above that water has three fundamental vibrations: $2A_1$ and B_2. Reference to the applicable character table, C_{2v}, shows that coordinate z transforms as A_1 and y as B_2. Hence, for the B_2 vibration, the asymmetric stretch, the integrand $\chi^0 y \chi^j$ is totally symmetric since $A_1 \cdot B_2 \cdot B_2 = A_1$. For the A_1 modes, the bend and the symmetric stretch, $\chi^0 z \chi^j$ is totally symmetric since $A_1 \cdot A_1 \cdot A_1 = A_1$. Hence all three fundamental vibrations of water are allowed. The B_2 vibration is said to be y-polarized and the two A_1 transition are z-polarized, that is, the absorption occurs when the y- and z-axes of the molecule coincide with the direction of polarization of the exciting light.

The actual spectrum of water taken as the vapor indeed shows three bands: $\nu_3 = 3756 \text{ cm}^{-1}$, $\nu_1 = 3652 \text{ cm}^{-1}$, and $\nu_2 = 1545 \text{ cm}^{-1}$ (these are shown schematically in Fig. 9.8). The questions of what the fundamental vibrations look like and which vibration can be assigned to which band remain to be answered. Intuitively it is reasonable to associate a vibration with each bond between atoms (two bonds); hence we expect two stretching vibrations. It is also reasonable to expect a bending vibration associated with the angle between the bonds (one in water). Furthermore, we might

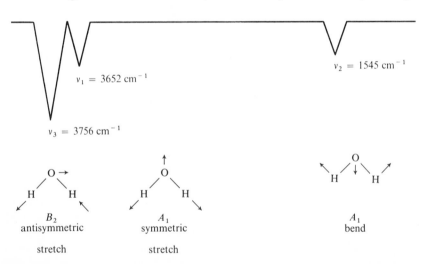

Fig. 9.8 Schematic infrared spectrum of water vapor. (Since the oxygen atom is eight times as heavy as the two hydrogen atoms, it will move a considerably smaller distance and in a direction opposite to the sum of the displacement vectors of the hydrogens to maintain the center of gravity stationary.)

expect that the two bond-stretching fundamentals would be relatively close in energy. The choice between stretching and bending modes can be made on the basis of the force constants for such motions; it always requires less energy to change angles (bending motions) than to change bond distances (stretching motions). Hence the low-frequency band at 1545 cm^{-1} is the bending mode. The bands at 3756 and 3652 cm^{-1} have been assigned to the asymmetric and symmetric stretching modes, respectively.

At this point it is of interest to consider briefly the subject of *Raman spectra* and to contrast the selection rules for infrared and Raman spectra. Although vibrational spectra are observed as either infrared or Raman, the origins of the two spectra are quite different. Infrared absorption originates in transitions between vibrational levels in the ground state of the molecules, and such transitions are observed as absorption spectra in the infrared region. Raman spectra, on the other hand, arise when monochromatic light in the ultraviolet or visible region, such as that emitted from a low-pressure mercury arc, is passed through the sample. Although the vast majority of the photons are simply scattered by the molecules in the sample, a few are able to exchange some energy with the scattering molecules; some photons gain energy, and some lose it. These energy gains and losses occur as quanta of the vibrational and rotational energy of the scattering molecules. In the Raman spectrum, one analyzes the photons which have lost some of their energy, and in the process have excited the sample molecules to higher rotational and vibrational states. Because of the electronic polarization induced in the molecule by the incident light of frequency v, the scattered light (*Rayleigh scattering*) will have a frequency of $v - v_i$, where v_i represents a vibrational frequency. The vibrational frequencies are thus seen as shifts (*Raman shifts*) from the original frequency. In order to be allowed, the polarizability of the molecule must be changed during the vibration.

The intensity of a Raman transition is determined by an integral analogous to that for infrared transitions except that the dipole moment vector is replaced by a *polarizability tensor*, P. For Raman scattering, it is necessary that at least one integral of the type

$$\int \psi_v^0 \, P \psi_v^j \, d\tau$$

be nonzero. The polarizability tensor, P, is a symmetric 3×3 matrix with components α_{xx}, α_{xy}, α_{xz}, etc. There are six distinct components; these are frequently indicated as α_{xx}, etc., or simply as x^2, y^2, z^2, xy, xz, and yz, since they transform like the squares and products of the Cartesian coordinates, and they (or their combinations, such as $x^2 - y^2$) are listed in the character tables opposite the symmetry species they generate (cf. Chapter 5). In order for a vibration to be Raman active, one of the six components of the integral given above must be nonvanishing, which means that there must be a change in the polarizability of the molecule when the transition occurs. All three vibrations of H_2O are Raman active as well as infrared active.

In considering the vibrations of a molecule with a center of symmetry, it will be recognized that, on inversion, all coordinates change sign. Consequently, all components of the dipole moment operator, M, occurring in the infrared intensity integral, $\int \psi_v^0 M \psi_v^j d\tau$, are u or ungerade. Then, transitions will be allowed only if the product $\chi^0 \chi^j$ is also u, since $u \times u = g$, and *only* the integrals of even functions have a finite value. However, we have seen that χ^0 is always totally symmetric and therefore is of a g species. Accordingly, when χ^j is g, the transition will be forbidden in the infrared, and only transitions of u vibrational modes are allowed.

On the other hand, consider the components of the quadratic functions of the Cartesian coordinates and their behavior on inversion. Binary products of two Cartesian coordinates, say $x \cdot y$ or z^2, do not change sign on inversion, since each coordinate separately changes sign and $-1 \times -1 = 1$. Therefore, all components of the polarizability tensor, P, in the intensity integral of the Raman transition are gerade. Now, since all χ_v^0 are also gerade, $\chi_v^0 \times P$ is always gerade, and only when χ^j is g will the Raman transition be allowed. This leads to the exclusion rule: In a molecule with a center of symmetry, no Raman-active vibration is also infrared active and no infrared-active vibration is also Raman active.

9.7 Tables for the Number of Normal Vibrations in Various Symmetry Species of Nondegenerate Point Groups

In considering the vibrations of atoms in molecules and the symmetry species to which they belong, it is helpful to classify the atoms in terms of sets of symmetry-equivalent atoms. Atoms are symmetry equivalent and hence are members of a set when they can be transformed into each other by a symmetry operation appropriate to the molecule (Chapter 5). In the nondegenerate point groups, specification of the displacement of one atom in a set in a given vibrational mode and identification of the symmetry species to which the vibration belongs, are sufficient to define the displacements of all other atoms in the set in that mode.

Consider the displacement vector of H_A along the $O-H_A$ bond, shown in Fig. 9.9a. If it represents a motion in a vibration of species A_1, this vector must transform with the character of $+1$ under C_2^z and σ_{xz}, that is, it must transform into a new vector on H_B with the same magnitude and direction, as shown in Fig. 9.9b. This means that the sets of displacement vectors on H_A and H_B shown in Figs. 9.9a and 9.9b belong to species A_1. To maintain the center of gravity fixed, obviously the O atom must move slightly away from the H atoms, so that the complete motion is as shown in Fig. 9.9c. The combination of all three displacement vectors represents *the symmetry co-ordinate* corresponding to this vibration. On the other hand, if the vibration represented by the displacement vector of Fig. 9.9a is of species B_2, the character under both C_2^z and σ_{xz} is -1, and transformation of the displacement vector of H_A leads to a displacement vector on H_B of the same magnitude and the opposite direction (multiplication of a vector by -1 reverses its direction), as shown in Fig. 9.9d. The sets of displacement vectors on H_A and H_B shown as Figs. 9.9a and 9.9d together represent the displacement vectors on these atoms in the B_2 vibration mode, which is obviously the anti-symmetric stretch. Together with the displacement vector of the O atom, required to maintain the center of gravity in place, they are *the* symmetry coordinate of this vibration (cf. Fig. 9.9e); and, since there is only one vibration of this symmetry type, they also are the normal coordinate.

In dealing with sets of atoms, it is of interest to ask how many atoms must belong to a given set. If the atom does not lie on any symmetry element (i.e., is a general atom), the number of atoms in the set is just equal to the order of the point group. If the atom lies on a symmetry element, this number is divided by the order of the element, where planes are of order 2, and axes C_p of order p. If the atom lies on several symmetry elements, examination usually shows readily how many are in the set.

We next wish to determine how many normal vibrations belong to each of the symmetry species. Let us again consider the water molecule, point group

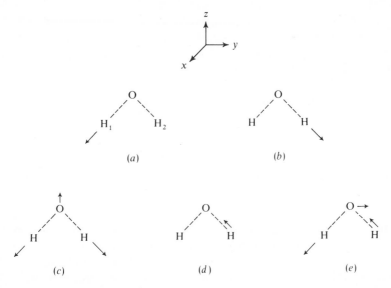

Fig. 9.9 Transformation of a vector on H_A (a): under C_2^z in species A_1 (b), and in species B_2 (d); the displacement vectors for vibrations of species A_1 (c) and of species B_2 (e).

C_{2v}, Fig. 9.7. All three atoms lie on symmetry elements. The two H atoms are members of a set and lie on σ_{yz}. The O atom is a member of a different set of one; it lies on all three symmetry elements. The O atom contributes 3 degrees of freedom, which can be thought of as motion in the x, y, and z directions, respectively. Motion in the z direction transforms into itself under all symmetry operations and hence belongs to species A_1; similarly motions in the x and in the y directions, respectively, transform as B_1 and B_2, as may be seen simply by looking at the behavior of the axes in the character table, or by transforming such a displacement vector and obtaining the characters, whereupon the assignment can be made to the proper symmetry species.

The vibrations contributed by the set of H atoms are a little trickier to determine; the set contributes a total of 6 degrees of freedom (d.f.)· each atom contributing 3. Two of these for each atom are motions in the yz-plane, the molecular plane, motion parallel to the y- and z-axes. Since these motions lie in the yz-plane, they must be symmetric with respect to it, and the degrees of freedom must belong to A_1 and/or B_2. Specification of a vector in the yz-plane on H_A still permits transformation into either of two vectors on H_B, depending whether the transformation is made according to the rules of A_1 or B_2, and thus the in-plane motion of the H atoms contributes 2 d.f. each to A_1 and B_2. The motion of the H atoms normal to the plane, that is, in the x direction, provides the last 2 d.f., one in B_1, the other in A_2, depending on whether the two H atoms move in the same or in opposite directions. The set of H atoms thus contributes a total of 6 d.f.: $2A_1 + 2B_2 + A_2 + B_1$. Adding to these the 3 d.f. contributed by the O atom gives $3A_1 + A_2 + 2B_1 + 3B_2$. Subtracting the six nongenuine vibrations, A_1, B_1, and B_2 for translation and A_2, B_1, and B_2 for rotation, leaves $2A_1 + B_2$ for the three genuine vibrations, as derived in an earlier section.

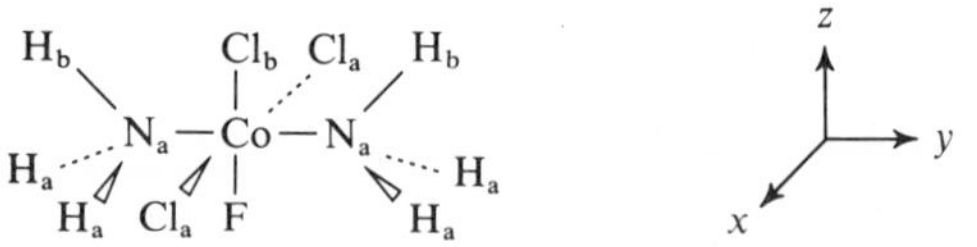

Fig. 9.10a A hypothetical molecule with C_{2v} symmetry in which the set of four H_a atoms does not lie on any symmetry element.

As a further example, let us examine another (hypothetical) molecule in point group C_{2v}, in which some atoms do not lie on any symmetry element, Fig. 9.10, so that we may develop some useful general formulas for determining the symmetry species of normal vibrations. It is now important to determine the number of sets having the various possible symmetry behaviors, and these numbers are usually denoted by a small m, with appropriate subscripts.

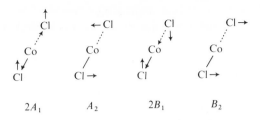

Fig. 9.10*b* Vibrations of the set m_{xz} (Cl$_a$) in the various symmetry species.

In this molecule, there is one, and only one, set of atoms (H$_a$) not on any symmetry element, and hence m (without subscript, denoting sets on no symmetry element) = 1. It is apparent that one representative atom in the set generates the three others by performing the operations of C_{2v}. Next we look for sets on the symmetry element σ_{xz} only, and on no other symmetry element, and call their number m_{xz}. There is only one set, Cl$_a$, and hence $m_{xz} = 1$. On σ_{yz} only, there are the sets of H$_b$ and N$_a$, and hence $m_{yz} = 2$. On all symmetry elements there are three sets, Co, Cl$_b$, and F, and their number, m_0, equals 3.

We now examine the degrees of freedom allowed these sets in the various symmetry species of point group C_{2v}, Table 9.3. In considering the degrees of freedom, we treat sets together as we did earlier. Thus, although there are four H$_a$ atoms, they all belong to the single set; and, since this set does not lie on any symmetry element, it contributes 3 d.f. in each of the four symmetry species of C_{2v}, for a total of 12 d.f. Now let us look at the Cl$_a$ set, m_{xz}, and examine the degrees of freedom allowed in species A_1. In order for the two Cl$_a$ atoms to move so that the movement is symmetric to all symmetry operations as required by A_1, the movement has to be in the xz-plane, and hence the set contributes 2 d.f. to species A_1. In A_2, the only motion permitted for this set is along a line in the y direction (1 d.f.), one Cl atom moving in the $+y$, the other in the $-y$, direction in order that the motion be antisymmetric with respect to both symmetry planes as required by A_2. In B_1 the Cl$_a$ atoms must move in the σ_{xz}-plane; hence the set contributes 2 d.f. to this species. In B_2 the Cl$_a$ set can move only in the y direction, but both Cl must move simultaneously in the same direction as required by the $+1$ character on σ_{yz} of B_2. Thus, to summarize, the set of Cl$_a$ atoms (on σ_{xz}) contributes $2A_1 + A_2 + 2B_1 + B_2$, for a total of 6 d.f. as sketched in Fig. 9.10*b*.

We can continue the process for the sets on σ_{yz}. There are $m_{yz} = 2$ such sets; a set of N atoms and a set of H atoms. Each set contributes $2A_1 + A_2 + B_1 + 2B_2$ for a total of 6 d.f., or a total of 12 d.f. for the four atoms. Thus, similar information can be secured for the $m_0 = 3$ sets of atoms on all symmetry elements. Each set contributes $A_1 + B_1 + B_2$ or a total of 3 d.f. for each of the three sets of single atoms, for a total of 9 d.f.

TABLE 9.6

Determination of the Number of Vibrations of Each Species for the Point Group C_{2v}

		Degrees of Freedom Contributed by Each Set of Nuclei			Number of Normal Vibrations		
Species	On No Symmetry Element	Only on $\sigma_v(xz)$	Only on $\sigma_v(yz)$	On C_2, $\sigma_v(xz)$, $\sigma_v(yz)$	Nongenuine T	R	Genuine
A_1	$3m$	$2m_{xz}$	$2m_{yz}$	$1m_o$	1		$3m + 2m_{xz} + 2m_{yz} + m_o - 1$
A_2	$3m$	$1m_{xz}$	$1m_{yz}$	0		1	$3m + m_{xz} + m_{yz} - 1$
B_1	$3m$	$2m_{xz}$	$1m_{yz}$	$1m_o$	1	1	$3m + 2m_{xz} + m_{yz} + m_o - 2$
B_2	$3m$	$1m_{xz}$	$2m_{yz}$	$1m_o$	1	1	$3m + m_{xz} + 2m_{yz} + m_o - 2$
Total	$12m$	$6m_{xz}$	$6m_{yz}$	$3m_o$	6		$12m + 6m_{xz} + 6m_{yz} + 3m_o - 6$

The foregoing information is summarized in Table 9.6. Here we show how the total number of 39 degrees of freedom of the molecule are distributed among the different symmetry species. Analogous, general equations can readily be obtained for other nondegenerate point groups and are given in standard textbooks. As a check on the correct application of these tables we can calculate the total number of atoms from the sum of the number of atoms in each set, summed over all sets. In the case for C_{2v}, the total number of atoms is always $N = 4m + 2m_{xz} + 2m_{yz} + m_0$, since each set in m has four atoms, each set in m_{xz} or m_{yz} has two, and each set in m_0 has only one. For the case we have been discussing, $m = 1$, $m_{xz} = 1$, $m_{yz} = 2$, $m_0 = 3$; and hence the total number of atoms is 13.

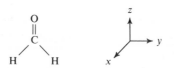

Fig. 9.11   Formaldehyde, C_{2v}.

As an exercise in the application of the formulas let us consider formaldehyde, Fig. 9.11. Here $m = 0$, $m_{xz} = 0$, $m_{yz} = 1$, $m_0 = 2$. To check on the correct assignment it is best to first calculate the total number of atoms by the formula to see whether the result corresponds to reality: $N = 4m + 2m_{xz} + 2m_{yz} + m_0 = 2 + 2 = 4$. We then proceed to calculate the number of vibrations in each species by using the formulas in the last column of Table 9.6, from which we see that the $3n - 6 = 12 - 6 = 6$ genuine vibrations belong to $3A_1 + B_1 + 2B_2$.

9.8   Normal Vibrations; Degenerate Point Groups

The number of vibrations in the various symmetry species of degenerate point groups can be determined by the method applied above to non-degenerate point groups. The method can be demonstrated by application to the ion CO_3^{2-} of D_{3h} symmetry; the character table for D_{3h} is shown in Table 9.7. The coordinate system on the ion and on each of the four atoms

TABLE 9.7
Symmetry Species and Characters for the Point Group D_{3h}

D_{3h}	I	$2C_3(z)$	$3C_2$	σ_h	$2S_3$	$3\sigma_v$	
A_1'	$+1$	$+1$	$+1$	$+1$	$+1$	$+1$	$\alpha_{xx} + \alpha_{yy}, \alpha_{zz}$
A_1''	$+1$	$+1$	$+1$	-1	-1	-1	...
A_2'	$+1$	$+1$	-1	$+1$	$+1$	-1	R_z
A_2''	$+1$	$+1$	-1	-1	-1	$+1$	M_z
E'	$+2$	-1	0	$+2$	-1	0	$M_x, M_y; \alpha_{xx} - \alpha_{yy}, \alpha_{xy}$
E''	$+2$	-1	0	-2	$+1$	0	$R_x, R_y; \alpha_{xy}, \alpha_{xz}$

is shown in Fig. 9.12. The reducible representation generated by operating on the twelve local coordinate axes, using one each of the classes of symmetry operations, is shown in Table 9.8. Each operation produces a 3×3 matrix for each of the four atoms. The way in which the fractions under the C_3 column are obtained is shown in detail in Fig. 9.13. The numbers that appear in Table 9.8 are the diagonal elements of each 3×3 submatrix. Thus the submatrix for operation I on the three coordinates of atom 1 is:

$$\begin{pmatrix} 1 & 0 & 0 \\ 0 & 1 & 0 \\ 0 & 0 & 1 \end{pmatrix}$$

Since we have no interest in the off-diagonal elements because they do not

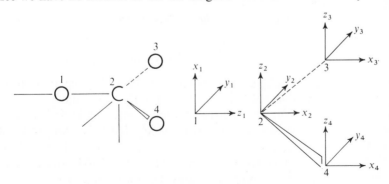

Fig. 9.12   The coordinate system for the carbonate ion.

TABLE 9.8
The Symmetry Species of the Vibrations in D_{3h}

D_{3h}	I	C_3	$C_2{}^x$	σ_h	S_3	$\sigma_v{}^x$
x_1	1	0	$+1$	$+1$	0	$+1$
y_1	1	0	-1	$+1$	0	-1
z_1	1	0	-1	-1	0	$+1$
x_2	1	$-\frac{1}{2}$	$+1$	$+1$	$-\frac{1}{2}$	$+1$
y_2	1	$-\frac{1}{2}$	-1	$+1$	$-\frac{1}{2}$	-1
z_2	1	$+1$	-1	-1	-1	$+1$
x_3	1	0	0	$+1$	0	0
y_3	1	0	0	$+1$	0	0
z_3	1	0	0	-1	0	0
x_4	1	0	0	$+1$	0	0
y_4	1	0	0	$+1$	0	0
z_4	1	0	0	-1	0	0
	$+12$	0	-2	$+4$	-2	$+2$

$$(A_1' + A_2' + 2A_2'' + 3E' + E'') - (A_2' + A_2'' + E' + E'') = A_1' + A_2'' + 2E'$$

contribute to the character, we simply record the diagonal elements. As usual, if the atom is shifted by the operation, the character for the submatrix is zero. The irreducible representations corresponding to the reducible representations are listed in the last row of Table 9.8, and after subtracting the nongenuine vibrations the symmetry species of the $3n - 6 = 6$ vibrations are shown.

Formulas for determining the number of vibrations in each symmetry species for the degenerate point groups can readily be derived by methods analogous to those shown above, and are available in standard textbooks. In the case of D_{3h}:

$$N = 12m + 6m_v + 6m_h + 3m_2 + 2m_3 + m_0$$

and, for example,

$$E' = 6m + 3m_v + 4m_h + 2m_2 + m_3 + m_0 - 1$$

where m and m_0 are the sets of atoms on no and on all symmetry elements, respectively; m_v and m_h are the sets lying on only one of the σ_v or the σ_h, respectively; and m_2 and m_3 are the sets lying on two-fold and three-fold axes but not on any other element with which that axis does not wholly coincide. For the $CO_3{}^{2-}$ ion, $m = 0$, $m_v = 0$, $m_h = 0$, $m_2 = 1$, $m_3 = 0$, and $m_0 = 1$. Of the $3n - 6$ vibrations, we expect $n - 1 = 3$ to be stretching and the remainder (3), to be bending vibrations. The formulas give, of course, the same result as that derived in Table 9.8, $A_1' + A_2'' + 2E'$. Reference to the

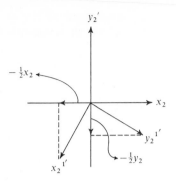

Fig. 9.13   The transformation of the coordinates on the carbon atom in $CO_3{}^{2-}$ on the C_3 operation; the full transformation matrix is

$$\begin{pmatrix} -\frac{1}{2} & -\frac{\sqrt{3}}{2} & 0 \\ \frac{\sqrt{3}}{2} & -\frac{1}{2} & 0 \\ 0 & 0 & 1 \end{pmatrix}$$

Only the diagonal elements are shown in Table 9.8.

character table (Table 9.7) shows that, since there is no component of the transition moment in species $A_1{}'$, this vibration is forbidden, and we expect only the $A_2{}''$ and the two E' vibrations to be observed in the infrared.

The exact movements of atoms that correspond to various species are rather tricky to generate and will be discussed in very general terms only.

9.9   Motions of the Atoms in the $3n - 6$ Vibrations

In ethylene, there are $3n - 6 = (3 \times 6) - 6 = 12$ genuine normal vibrations. The formula for calculating the number of normal vibrations in each of the symmetry species is given in Table 9.9, and then all are applied to the case of ethylene. If we now ask about the exact motions of the atoms in these twelve normal vibrations, we are in some trouble. However, we can fairly readily write down symmetry vibrations or, better, symmetry coordinates. The symmetry coordinate represents how far along the path of a normal vibration the atoms in a molecule have moved, and consequently are represented by an arrow showing the direction and magnitude of the motion of each atom. Thus, a symmetry coordinate consists of the assembly of displacement vectors (shown in Fig. 9.14), one such vector for each atom. These displacement vectors, and consequently the symmetry coordinates, transform as the various symmetry species of the molecule.

Aside from the considerations explicitly dealt with in what follows, it is necessary to retain the center of gravity and three principal (rotational) axes

TABLE 9.9

The Number of Normal Vibrations in a Molecule Belonging to D_{2h}, Applied to Ethylene

Species	Formula	Result
A_g	$3m + 2m_{xy} + 2m_{xz} + 2m_{yz} + m_{2x} + m_{2y} + m_{2z}$	3
A_u	$3m + m_{xy} + m_{xz} + m_{yz}$	1
B_{1g}	$3m + 2m_{xy} + m_{xz} + 2m_{yz} + m_{2x} + m_{2y} - 1$	0
B_{1u}	$3m + m_{xy} + 2m_{xz} + m_{yz} + m_{2x} + m_{2y} + m_{2z} + m_o - 1$	2
B_{2g}	$3m + m_{xy} + 2m_{xz} + m_{yz} + m_{2x} + m_{2z} - 1$	2
B_{2u}	$3m + 2m_{xy} + m_{xz} + 2m_{yz} + m_{2x} + m_{2y} + m_{2z} + m_o - 1$	1
B_{3g}	$3m + m_{xy} + m_{xz} + 2m_{yz} + m_{2y} + m_{2z} - 1$	1
B_{3u}	$3m + 2m_{xy} + 2m_{xz} + m_{yz} + m_{2x} + m_{2y} + m_{2z} + m_o - 1$	2
$N \quad =$	$8m + 4m_{xy} + 4m_{xz} + 4m_{yz} + 2m_{2x} + 2m_{2y} + 2m_{2z} + m_o$	6
$m_{xz} \quad = 1$ (4H's)		
$m_{2z} \quad = 1$ (2C's)		
$m = m_{xy} = m_{yz} = m_{2x} = m_{2y} = m_o = 0$		

unchanged in each genuine vibrational symmetry coordinate. This is achieved, for example, in v_5 by the small displacement vector assigned to the C atoms.

First, we know that the molecule has five bonds, and hence we may expect five stretching vibrations. The first of these, for the C-C bond, is immediately obvious (v_2 in Fig. 9.14) and belongs to A_g. Of the four C—H stretches, obviously one belongs to A_g, where all C-H bonds stretch together (v_1 of Fig. 9.14). The other three must involve the simultaneous stretching of two bonds and contraction of two others. If we say that C^1—H^1 always stretches, there are only three such combinations, one each in which one of C^1—H^2, C^2—H^3, and C^2—H^4 also stretches and the other two contract (motions in which C^1—H^1 contracts are equivalent to the above, but in a different phase of the motion). These three motions, as can be seen from Fig. 9.14, transform as B_{3g} (v_{10}), B_{2u} (v_8), and B_{1u} (v_5), respectively.

Having thus fixed the five stretching vibrations, we must now find seven more modes, all of which must involve some deformations of angles. There are six angles in the plane, two $\measuredangle$ HCH and four $\measuredangle$ HCC. Of these, however, only four are independent, two at each C atom, if the molecule is to remain planar; thus we expect four bending vibrations in the plane. Two of these obviously could be the bending modes of the $\measuredangle$ HCH; one when these two angles change in phase, one when the motion is out of phase; these two vibrations are v_3 (A_g) and v_6 (B_{1u}) (cf. Fig. 9.14). The other two in-plane vibrations must involve the changes of the $\measuredangle$ HCC, and they correspond to *wagging* of the CH_2 groups. Again the motions may occur in phase, v_{11} (B_{3g}), or out of phase, v_9 (B_{2u}). This leaves three motions which must, of necessity, be out of plane. The simplest of these is the *torsional* motion of

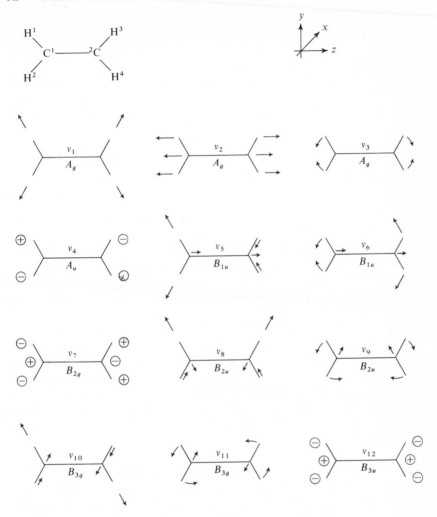

Fig. 9.14   The twelve symmetry coordinates of ethylene.

one CH_2 group with respect to the other, v_4 (A_u). Finally, we have the two motions in which the CH_2 planes are tilted with respect to the molecular plane, again in and out of phase, v_{12} (B_{3u}) and v_7 (B_{2g}).

Each of the three out-of-plane motions belongs to a symmetry species to which no other vibration belongs, and hence are normal vibrations. The remaining symmetry coordinates belong to species in pairs (or even three in the case of A_g), and the normal coordinates should be some mixture of the symmetry coordinates. Determination of the exact mixing is a sensitive

TABLE 9.10
The Infrared Spectrum of Ethylene

Symmetry Species	Active in*	C—C Stretch	C—H Stretch	HCH Bend	Wag	Out-of-Plane Bend	Observed, cm⁻¹	in*
A_g	R	1	1	1			3019 (CH), 1623, 1342	R
A_u	...					1	Forbidden	...
B_{1g}	R							
B_{1u}	IR	1	1				2990 (CH), 1444	IR
B_{2g}	R		1		1		3272 (CH)	R
B_{2u}	IR					1	949	IR
B_{3g}	R					1	943	R
B_{3u}	IR	1			1		3106 (CH), 995	IR

* R = Raman; IR = infrared.

function of the various force constants, that is, the interaction of all atoms, and is achieved by solving sets of secular equations, most commonly by Wilson's F and G matrix method.

The symmetry coordinates alone provide an approximate idea of the normal modes, particularly when, as is the case here, one would require the mixing of stretching and bending modes, which are energetically quite distinct. Thus we see in Table 9.10 that four fundamentals occur near 3000 cm⁻¹, all of which can be assigned to nearly pure C—H stretching modes. The two out-of-plane vibrations observed occur near 950 cm⁻¹; the third is forbidden in both infrared and Raman, and information can at best be obtained from combination bands or from the vibrational structure of electronic emission spectra. Table 9.10 summarizes the vibrational spectrum of ethylene.

9.10   Vibrational Spectra of Metal Carbonyl Complexes

There are many carbonyl complexes of the transition metals, and analysis of their vibrational spectra sheds considerable light on the nature of the bonding in such complexes. The carbonyl stretching frequencies usually appear as strong bands in the infrared spectra, and many studies of these bands have been made in order to determine the extent of $d\pi$-$p\pi$ interaction.

The frequency of the stretching vibrations of a CO group in a carbonyl complex depends on the force constant of the particular C-O bond in question and on the interactions of the stretching vibrations of the carbonyl group with the stretching vibrations of other groups in the molecule. The force constant which is calculated from the frequency:

$$v = \tfrac{1}{2}\pi \sqrt{\frac{k}{\mu}} \qquad (9.11)$$

is thus not the force constant of the C-O bond only, but represents a total force constant which is a sum of the bond force constant plus the interaction force constant. Disentangling these contributions to k can be achieved rather successfully in octahedral and square planar complexes, where *cis* and *trans* effects can be evaluated, without recourse to detailed coordinate analysis, but simplifying assumptions cannot be so readily made in other cases, for example, in tetrahedral symmetry. In what follows we shall illustrate how the symmetry and number of vibrational modes for the C-O bonds in octahedral carbonyls may be determined by arguments we have used earlier in this book, particularly in Chapter 6.

For the octahedral carbonyl we can consider the six ligand carbonyls, numbered as shown previously in Fig. 6.5a, and use the coordinate system and operations shown in Fig. 6.5b. Although the compound $M(CO)_6$ has 39 degrees of freedom, we are interested here only in the CO group and we handle the problem as though we are generating group orbitals from the six CO groups, treating the CO as a single group. Step 1 is to generate the reducible representation, using Table 6.1 for point group O, transforming each of the ligand CO's as in Table 6.3a to obtain the reducible representation, which we find corresponds to $A_1 + E + T_1$ as in Table 6.3c. Now ligand 1 is transformed under the symmetry operations of the point group into new ligands, Table 6.4, and these are multiplied by the characters of the reducible representations. The results are the same as the six group orbitals:

$$
\begin{aligned}
A_1 &= 1 + 2 + 3 + 4 + 5 + 6 \\
E &= 1 - 2 + 3 - 4 \\
&\quad -1 - 2 - 3 - 4 + 2(5) + 2(6) \\
T_1 &= 1 - 3 \\
&\quad\; 2 - 4 \\
&\quad\; 5 - 6
\end{aligned}
$$

If we consider a plus sign as a stretch and a minus sign as a contraction of the C-O bond, the motions can be indicated as shown in Fig. 9.15. The g and u characters in O_h are obvious from inspection of the direction of the arrows. Reference to the character table for O_h shows that only the T_{1u} vibration is infrared active and that the other two vibrations are Raman active. Accordingly we expect one strong band in the infrared for the carbonyl stretching frequency in $M(CO)_6$ compounds, a triply degenerate T_{1u} vibration, and two vibrations in the Raman spectrum, and this is what is actually found.

We might ask how we can determine the relative frequencies of the three bands. To do this we must inquire further into the nature of the force constants.

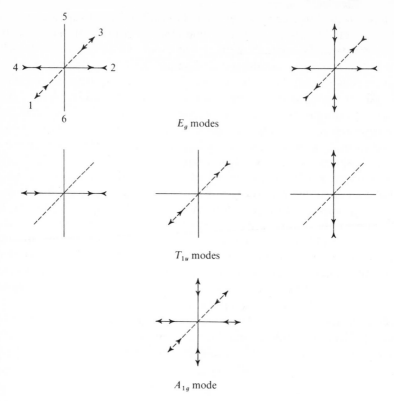

E_g modes

T_{1u} modes

A_{1g} mode

Fig. 9.15   The six C—O stretching vibrations in M(CO)$_6$. (Only the T_{1u} antisymmetric stretching mode is infrared active, and hence only one band is observed.)

We may rewrite equation 9.11 as

$$k = 4c^2\mu v^2\pi^2 \qquad (9.12)$$

The force constant for the C-O bond, k, as explained above, may be considered to be made up of two parts. The first part, which we will call k_1, is the force constant of the particular C-O bond in question; it increases with increasing extent of π bonding between the C and the O atoms. The second factor, which we will call k_i, is the interaction of the stretching vibration of the CO group in question with the stretching vibration of other carbonyl groups present in the molecule. Hence

$$k = k_1 + k_i \qquad (9.13)$$

where k is the force constant calculated from the observed stretching frequency according to equation 9.12, and other interactions are neglected.

We can make the following qualitative observations with respect to k's in equation 9.13:

1. All k_i involving the stretching interaction of any particular CO with other CO's are positive, that is, such interaction strengthens other C-O bonds. The stretching of a C-O bond in a carbonyl increases the (metal) $d \rightarrow \pi^*(CO)$ interaction at that CO so that the metal d orbitals are less available for other CO's. Hence the bonding interaction between the C and the O atoms in other CO's is increased.

2. The k_i of carbonyl groups *trans* to each other is approximately twice that of carbonyls *cis* to each other, since the *trans* CO groups compete for the same set of two metal d orbitals, whereas *cis* CO groups compete for different d orbitals.

If we now examine the A_{1g} vibration, Fig. 9.15, we see that k for any particular CO group will be k_1 plus the interaction term for one *trans* and four *cis* groups, since they all stretch simultaneously:

$$k = k_1 + k_t + 4k_c$$

From the discussion in observation 2 above we can assume that $k_t = 2k_c$, and hence

$$k = k_1 + 6k_c \quad \text{for } A_{1g}$$

In one of the E_g vibrations it will be noted that, if we focus on a CO group undergoing a stretching of the C—O, the *trans* CO is also stretching but the two *cis* CO groups are contracting, that is, they are undergoing a negative stretch, so that

$$k = k_1 + k_t - 2k_c$$

and, since $k_t = 2k_c$,

$$k = k_1 \quad \text{for } E_g$$

Similar analysis of the other E_g vibration leads to the same conclusion.

Using the same type of reasoning for the T_{1u} vibration, we obtain

$$k = k_1 - 2k_c \quad \text{for } T_{1u}$$

Accordingly we would expect the observed stretching frequencies to decrease in the order $A_{1g} > E_g > T_{1u}$. The observed frequencies are 2062 cm^{-1} (A_{1g}), 2020 cm^{-1} (E_g), and 2000 cm^{-1} (T_{1u}). Quantitatively, the above arguments suggest that the frequency ratio $A_{1g} - E_g/E_g - T_{1u}$ should be $6k_c/2k_c = 3:1$; experimentally, from the above values this ratio is 42/20, or about 2:1. The intensities of the bands and other considerations verify the correct assignment.

The analysis of the spectrum of $M(CO)_6$ can be extended to compounds in which one or more of the CO groups are replaced by other ligands, L, such as phosphines, and valuable information as to the location of the L groups may be deduced from the infrared spectra of the resulting compounds. Much work of this nature now appears in the literature.

PROBLEMS

9.1 The stretching frequency for F-Cl is $\bar{\nu} = 313.5$ cm^{-1}. What is the force constant for the F-Cl bond (F = 19, Cl = 35). (b) Assuming the same force constant as in (a), calculate the stretching frequency for F-Cl where F = 19 and Cl = 37.

9.2 Show that the vibrations of CO_3^{2-}, drawn below, belong jointly to species E' in D_{3h} (the direction, relative magnitude, and angles of the displacements are indicated).

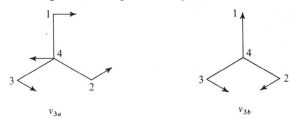

9.3 Determine the number of normal modes of vibration of tetrahedral Ni(CO)$_4$ in each of the symmetry species of T_d, treating each CO as a single group on the bond axis.

9.4 Determine the symmetry species to which the five carbonyl stretches in the C_{4v} compound ClMn(CO)$_5$ belong. How many carbonyl bands do you expect to see in the infrared spectrum of this compound?

9.5 Using the equations for determining the number of normal vibrations in various symmetry species, find the number and symmetry of the vibrations in acetone:

$$
\begin{array}{c}
\text{O} \\
\parallel \\
\text{C} \\
\text{H} \diagdown \quad \diagup \quad \diagdown \quad \diagup \text{H} \\
\text{C} \qquad \text{C} \\
\text{H} \overset{}{\text{H}} \qquad \text{H} \text{H}
\end{array}
$$

9.6 Repeat problem 9.5 for (a) naphthalene, (b) Ni(CO)$_4$, and (c) (CO$_3$)$^{2-}$. (Formulas for D_{2h} are given in Table 9.9; those for other point groups may be found in H. H. Jaffé and M. Orchin, *Symmetry in Chemistry*, John Wiley and Sons, 1965, Appendix 2.) Determine which of the vibrations are infrared active and which are Raman active.

GENERAL REFERENCES

1. E. B. Wilson, Jr., J. C. Decius, and P. C. Cross, *Molecular Vibrations*, McGraw-Hill Book Company, New York, 1955.
2. R. P. Bauman, *Absorption Spectroscopy*, John Wiley and Sons, New York, 1962.
3. G. Herzberg, *Molecular Spectra and Molecular Structure: Infrared and Raman Spectra of Polyatomic Molecules*, Vol. 2, D. Van Nostrand Co., Princeton, N.J., 1966.
4. G. W. King, *Spectroscopy and Molecular Structure*, Holt, Rinehart & Winston, New York, 1964.
5. W. J. Potts, Jr., *Chemical Infrared Spectroscopy*, Vol. 1, John Wiley and Sons, New York, 1963.
6. C. E. Meloan, *Elementary Infrared Spectroscopy*, The Macmillan Company, New York, 1963.

10 The Hückel molecular orbital treatment and symmetry simplifications

10.1 Introduction

In classical mechanics the total energy E of a molecule is the sum of its kinetic energy, T, and its potential energy, V:

$$E = T + V = H$$

The total energy is frequently given the symbol H in honor of Lord Hamilton, the English physicist, who made important contributions to the mathematics of classical mechanics. In wave mechanics, however, the fundamental equation involving the energy of a molecule is:

$$H\Psi = E\Psi \tag{10.1}$$

Here H is an energy operator, and $H\Psi$ is the result of the operation of H on Ψ. The value of E cannot be calculated directly from this equation for molecular systems, because it is not possible to find a function Ψ for which $H\Psi$ is some constant multiple of Ψ so that $H\Psi/\Psi$ is a constant and not a variable function of the position of the electrons. Accordingly, both sides of equation 10.1 are multiplied by Ψ and then integrated over all space; this process corresponds to averaging $H\Psi$ over all space:

$$\int \Psi H \Psi \, d\tau = \int \Psi E \Psi \, d\tau = E \int \Psi \Psi \, d\tau$$

$$E = \frac{\int \Psi H \Psi \, d\tau}{\int \Psi \Psi \, d\tau} \tag{10.2}$$

Equations 10.1 and 10.2 apply to the energy of the complete molecule and all of its electrons. For most spectroscopic purposes we consider only electronic motion (Born-Oppenheimer approximation), and then Ψ is a function of the coordinates of all the electrons. As a crude approximation we assume that Ψ can be factored into a product of many ψ, each the wave function of a single electron (an orbital) (cf. also Chapter 12).

10.2   Approximate Solutions of the Schrödinger Equation

Equation 10.2 provides a basis for the calculation of the energy of molecular systems. However, a numerical evaluation of the equation requires a knowledge of the functions ψ, that is, the molecular orbitals. In accordance with a suggestion by Mulliken, it has become customary to express the MO's as a *linear combination of atomic orbitals* (LCAO):

$$\psi_j = \sum_r c_{jr}\phi_r \tag{10.3}$$

This means that the molecular orbital ψ_j is expressed as a weighted sum of all atomic orbitals ϕ_r, the weights being given by the coefficients c_{jr}, and the sum extended over all atoms (or better all AO's) which are part of the system. Let us illustrate with a homonuclear diatomic molecule such as H_2, where the two MO's are of the forms:

$$\psi_1 = c_{11}\phi_1 + c_{12}\phi_2 \tag{10.4}$$

$$\psi_2 = c_{21}\phi_2 + c_{22}\phi_2 \tag{10.4a}$$

Substituting the value of ψ_1 for ψ in equation 10.2 gives:

$$E = \frac{\int [(c_{11}\phi_1 + c_{12}\phi_2)H(c_{11}\phi_1 + c_{12}\phi_2)]\,d\tau}{\int [(c_{11}\phi_1 + c_{12}\phi_2)(c_{11}\phi_1 + c_{12}\phi_2)]\,d\tau}$$

$$= \frac{\int [(c_{11}\phi_1 Hc_{11}\phi_1 + c_{11}\phi_1 Hc_{12}\phi_2 + c_{12}\phi_2 Hc_{11}\phi_1 + c_{12}\phi_2 Hc_{12}\phi_2)]\,d\tau}{\int [(c_{11}\phi_1 c_{11}\phi_1 + c_{12}\phi_2 c_{11}\phi_1 + c_{11}\phi_1 c_{12}\phi_2 + c_{12}\phi_2 c_{12}\phi_2)]\,d\tau}$$

Now, since c_{11} and c_{12} are simply constants to be evaluated, we can take them out of the integral, and we can break up the integrals as follows:

$$E = \frac{c_{11}^2\int\phi_1 H\phi_1\,d\tau + c_{11}c_{12}\int\phi_1 H\phi_2\,d\tau + c_{11}c_{12}\int\phi_2 H\phi_1\,d\tau + c_{12}^2\int\phi_2 H\phi_2\,d\tau}{c_{11}^2\int\phi_1\phi_1\,d\tau + c_{11}c_{12}\int\phi_1\phi_2\,d\tau + c_{11}c_{12}\int\phi_2\phi_1\,d\tau + c_{12}^2\int\phi_2\phi_2\,d\tau} \tag{10.5}$$

In all cases of physical reality, $\int \phi_1 H\phi_2\,d\tau = \int \phi_2 H\phi_1\,d\tau$, and when this occurs the operator H is said to be Hermetian.

Examination of equation 10.5 shows that there are two kinds of integrals: those in the numerator are of the type $\int \phi_i H\phi_j\,d\tau$ and are called the *matrix elements* (sometimes also called the H integrals) and are referred to as H_{ij}; those in the denominator are of the type $\int \phi_i\phi_j\,d\tau$, the *overlap integrals*, usually denoted as S_{ij}.

Let us consider first the overlap integral $S_{ij} = \int \phi_i\phi_j\,d\tau$. If $i = j$, then $S_{ii} = \int \phi_i\phi_i\,d\tau = 1$ for normalized AO's; when $i \neq j$, $1 \geqslant |S_{ij}| \geqslant 0$. The term S_{ij} is a dimensionless number. If $S_{ij} = 0$, then ϕ_j and ϕ_i are orthogonal. If the functions are not orthogonal, the overlap integral will be some number other than 0 but between -1 and $+1$. In general, at or near bonding distances, S_{ij} is positive, and the closer the centers of the ϕ functions the larger S_{ij}.

Thus this integral is a measure of how much the orbitals i and j overlap: hence the name *overlap integral*. In the most crude or "zeroth-order" approximation of the LCAO method, S_{ij} ($i \neq j$) is neglected, that is, is taken equal to zero.

The first type of matrix element that we will describe is the one in which $i = j$ and $H_{ii} = \int \phi_i H \phi_i \, d\tau$, the *Coulomb integral*, commonly designated as α_i. Here H_{ii} may be regarded as the Coulomb energy of an electron, and α_i is thus associated with the ionization potential, that is, the energy required to remove an electron from the atom. Hence α_i must be a function of the nuclear charge and the type of orbital. For π orbitals, for example, $\alpha_C = -11.16$ eV, $\alpha_N = -11.96$ eV, and $\alpha_O = -17.70$ eV (cf. Chapter 12 for further discussion). Although α_i is affected relatively little by neighboring atoms, procedures are available in higher-order approximations for taking this effect into account.

The other type of matrix element arises when $i \neq j$, $H_{ij} = \int \phi_i H \phi_j \, d\tau$. This integral, called the *resonance integral* and given the symbol β_{ij}, is a function of the atomic number, the types of orbitals, and the degree of overlap. Since the integral depends on the overlap, β is also a function of the internuclear distance and the angle of the orbitals with respect to the internuclear axis. The resonance integrals, β_{ij}, are usually taken proportional to the overlap integrals, S_{ij}.

Returning now to equation 10.5 and substituting the above equivalents for the integrals in that equation, we obtain:

$$E = \frac{c_{11}^2 H_{11} + c_{11}c_{12}H_{12} + c_{11}c_{12}H_{21} + c_{12}^2 H_{22}}{c_{11}^2 S_{11} + c_{11}c_{12}S_{12} + c_{11}c_{12}S_{21} + c_{12}^2 S_{22}} \tag{10.6}$$

and, since $H_{21} = H_{12}$ and $S_{12} = S_{21}$,

$$E = \frac{c_{11}^2 H_{11} + 2c_{11}c_{12}H_{12} + c_{12}^2 H_{22}}{c_{11}^2 S_{11} + 2c_{11}c_{12}S_{12} + c_{12}^2 S_{22}} \tag{10.7}$$

Since the coefficients c_{jr} are unknown, the problem of evaluating E now involves the problem of evaluating the c's. Here, use is made of the *variational principle*, which states that the value of E obtained from an approximate function cannot be smaller than the value of E given by the correct ψ. Consequently, the lowest value of E obtained from any function of the type used is the best approximation to the correct E obtainable by this type of function. Minimization of the energy is achieved in standard fashion by setting the partial derivatives of E with respect to each of the c's equal to zero:

$$\left(\frac{\partial E}{\partial c_{11}} \right)_{c_{12}} = 0, \quad \left(\frac{\partial E}{\partial c_{12}} \right)_{c_{11}} = 0$$

Rearranging equation (10.7), we obtain:

$$E(c_{11}^2 S_{11} + 2c_{11}c_{12}S_{12} + c_{12}^2 S_{22}) = c_{11}^2 H_{11} + 2c_{11}c_{12}H_{12} + c_{12}^2 H_{22}$$

and differentiating implicitly with respect to E, c_{11}, and c_{12} gives:

$$E(2c_{11}S_{11}dc_{11} + 2c_{11}S_{12}dc_{12} + 2c_{12}S_{12}dc_{11} + 2c_{12}S_{22}dc_{12}) +$$
$$dE(c_{11}^2 S_{11} + 2c_{11}c_{12}S_{12} + c_{12}^2 S_{22})$$
$$= 2c_{11}H_{11}dc_{11} + 2c_{11}H_{12}dc_{12} + 2c_{12}H_{12}dc_{11} + 2c_{12}H_{22}dc_{12}$$

With c_{12} constant, $dc_{12} = 0$:

$$(c_{11}^2 S_{11} + 2c_{11}c_{12}S_{12} + c_{12}^2 S_{22})\partial E$$
$$= 2c_{11}H_{11}\partial c_{11} + 2c_{12}H_{12}\partial c_{11} - E(2c_{11}S_{11} + 2c_{12}S_{12})\partial c_{11}$$

and with c_{11} constant, $dc_{11} = 0$:

$$(c_{11}^2 S_{11} + 2c_{11}c_{12}S_{12} + c_{12}^2 S_{22})\partial E$$
$$= 2c_{11}H_{12}\partial c_{12} + 2c_{12}H_{22}\partial c_{12} - E(2c_{11}S_{12} + 2c_{12}S_{22})\partial c_{12}$$

Setting $\partial E/\partial c_{11}$ and $\partial E/\partial c_{12}$ equal to zero and collecting terms, we obtain

$$c_{11}(H_{11} - ES_{11}) + c_{12}(H_{12} - ES_{12}) = 0$$
$$c_{11}(H_{21} - ES_{21}) + c_{12}(H_{22} - ES_{22}) = 0 \qquad (10.8)$$

Equation 10.8 is a set of two homogeneous linear equations in two unknowns, called the *secular equations*. One solution is $c_{11} = c_{12} = 0$, but this solution is unacceptable, since then $\psi = 0$. Other solutions exist only for certain values of E, which can be found by solving the determinant, called the *secular determinant*, of the multipliers of the c's in equation 10.8:

$$\begin{vmatrix} H_{11} - ES_{11} & H_{12} - ES_{12} \\ H_{21} - ES_{21} & H_{22} - ES_{22} \end{vmatrix} = 0 \qquad (10.9)$$

In the zero-order approximation the calculations are very much simplified since we assume that $S_{12} = S_{21} = 0$. Substituting these values for the overlap integral between neighboring atoms, replacing $H_{11} = H_{22}$ by α, the Coulomb integral, and $H_{12} = H_{21}$ by β, the resonance integral, and remembering that $S_{11} = S_{22} = 1$, we obtain as the secular determinant:

$$\begin{vmatrix} \alpha - E & \beta \\ \beta & \alpha - E \end{vmatrix} = 0 \qquad (10.10)$$

Dividing through by β gives:

$$\begin{vmatrix} \dfrac{\alpha - E}{\beta} & 1 \\ 1 & \dfrac{\alpha - E}{\beta} \end{vmatrix} = 0 \tag{10.11}$$

If we now let

$$x = \frac{\alpha - E}{\beta} \tag{10.12}$$

we obtain:

$$\begin{vmatrix} x & 1 \\ 1 & x \end{vmatrix} = 0 \tag{10.13}$$

and, solving, $x^2 - 1 = 0$, whence $x = \pm 1$. For $x = 1$ from equation 10.10 we get $E = \alpha - \beta$, and for $x = -1$, $E = \alpha + \beta$.

This procedure has actually given us two values of E. Had we carried out the same algebraic calculations represented by equations 10.4–10.9 on the second orbital, $\psi_2 = c_{21}\phi_1 + c_{22}\phi_2$, we would have obtained the identical determinant 10.9. Thus the single calculation always gives us the energies of all the orbitals.

We have now determined the energies of the two MO's of the H_2 molecule in terms of α and β. The energy $\alpha + \beta$ corresponds to the lower-energy orbital since β, by its nature, is a negative quantity (a negative energy refers to an attractive force between an electron and the nuclei, while a positive energy refers to repulsion). In order to obtain these energies we must have the values of α and β for the hydrogen molecule.

The secular determinant, equation 10.9, represents the energy equation for any MO made up of two equivalent AO's. In the most widely used application of crude quantum-mechanical arguments we are concerned only with the π electrons of unsaturated and aromatic systems. In these applications it is assumed that the σ bond framework is constant, and that the π electrons can be treated completely separately. The π electron system of ethylene is made up of just the two $2p\pi$ orbitals of the two carbon atoms, and equation 10.9 applies to this system. Thus, if we wish to calculate the energies of the two MO's generated from the two $p\pi$ carbon atoms in ethylene, we solve equation 10.9 by simplifying to equation 10.13 and end up with the solutions:

$$E(\psi_b) = \alpha + \beta$$
$$E(\psi_a) = \alpha - \beta$$

In the ethylene molecule, both of the C atoms are sp^2 hybridized and the orbitals of concern are pure $p\pi$ orbitals, for which:

$$\alpha = -11.16 \text{ eV}, \quad \text{and} \quad \beta = -2.32 \text{ eV}$$

Thus the energy (eigenvalue) for the bonding MO in ethylene is $\alpha + \beta = -13.48$ eV, and the eigenvalue corresponding to the higher-energy antibonding MO is $\alpha - \beta = -9.84$ eV. It is customary, especially in dealing with conjugated $p\pi$ systems, to give the energies of the MO's in units of β because we are generally interested in how much more or less stable the orbital is relative to the energy of the isolated atom. The MOED is accordingly sketched with $\alpha = 0$ and the bonding and antibonding MO's represented by horizontal lines below and above the lines representing the isolated atoms. For ethylene (Fig. 10.1) the bonding MO is 2.32 eV more stable, and the antibonding orbital 2.32 eV less stable, than the isolated carbon $p\pi$ orbital.

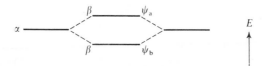

Fig. 10.1   The molecular orbital energy diagram for ethylene.

In the ground state both electrons of ethylene are in ψ_b. Promotion of an electron from ψ_b to ψ_a would require 4.64 eV = 107kcal/mole corresponding to 286 nm (equation 1.2). Actually the $\pi \rightarrow \pi^*$ transition for ethylene occurs at about 185 nm, and this discrepancy brings up several new problems. One of these relates to the proper selection of β values. There are several scales of β values, and if a particular scale is adhered to, results with internal consistency are achieved. For spectroscopic work, analysis of a single spectrum can lead to an empirical value of β which is called the *spectroscopic β* and which is generally different from the conventional β obtained from, say, thermochemical data. Another problem is that the energy level diagram does not distinguish between singlet and triplet states but rather gives a "center of gravity" of singlet and triplet (see Chapter 11) energies. Since singlet-triplet separations can be quite large and the observed $\pi \rightarrow \pi^*$ transition is a singlet, the actual energy may be considerably higher than that indicated by the 2β energy difference between the ground and excited states.

10.3   Constructing and Solving the Secular Determinants of Conjugated Systems; the Hückel Molecular Orbital Treatment (HMO)

We have thus far set up the secular determinant for a system of two atoms, such as H_2 or C=C, and have obtained a 2 × 2 determinant containing four terms. In general, the secular determinant contains a total number of terms which is the square of the number of AO's involved in the conjugated system. If we combine four AO's in butadiene, we will generate a 4 × 4 secular determinant containing sixteen terms.

The general form of a determinant is:

$$
\begin{vmatrix}
a_{11} & a_{12} & a_{13} & \cdots & a_{1n} \\
a_{21} & a_{22} & a_{23} & \cdots & a_{2n} \\
a_{31} & a_{32} & a_{33} & \cdots & a_{3n} \\
\vdots & \vdots & \vdots & \vdots & \vdots \\
a_{n1} & a_{n2} & a_{n3} & \cdots & a_{nn}
\end{vmatrix} = 0
$$

Each a is called an element, and the subscripts refer to the position, that is, the row and column, respectively, of the element. Thus element a_{32} is the element in the third row and the second column.

The secular determinants for the $p\pi$ systems of butadiene can readily be set up. Starting at one end of the conjugated system, we number consecutively the atoms in the $p\pi$ system. The first element, a_{11}, represents carbon 1 and is given the value $x = (\alpha - E)/\beta$. The next element in the first row, a_{12}, represents carbon atom 1 bonded to 2 and is given the value 1. Resonance integrals between nonbonded atoms are usually neglected; and, since in butadiene carbon 1 is not directly bonded to 3 and 4, elements a_{13} and a_{14} are given the values 0. Carbon atom 2 is bonded to 1 and to 3 but not to 4, and hence elements $a_{21} = a_{23} = 1$ and $a_{24} = 0$; the element a_{22} (a diagonal element) is again x. The justification for these substitutions is straightforward.

In summary, the elements of the diagonal ($a_{11}, a_{22}, a_{33}, \ldots$) in the secular determinant are designated as x, the elements with subscripts of nearest-neighbor numbers ($a_{12}, a_{21}, a_{23}, \ldots$) are given the value 1 (if all bonds are assumed equivalent, i.e., all β equal), and other elements are 0. Thus:

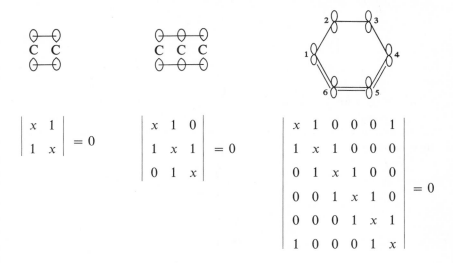

$$
\begin{vmatrix}
x & 1 \\
1 & x
\end{vmatrix} = 0
$$

$$
\begin{vmatrix}
x & 1 & 0 \\
1 & x & 1 \\
0 & 1 & x
\end{vmatrix} = 0
$$

$$
\begin{vmatrix}
x & 1 & 0 & 0 & 0 & 1 \\
1 & x & 1 & 0 & 0 & 0 \\
0 & 1 & x & 1 & 0 & 0 \\
0 & 0 & 1 & x & 1 & 0 \\
0 & 0 & 0 & 1 & x & 1 \\
1 & 0 & 0 & 0 & 1 & x
\end{vmatrix} = 0
$$

After the determinant is set up, the next step is to solve the determinant for x and then substitute $(\alpha - E)/\beta$ for x, where α is the Coulomb integral and β the resonance integral.

Such an $n \times n$ determinant can always be expanded into a polynomial equation of nth order, and the nature of the equation is such that it always has n real solutions (roots). Since the advent of computers, however, programs which "diagonalize" the matrix without expansion into an nth-order equation are readily available, and thus the eigenvalues are calculated rapidly.

A general procedure for the evaluation of an $n \times n$ determinant involves the use of minors, a minor being what remains of the determinant after removal of one row and one column. The steps are as follows:

1. Select a convenient row (or column) and write the value of the determinant as the sum of the products of each element of the row by its minor, giving each product a plus or minus sign, depending on whether the sum of the number of the row and the number of the column of the element is even or odd, respectively.

2. Expand each determinant obtained in step 1 by the same method. Continue until no determinants remain.

In the example below, using the secular determinant of the allyl radical, we arbitrarily choose the first row for our development:

$$\begin{vmatrix} x & 1 & 0 \\ 1 & x & 1 \\ 0 & 1 & x \end{vmatrix} = x \begin{vmatrix} x & 1 \\ 1 & x \end{vmatrix} - 1 \begin{vmatrix} 1 & 1 \\ 0 & x \end{vmatrix} + 0 \begin{vmatrix} 1 & x \\ 0 & 1 \end{vmatrix}$$

$$= x[x^2 - 1] - 1[x] = x^3 - x - x = x^3 - 2x = 0$$
$$\therefore \quad x = 0 \quad \text{or} \quad x^2 - 2 = 0, x = \pm\sqrt{2}$$

The energies for the three MO's of the allyl system are, therefore, $E_1 = \alpha + \sqrt{2}\,\beta$, $E_2 = \alpha$ (a nonbonding orbital), and $E_3 = \alpha - \sqrt{2}\,\beta$.

For butadiene, we require a 4×4 determinant which can be solved by the procedure of minors:

$$\begin{vmatrix} x & 1 & 0 & 0 \\ 1 & x & 1 & 0 \\ 0 & 1 & x & 1 \\ 0 & 0 & 1 & x \end{vmatrix}$$

$$= x \begin{vmatrix} x & 1 & 0 \\ 1 & x & 1 \\ 0 & 1 & x \end{vmatrix} - 1 \begin{vmatrix} 1 & 1 & 0 \\ 0 & x & 1 \\ 0 & 1 & x \end{vmatrix} + 0 \begin{vmatrix} 1 & x & 0 \\ 0 & 1 & 1 \\ 0 & 0 & x \end{vmatrix} - 0 \begin{vmatrix} 1 & x & 1 \\ 0 & 1 & x \\ 0 & 0 & 1 \end{vmatrix}$$

$$= x \left[x \begin{vmatrix} x & 1 \\ 1 & x \end{vmatrix} - 1 \begin{vmatrix} 1 & 1 \\ 0 & x \end{vmatrix} + 0 \begin{vmatrix} 1 & x \\ 0 & 1 \end{vmatrix} \right]$$

$$- 1 \left[1 \begin{vmatrix} x & 1 \\ 1 & x \end{vmatrix} - 1 \begin{vmatrix} 0 & 1 \\ 0 & x \end{vmatrix} + 0 \begin{vmatrix} 0 & x \\ 0 & 1 \end{vmatrix} \right]$$

$$= x[x(x^2 - 1) - 1(x)] - 1[1(x^2 - 1)]$$
$$= x(x^3 - x - x) - x^2 + 1 = x^4 - 3x^2 + 1 = 0$$

To solve this quadratic equation in x^2, we let $y = x^2$, whence $y^2 - 3y + 1 = 0$ and

$$y = \frac{3 \pm \sqrt{5}}{2} = \frac{1 \pm 2\sqrt{5} + 5}{4} = \left(\frac{1 \pm \sqrt{5}}{2} \right)^2 = x^2$$

$$x = \pm \frac{1 \pm \sqrt{5}}{2}$$

Since $x = (\alpha - E)/\beta$,

$$E_1 = \alpha + \left(\frac{\sqrt{5} + 1}{2} \right) \beta = \alpha + 1.618\beta$$

$$E_2 = \alpha + \left(\frac{\sqrt{5} - 1}{2} \right) \beta = \alpha + 0.618\beta$$

$$E_3 = \alpha - \left(\frac{\sqrt{5} - 1}{2} \right) \beta = \alpha - 0.618\beta$$

$$E_4 = \alpha - \left(\frac{\sqrt{5} + 1}{2} \right) \beta = \alpha - 1.618\beta$$

Hence the four MO's of butadiene can be arranged in the MOED (energies in units of $-\beta$), Fig. 10.2; and in the ground state the four π electrons are placed in two bonding MO's.

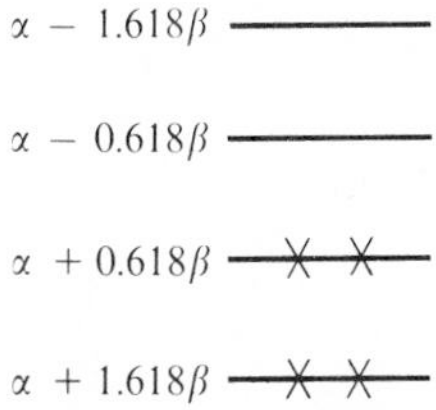

Fig. 10.2   The energy levels of the molecular orbitals of butadiene.

10.4 Determination of the Coefficients

In Chapter 4 we pointed out that the value of the amplitude of the FEM wave function for linear polyenes, $\psi_j = \sqrt{2/a}\,\sin(n_j\pi x/a)$, reproduces the coefficients c_{jr} of the HMO wave function:

$$\psi_j = c_{j1}\phi_1 + c_{j2}\phi_2 + c_{j3}\phi_3 + \dots$$

In general the coefficients tell us how much each AO contributes to the MO.

The ratio of the coefficients can be calculated directly from the secular equations, and then the values can be obtained from the normalization condition:

$$\sum_r c^2{}_{jr} = 1$$

Thus, for the π system of ethylene, the secular determinant is

$$\begin{vmatrix} x & 1 \\ 1 & x \end{vmatrix} = 0, \quad x = \pm 1$$

which corresponds to the homogeneous simultaneous equations:

$$c_1 x + c_2 = 0$$
$$c_1 + c_2 x = 0$$

For ψ_1, the bonding MO, $x = -1$, whence $c_1(-1) + c_2 = 0$ and $c_1 = c_2$:

$$c_1{}^2 + c_2{}^2 = 1, \quad 2c_1{}^2 = 1, \quad \text{and} \quad c_1 = c_2 = \frac{1}{\sqrt{2}}$$

Hence:

$$\boxed{\psi_1 = \frac{1}{\sqrt{2}}\phi_1 + \frac{1}{\sqrt{2}}\phi_2} \qquad E = \alpha + \beta$$

For ψ_2, $x = +1$ and, since $c_1 x + c_2 = 0$, $c_1 = -c_2$. Again, $c_1{}^2 + c_2{}^2 = 2c_1{}^2 = 1$, $c_1 = -c_2 = 1/\sqrt{2}$. Hence:

$$\boxed{\psi_2 = \frac{1}{\sqrt{2}}\phi_1 - \frac{1}{\sqrt{2}}\phi_2} \qquad E = \alpha - \beta$$

For allyl, the secular equations are:

$$\begin{aligned} &\text{(a)} && c_1 x + c_2 && && = 0 \\ &\text{(b)} && c_1 + c_2 x + c_3 && = 0 \\ &\text{(c)} && c_2 + c_3 x && = 0 \end{aligned}$$

For ψ_1, $x = -\sqrt{2}$ and, from (a), $-\sqrt{2}\, c_1 = -c_2$ or $c_1 = c_2/\sqrt{2}$. From (c), $c_2 = \sqrt{2}\, c_3$ or $c_3 = c_2/\sqrt{2}$.

Now we have c_1 and c_3 expressed in terms of c_2.

$$c_1{}^2 + c_2{}^2 + c_3{}^2 = \left(\frac{c_2}{\sqrt{2}}\right)^2 + c_2{}^2 + \left(\frac{c_2}{\sqrt{2}}\right)^2 = 1\,;$$

$$2c_2{}^2 = 1, \quad \text{and} \quad c_2 = \frac{1}{\sqrt{2}}$$

$$c_1 = \frac{c_2}{\sqrt{2}} = \frac{1}{\sqrt{2}} \cdot \frac{1}{\sqrt{2}} \cdot \sqrt{2} = \frac{1}{2} = c_3$$

$$\therefore \quad \boxed{\psi_1 = \tfrac{1}{2}\phi_1 + \frac{1}{\sqrt{2}}\phi_2 + \tfrac{1}{2}\phi_3} \qquad E = \alpha + \sqrt{2}\beta$$

For ψ_2, $x = 0$, $c_2 = 0$, and $c_1 = -c_3$.

$$\therefore \quad c_1 = \frac{1}{\sqrt{2}}, \; c_2 = 0, \; c_3 = \frac{-1}{\sqrt{2}}$$

$$\boxed{\psi_2 = \frac{1}{\sqrt{2}}\phi_1 - \frac{1}{\sqrt{2}}\phi_3} \qquad E = \alpha$$

Similarly, for ψ_3:

$$\boxed{\psi_3 = \tfrac{1}{2}\phi_1 - \frac{1}{\sqrt{2}}\phi_2 + \tfrac{1}{2}\phi_3} \qquad E = \alpha - \sqrt{2}\,\beta$$

As a final example, let us determine the coefficients for each of the AO's in one of the four MO's of butadiene, say ψ_4, where $x = 1.618$. The four linear equations corresponding to the secular determinants are:

(a) $c_1 x + c_2 \qquad\qquad\qquad = 0$

(b) $c_1 \;\; + c_2 x + c_3 \qquad\quad = 0$

(c) $\qquad\quad c_2 \;\; + c_3 x + c_4 = 0$

(d) $\qquad\qquad\qquad c_3 \;\; + c_4 x = 0$

Let us now determine all the coefficients in terms of c_1.

From (a), $c_2 = -c_1 x = -1.618 c_1$.

From (b), $c_3 = -c_2 x - c_1 = -c_1(x^2 - 1) = 1.618 c_1$.

From (c), $c_4 = -c_3 x - c_2 = -c_1[x(x^2 - 1) - x]$
$$= -c_1[x(x^2 - 2)] = -c_1.$$

$$c_1{}^2 + c_2{}^2 + c_3{}^2 + c_4{}^2 = 2(c_1{}^2 + 1.618^2 c_1{}^2) = 7.24 c_1{}^2 = 1$$

$$\therefore \quad c_1 = 0.372, \quad c_2 = -0.602, \quad c_3 = 0.602, \quad c_4 = -0.372$$

$$\boxed{\psi_4 = 0.372\phi_1 - 0.602\phi_2 + 0.602\phi_3 - 0.372\phi_4} \qquad E = \alpha + 1.618\beta$$

10.5   Delocalization Energies

We have calculated the energies of the MO's of ethylene and butadiene. In the ground state, two electrons are in ψ_1 of ethylene and the total energy is $2(\alpha + \beta)$. In butadiene, with two electrons in each of ψ_1 and ψ_2, the total energy of the ground state is $2(\alpha + 1.618\beta) + 2(\alpha + 0.618\beta) = 4\alpha + 4.47\beta$, and this energy is numerically larger (actually more negative) than that of two ethylenes by 0.47β. This is the calculated energy released when the π systems of two isolated ethylenes, each with a pair of π electrons, combine (theoretically) to form the conjugated diene. The four electrons are then delocalized (in two MO's) over four atoms, and the energy released by this delocalization is called the *delocalization energy*. If we had worked out the energies of the six benzene MO's, we would have obtained energies for ψ_1, ψ_2, and ψ_3 (the MO's occupied in the ground state) of $\alpha + 2\beta$, $\alpha + \beta$, $\alpha + \beta$, respectively. The total energy would then be $6\alpha + 8\beta$. Three isolated double bonds of ethylene would contribute $6\alpha + 6\beta$ energy, and hence benzene is more stable than three noninteracting ethylenes by 2β, the delocalization energy (DE).

Fig. 10.3   The molecular orbital energy level diagram for allyl.

It is of interest to calculate the DE for the allyl system, Fig. 10.3. In the cation, there are two electrons in ψ_1 and the DE is $2(\alpha + \sqrt{2}\beta) - 2(\alpha + \beta) = 0.83\beta$. In the radical, with one electron in ψ_2, π energy $2(\alpha + \sqrt{2}\beta) + \alpha = 3\alpha + 2\sqrt{2}\,\beta$. From this value we need to subtract $(2(\alpha + \beta))$, the energy of two electrons in an isolated double bond plus the energy of a single electron in a $p\pi$ orbital, which we have defined by α. Hence $3\alpha + 2\sqrt{2}\beta - [2(\alpha + \beta) - \alpha] = 0.83\beta$. The DE for the anion with two electrons in ψ_2 will similarly come out to be 0.83β, and thus in the HMO approximation the allyl cation, radical, and anion all have the same DE and are equally stable.

This may seem peculiar at first, since we expect that four $p\pi$ electrons delocalized over three atoms should lead to a more stable molecule than two $p\pi$ electrons similarly delocalized. However, additional electrons after the two in ψ_1, a strongly bonding MO, go into ψ_2, the occupation of which does not lead to any delocalization energy because it is the nonbonding MO. Of course, the total energy of the three-electron system is greater than that of a two-electron system, but it is the DE and not the total energy that we are concerned with here. Actually the stability sequence is opposite to that expected. In feeding one or two electrons into the carbonium ion to form the free radical and the anion, respectively, we add additional electrons and hence increase the electron-electron repulsions. Since these repulsions were previously neglected, the stability sequence should be cation > radical > anion.

At this point we can profitably consider briefly the topics of resonance energies and aromaticity because they are closely related to delocalization and delocalization energies.

10.6   Resonance Energies and Aromaticity

We are familiar with the notion that the structure of a molecule possessing a conjugated system of multiple bonds can be expressed by several resonance structures, and the closer these structures are in energy, the greater the *resonance energy* (RE). Thus, because benzene can be expressed equally well by two equivalent resonance structures (the two Kekulé structures), whereas 1,3,5-hexatriene cannot, we assume, correctly, that the former possesses considerably more resonance energy than the acyclic triene. The RE can be determined by measuring the heat of hydrogenation (or the heat of combustion) and comparing this experimental value with a value calculated for a single reference structure. Thus the hydrogenation of benzene liberates 49.8 kcal/mole, whereas the heat liberated by the hydrogenation of cyclohexene is 28.6 kcal/mole. A single Kekulé structure of benzene, that is, a six-membered ring with three noninteracting double bonds, might be expected to liberate $28.6 \times 3 = 85.8$ kcal/mole on hydrogenation, and hence the empirical RE of benzene is 36 kcal/mole. The difference between the observed heat of combustion of benzene (789.1 kcal/mole) and that calculated for a single Kekulé structure (825.1 kcal/mole) gives the same value for the RE.

In determining the delocalization energy (DE) of benzene in Section 10.5, we used the total π energy of three isolated double bonds as a reference, each double bond contributing 2β to the total π energy. The set of three double bonds corresponds to a single Kekulé structure. The delocalized structure used as a model for the HMO calculation corresponds to the resonance hybrid of two Kekulé structures (plus others which we neglect), and thus the DE in MO theory corresponds to the RE in valence bond theory. Accordingly, it is not unexpected that there is a good correlation between the DE and the RE for many aromatic hydrocarbons.

In the preceding discussion of benzene, the reference structure for calculating both DE and RE is a single Kekulé structure, in which all the C-C bond distances are equal. However, if we think of the six-membered ring as having three noninteracting double bonds, there is considerable validity in considering the reference compound as 1,3,5-cyclohexatriene, with alternating single and double bonds. With this as a reference structure we would first require some deformation energy to convert this model into the Kekulé model, that is, we would need to compress the single bonds and stretch the double bonds to obtain the Kekulé structure, in which, as mentioned above, all C-C bond distances are equal. This distortion energy has been estimated to be 27 kcal/mole, that is, the Kekulé structure is 27 kcal/mole less stable (higher in energy) than the structure (1,3,5-cyclohexatriene) in which there are alternating single and double bonds. If we use 1,3,5-cyclohexatriene as a reference structure for the RE calculation, the resonance hybrid is required to be 36 kcal/mole more stable (lower in energy) than the reference. Hence the total energy difference between the Kekulé structure and the hybrid (or delocalized) structure now becomes the sum of the distortion energy and the empirical resonance energy, or $27 + 36 = 63$ kcal/mole.

The energy difference between a Kekulé structure and the resonance hybrid structure that includes the distortion energy is called the *vertical resonance energy* in valence bond theory and the *vertical delocalization energy* in MO theory. The use of the word vertical implies a comparison between structures of different energy during which the geometry of each structure is held stationary, a concept that is expressed in the Franck-Condon principle (cf. Chapter 8). However, unless otherwise specified, the RE of hydrocarbons is usually given as the empirical and not the vertical resonance energy, that is, for benzene, 36 and not 63 kcal/mole.

One of the most fundamental of all the principles governing the structure of ring compounds is the concept of the *aromatic sextet*. The presence of six π electrons in a ring confers on it peculiar stability and reactivity. The tendencies of the cyclopentadienyl radical to take up an electron to give $C_5H_5^-$, and of the cycloheptatrienyl radical to give up an electron to form $C_7H_7^+$, are dramatic examples of this phenomenon. As with benzene, these isoelectronic systems have three bonding orbitals, two degenerate ones and a single lowest-energy level (cf. Chapter 4). Since each orbital accommodates two electrons, the stable number of electrons is $4n + 2$, where n is any integer. When $n = 0$, the "aromatic" system should consist of two electrons, and in a remarkable confirmation of this predicted stability it has been found that substituted cyclopropenyl cations are relatively stable.

On the basis of the Hückel rule of $4n + 2$, it might be expected that cyclobutadiene, with $4n$ ($n = 1$) electrons, would not be stable. Simple HMO calculations show no resonance stabilization, and calculations which consider electron repulsion terms indicate that the system is less stable than one with two isolated double bonds. Also, some chemical evidence points to the possibility that in cyclobutadiene there is actually resonance destabilization. The term *antiaromatic* has been proposed for such π systems.

Now we turn to other important properties of molecules that we can calculate from the MO wave functions.

10.7   Electron Densities

In the LCAO (HMO) approximation, $c_r{}^2$ is the probability that an electron in an MO is associated with atomic orbital ϕ_r, or $c_r{}^2$ is the electron density at atom r in the MO.

The total electron density at an atom r, q_r, is the sum of the electron densities at that atom contributed by each electron or by each occupied orbital:

$$q_r = \sum_j n_j c_{jr}^2$$

Here c_{jr} is the coefficient of atom r in the jth molecular orbital, which is occupied by n_j electrons. The sum is taken over all the MO's. Thus, in butadiene, ψ_1 and ψ_2 are each occupied by two electrons, so that $n_1 = n_2 = 2$, and ψ_3 and ψ_4 are vacant, $n_3 = n_4 = 0$. Hence

$$q_1 = 2c_{11}^2 + 2c_{21}^2 = 2(0.371)^2 + 2(0.600)^2 = 2(0.140) + 2(0.36) = 1.00$$
$$q_2 = 2c_{12}^2 + 2c_{22}^2 = 2(0.6)^2 + 2(0.371) = 1.00$$

By symmetry, atoms 3 and 4 give the same answer; hence the electron density is unity at each atom. The sum of the electron densities over all the atoms must equal the total number of electrons; thus $q_1 + q_2 + q_3 + q_4 = 4$, the number of π electrons in butadiene.

The electron density at each carbon atom is unity, not only in butadiene but also in all *alternant hydrocarbons*, that is, compounds containing only conjugated carbon atoms and even-numbered rings, having all Coulomb integrals equal, and having as many π electrons as atoms in the conjugated system. This last requirement is met, for example, in the allyl radical, C_3H_5, but not in the anion or cation derived therefrom by addition or removal of an electron, respectively. Thus, for the allyl cation, with two electrons in ψ_1, $n_1 = 2$, $n_2 = n_3 = 0$, and $c_{11} = \frac{1}{2}$, $c_{12} = 1/\sqrt{2}$, $c_{13} = \frac{1}{2}$.

$$\psi_1 = \tfrac{1}{2}\phi_1 + 1/\sqrt{2}\phi_2 + \tfrac{1}{2}\phi_3 \qquad q_1 = 2c_{11}^2 = 2\cdot(\tfrac{1}{2})^2 = \tfrac{1}{2}$$
$$\psi_2 = 1/\sqrt{2}\phi_1 - 1/\sqrt{2}\phi_3 \qquad q_2 = 2c_{12}^2 = 2\cdot(\tfrac{1}{2})^2 = 1$$
$$\psi_3 = \tfrac{1}{2}\phi_1 - 1/\sqrt{2}\phi_2 - \tfrac{1}{2}\phi_3 \qquad q_3 = 2c_{13}^2 = 2\cdot(\tfrac{1}{2})^2 = \tfrac{1}{2}$$

The electron densities can be used to calculate very simply another property of interest, called the *charge density*, ζ (zeta).. For hydrocarbons, the charge density is defined as $1 - q_r$. In the allyl system the charge densities are thus:

Charge Density	Cation	Radical	Anion
$1 - q_1$	$\tfrac{1}{2}$	0	$-\tfrac{1}{2}$
$1 - q_2$	0	0	0
$1 - q_3$	$\tfrac{1}{2}$	0	$-\tfrac{1}{2}$

10.8   The Bond Order, p

The bond order, p, is a measure of the amount of π bond character in a particular bond:

$$p_{rs} = \sum_j n_j c_{jr} c_{js}$$

where n_j is the number of electrons in the jth molecular orbital, c_{jr} is the coefficient of atom r in the jth molecular orbital, c_{js} is the coefficient of atom s bonded to atom r, and p_{rs} has meaning only for adjacent atoms. This definition is exactly analogous to the one given in Chapter 4, except that the coefficients used in this chapter replace the amplitudes used there.

The product of the (normalized) coefficients of adjacent atoms may be construed as a bond-electron density. When both coefficients are large and are of like sign, the product is large and corresponds to substantial electronic cement binding the atoms. When one of the coefficients is zero, indicative of a node at an atom, the partial bond order vanishes in accordance with the nonbonding character. If the coefficients of adjacent atoms are of opposite sign, indicative of a node between atoms, we have a negative bond order in agreement with the antibonding situation.

As an example let us calculate p_{rs} for butadiene. The wave functions are:

$$\psi_1 = 0.372\phi_1 + 0.602\phi_2 + 0.602\phi_3 + 0.372\phi_4$$
$$\psi_2 = 0.602\phi_1 + 0.372\phi_2 - 0.372\phi_3 - 0.602\phi_4$$
$$\psi_3 = 0.602\phi_1 - 0.372\phi_2 - 0.372\phi_3 + 0.602\phi_4$$
$$\psi_4 = 0.372\phi_1 - 0.602\phi_2 + 0.602\phi_3 - 0.372\phi_4$$

For the ground state:

$$p_{12} = 2c_{11}c_{12} + 2c_{21}c_{22} = 2(0.372 \times 0.602) + 2(0.602 \times 0.372) \quad = 0.894$$

$$p_{23} = 2c_{12}c_{13} + 2c_{22}c_{23} = 2(0.602 \times 0.602) + 2(0.372 \times -0.372) = 0.447$$

By symmetry, $p_{34} = p_{12}$.

Thus we see that the essential double bonds, 1-2 and 3-4, have nearly completely double-bond character, and hence the name. But, because of the delocalization, the essential single bond 2-3 also has substantial double-bond character, though much less than the others. This double-bond character explains the tendency of a molecule to remain planar, as well as the small but real energy barrier between the two "isomers," *s-cis* and *s-trans* (cf. Section 4.3):

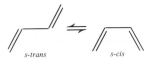

s-trans   *s-cis*

An excited state is generated by exciting one electron from one of the occupied orbitals to one of the vacant (virtual) ones. The first excited state of butadiene can be described as $\psi_1{}^2\psi_2\psi_3$, with $n_1 = 2$, $n_2 = n_3 = 1$, $n_4 = 0$, and hence bond orders:

$$p_{12} = 2c_{11}c_{12} + c_{21}c_{22} + c_{31}c_{32} = 0.445$$
$$p_{23} = 2c_{12}c_{13} + c_{22}c_{23} + c_{32}c_{33} = 0.72$$

Here, the bond order of the 1-2 bond has drastically decreased upon excitation, whereas that of the 2-3 bond has increased substantially. This increase is responsible for a high energy barrier to rotation around the 2-3 bond in the excited state, and for the fact that a molecule in the *s-cis* conformation, when excited, will remain in that conformation.

10.9   The Free Valence Number, F

The quantity F_r, called the *free valence number* of atom r, measures the extent to which the maximum valence of that atom is *not* satisfied by bonds, or the amount of residual valency that is on the atom. It is defined by:

$$F_r = F_{\max} - \sum_s p_{rs}$$

where the summation extends over all atoms adjacent to r, and $F_{\max}$ is a number assigned the value of $\sqrt{3} = 1.732$ for reasons which will be elaborated below.

As an exercise let us calculate F_2 in butadiene in the ground state:

$$F_2 = 1.732 - (p_{12} + p_{23}) = 1.732 - (0.894 + 0.447) = 0.391$$

Information about bond orders and free valence numbers is summarized in molecular diagrams in which the p_{rs} is written on each bond and the F_r at the end of a short arrow starting at an atom of each type. The molecular diagram for butadiene is:

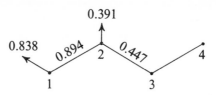

Since by symmetry $p_{12} = p_{34}$, $F_1 = F_4$, and $F_2 = F_3$, only one value for each pair is necessary.

The number 1.732 for the maximum free valence number is the sum of the p_{rs} for the central carbon atom in trimethylenemethane. In this molecule each methylene group can focus all its bonding power onto the central C atom, which is thus involved in $p\pi$ bonding with three other $p\pi$ orbitals. Hence it is satisfied to the greatest possible extent by π bonding to other atoms, and its residual free valency should be minimal (zero).

Let us proceed to make the necessary calculations:

$$
\begin{array}{c}
C_1 \\
| \\
C_4 \\
{}_2C \quad\quad C_3
\end{array}
$$

We have a $p\pi$ orbital on each of these C atoms perpendicular to the plane of the four atoms, and so we set up our secular determinant:

$$
\begin{vmatrix}
x & 0 & 0 & 1 \\
0 & x & 0 & 1 \\
0 & 0 & x & 1 \\
1 & 1 & 1 & x
\end{vmatrix} = 0
$$

Solving, we obtain $x^2(x^2 - 3) = 0$. (This can be done by the method of minors very quickly.) The four values for x are:

$$
x = +\sqrt{3}, \quad x = -\sqrt{3}, \quad x = 0, \quad x = 0
$$

ψ_4 ———————— $\alpha - \sqrt{3}\beta$

$\psi_{2,3}$ ═══════════ α

ψ_1 ———————— $\alpha + \sqrt{3}\beta$

Fig. 10.4   The molecular orbital energy diagram of trimethylenemethane.

The MOED for the system is shown in Fig. 10.4. After solving for the coefficients in the usual manner, we obtain the four final wave functions:

$$\psi_1 = \frac{1}{\sqrt{6}}(\phi_1 + \phi_2 + \phi_3 + \sqrt{3}\phi_4)$$

$$\psi_2 = \frac{1}{\sqrt{2}}(\phi_1 - \phi_3)$$

$$\psi_3 = \frac{1}{\sqrt{6}}(\phi_1 - 2\phi_2 + \phi_3)$$

$$\psi_4 = \frac{1}{\sqrt{6}}(\phi_1 + \phi_2 + \phi_3 - \sqrt{3}\phi_4)$$

From these we may calculate the bond order:

$$p_{14} = p_{24} = p_{34} = 2\left(\frac{1}{\sqrt{6}} \times \frac{1}{\sqrt{2}}\right) + 1\left(\frac{1}{\sqrt{2}} \times 0\right) + 1\left(\frac{1}{\sqrt{6}} \times 0\right) = \frac{1}{\sqrt{3}}$$

The sum of the bond order around C_4, the central atom, is thus $3 \times 1/\sqrt{3} = \sqrt{3} = 1.732$, and this value is taken as F_{max}. The value of F_4 is thus zero; there is no residual valency on carbon atom 4.

10.10   Symmetry Simplifications of the Hückel Molecular Orbital Method

The HMO method outlined in the preceding section involved: (1) setting up the secular determinant, (2) expanding the determinant, (3) determining the roots of the resulting polynomial, and, finally, (4) determining and then normalizing the coefficients. Without the aid of computers, the second and third steps of the procedure are exceedingly tedious for molecules (determinants) of any significant size. For naphthalene, $C_{10}H_8$, for example, a 10×10 secular determinant must be set up and expanded into a tenth-order polynomial, or a fifth-order polynomial in x^2. The numerical solution of such a polynomial by the Newton-Raphson method is computationally tedious, and the expansion of the determinant almost prohibitive. The problem can be greatly simplified, however, by the use of symmetry, provided the molecule belongs to a point group having any symmetry elements in addition to the plane of the molecule. The only thing required is the knowledge that each MO transforms as one of the irreducible representations of the proper point group.

Let us take naphthalene, Fig. 10.5, which belongs to point group D_{2h}, as an example; it has ten π electrons on ten atomic centers, and we wish to see whether we can find a way to simplify the 10×10 secular determinant. First we recognize that the molecular plane σ_{yz} is a plane of symmetry; all π orbitals are antisymmetric with respect to this plane, and hence all π

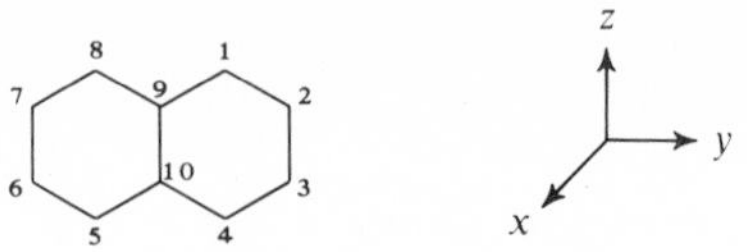

Fig. 10.5   Orientation of naphthalene

molecular orbitals must belong to representations having this property. In point group D_{2h}, Table A2.3, we see that only a_u, b_{1g}, b_{2g}, and b_{3u} qualify. Next, we see that the molecule has two additional planes of symmetry, one (σ_{xz}) passing through atoms 9 and 10 and the other (σ_{xy}) bisecting the 2-3, 9-10, and 6-7 bonds. These three symmetry elements suffice, since their combinations generate all other elements. Now, under σ_{xz}, atoms 2 and 7, 3 and 6, 1 and 8, and 4 and 5 are equivalent in pairs, and, under σ_{xy}, atoms 1 and 4, 5 and 8, 2 and 3, 6 and 7, and 9 and 10 are equivalent in pairs. Consequently, atoms 1, 4, 5, and 8, atoms 2, 3, 6, and 7, and atoms 9 and 10 form three different sets of symmetry-equivalent atoms.

As a consequence of these equivalences and the necessity that the MO's transform like the proper irreducible representation, we can conclude that in, for example, species b_{3u}, which is symmetric with respect to both σ_{xz} and σ_{xy}:

$$c_{j1} = c_{j4} = c_{j5} = c_{j8}$$
$$c_{j2} = c_{j3} = c_{j6} = c_{j7}$$
$$c_{j9} = c_{j10}$$

Hence an MO belonging to b_{3u} must have the form

$$\psi_j(b_{3u}) = \frac{c_{j1}}{2}(\phi_1 + \phi_4 + \phi_5 + \phi_8) + \frac{c_{j2}}{2}(\phi_2 + \phi_3 + \phi_6 + \phi_7)$$
$$+ \frac{c_{j9}}{\sqrt{2}}(\phi_9 + \phi_{10})$$

where normalizing factors of $\frac{1}{2}$ and $1/\sqrt{2}$ are introduced for computational convenience. This orbital has three undetermined constants, c_{jr}; consequently there are three such MO's, each of b_{3u} symmetry, and we have thus shown that a 3×3 determinant can be factored out of the original 10×10. In effect the function $\chi_1 = \frac{1}{2}(\phi_1 + \phi_4 + \phi_5 + \phi_8)$ is a symmetry orbital in the sense in which these were introduced in Chapter 6. The functions $\chi_2 = \frac{1}{2}(\phi_2 + \phi_3 + \phi_6 + \phi_7)$ and $\chi_3 = 1/\sqrt{2}(\phi_9 + \phi_{10})$ are also such symmetry orbitals, with the difference, however, that they include overlapping and interacting atomic orbitals ϕ_2 and ϕ_3, ϕ_6 and ϕ_7, and ϕ_9 and ϕ_{10}; consequently the handling is made more difficult.

Let us now set up the 3×3 secular determinant in terms of our group orbitals χ_1, χ_2, and χ_3. The elements of the determinant are given by

$$H_{11} = \int \chi_1 H \chi_1 \, d\tau$$
$$= \tfrac{1}{4} \int (\phi_1 + \phi_4 + \phi_5 + \phi_8) H (\phi_1 + \phi_4 + \phi_5 + \phi_8) \, d\tau$$
$$= \tfrac{1}{4}(\int \phi_1 H \phi_1 \, d\tau + \int \phi_4 H \phi_4 \, d\tau + \int \phi_5 H \phi_5 \, d\tau + \int \phi_8 H \phi_8 \, d\tau$$
$$+ \; 12 \text{ cross terms})$$

In this case each of the terms $\int \phi_r H \phi_r \, d\tau = \alpha$, and each of the cross terms $\int \phi_r H \phi_s \, d\tau = 0$, since no two centers in this set are adjacent. Hence

$$H_{11} = \tfrac{1}{4} \cdot 4\alpha = \alpha$$

Next

$$H_{22} = \int \chi_2 H \chi_2 \, d\tau$$
$$= \tfrac{1}{4} \int (\phi_2 + \phi_3 + \phi_6 + \phi_7) H (\phi_2 + \phi_3 + \phi_6 + \phi_7) \, d\tau$$

When this integral is expanded, we again obtain 4α, but in addition, out of the twelve cross terms, four do not vanish:

$$\int \phi_2 H \phi_3 \, d\tau, \quad \int \phi_3 H \phi_2 \, d\tau, \quad \int \phi_6 H \phi_7 \, d\tau, \quad \text{and} \quad \int \phi_7 H \phi_6 \, d\tau$$

because atoms 2 and 3 are bonded, as are 6 and 7, and consequently each of these integrals is β. Hence

$$H_{22} = \alpha + \beta$$

Similarly,

$$H_{33} = \int \chi_3 H \chi_3 \, d\tau = \tfrac{1}{2} \int (\phi_9 + \phi_{10}) H (\phi_9 + \phi_{10}) \, d\tau = \alpha + \beta$$
$$H_{12} = \tfrac{1}{4}(\int \phi_1 H \phi_2 \, d\tau + \int \phi_4 H \phi_3 \, d\tau + \int \phi_5 H \phi_6 \, d\tau + \int \phi_8 H \phi_7 \, d\tau$$
$$+ \; 12 \text{ vanishing terms})$$
$$= \beta$$

Similarly we find:

$$H_{13} = \sqrt{2}\beta, \quad H_{23} = 0$$

Thus, the determinant is:

$$\begin{vmatrix} H_{11} & H_{12} & H_{13} \\ H_{21} & H_{22} & H_{23} \\ H_{31} & H_{32} & H_{33} \end{vmatrix} = \begin{vmatrix} \alpha - E & \beta & \sqrt{2}\beta \\ \beta & \alpha + \beta - E & 0 \\ \sqrt{2}\beta & 0 & \alpha + \beta - E \end{vmatrix}$$

If, as usual, we divide through by β and let $x = (\alpha - E)/\beta$ to obtain:

$$\begin{vmatrix} x & 1 & \sqrt{2} \\ 1 & x + 1 & 0 \\ \sqrt{2} & 0 & x + 1 \end{vmatrix} = 0$$

the secular equation expands to (using the method of minors on the terms of the last row):

$$(x + 1)\left[x(x + 1) - 1\right] - 2(x + 1)$$
$$= (x + 1)(x^2 + x - 1) - 2(x + 1)$$
$$= (x + 1)(x^2 + x - 3) = 0$$

Hence $x = 1$, or $x^2 + x - 3 = 0$, $x = -2.303$ and $+1.303$. Substituting these values of x back into the secular equations, we obtain:

$$c_{j1}x + c_{j2} + \sqrt{2}c_{j3} = 0$$
$$c_{j1} + c_{j2}(x + 1) \quad\quad = 0$$
$$\sqrt{2}c_{j1} + c_{j3}(x + 1) = 0$$

and normalizing gives:

$$\psi_1 = 0.30(\phi_1 + \phi_4 + \phi_5 + \phi_8) + 0.23(\phi_2 + \phi_3 + \phi_6 + \phi_7)$$
$$+ 0.46(\phi_9 + \phi_{10})$$
$$\psi_2 = 0.41(\phi_2 + \phi_3 + \phi_6 + \phi_7) - 0.41(\phi_9 + \phi_{10})$$
$$\psi_3 = 0.40(\phi_1 + \phi_4 + \phi_5 + \phi_8) - 0.17(\phi_2 + \phi_3 + \phi_6 + \phi_7)$$
$$- 0.34(\phi_9 + \phi_{10})$$

If we now proceed to symmetry species b_{2g}, which is antisymmetric to σ_{xy}, we find:

$$c_{j1} = -c_{j4} = -c_{j5} = c_{j8}$$
$$c_{j2} = -c_{j3} = -c_{j6} = c_{j7}$$
$$c_{j9} = -c_{j10}$$

Thus we have again three MO's in species b_{2g}. Following through in the same way, we obtain the determinant

$$\begin{vmatrix} x & 1 & \sqrt{2} \\ 1 & x - 1 & 0 \\ \sqrt{2} & 0 & x - 1 \end{vmatrix} = 0$$

based on the symmetry orbitals:

$$\chi_4 = \tfrac{1}{2}(\phi_1 - \phi_4 - \phi_5 + \phi_8)$$
$$\chi_5 = \tfrac{1}{2}(\phi_2 - \phi_3 - \phi_6 + \phi_7)$$
$$\chi_6 = \frac{1}{\sqrt{2}}(\phi_9 - \phi_{10})$$

and the final wave functions:

$$\psi_4 = 0.40(\phi_1 - \phi_4 - \phi_5 + \phi_8) + 0.17(\phi_2 - \phi_3 - \phi_6 + \phi_7)$$
$$+ 0.35(\phi_9 - \phi_{10})$$

$$\psi_5 = 0.41(\phi_2 - \phi_3 - \phi_6 + \phi_7) - 0.41(\phi_9 - \phi_{10})$$
$$\psi_6 = 0.30(\phi_1 - \phi_4 - \phi_5 + \phi_8) - 0.23(\phi_2 - \phi_3 - \phi_6 + \phi_7)$$
$$- 0.46(\phi_9 - \phi_{10})$$

In symmetry species b_{1g}, which is antisymmetric to σ_{xz}, the relations between the coefficients are as follows:

$$c_1 = c_4 = -c_5 = -c_8$$
$$c_2 = c_3 = -c_6 = -c_7$$
$$c_9 = c_{10} = 0$$

Here, only two c's are undetermined since c_9 and c_{10} are zero, and we obtain the 2×2 determinant:

$$\begin{vmatrix} x & 1 \\ 1 & x + 1 \end{vmatrix} = 0$$

based on

$$\chi_7 = \tfrac{1}{2}(\phi_1 + \phi_4 - \phi_5 - \phi_8)$$
$$\chi_8 = \tfrac{1}{2}(\phi_2 + \phi_3 - \phi_6 - \phi_7)$$

giving the MO's

$$\psi_7 = 0.26(\phi_1 + \phi_4 - \phi_5 - \phi_8) + 0.43(\phi_2 + \phi_3 - \phi_6 - \phi_7)$$
$$\psi_8 = 0.43(\phi_1 + \phi_4 - \phi_5 - \phi_8) - 0.26(\phi_2 + \phi_3 - \phi_6 - \phi_7)$$

Finally, for the fourth and last symmetry species (a_u), having the character -1 for the σ coincident with the plane of the molecule (and there obviously can be no π molecular orbitals in any of the other species),

$$c_1 = -c_4 = c_5 = -c_8$$
$$c_2 = -c_3 = c_6 = -c_7$$
$$c_9 = c_{10} = 0$$

and again two MO's follow from the determinant

$$\begin{vmatrix} x & 1 \\ 1 & x - 1 \end{vmatrix} = 0$$

based on

$$\chi_9 = \tfrac{1}{2}(\phi_1 - \phi_4 + \phi_5 - \phi_8)$$
$$\chi_{10} = \tfrac{1}{2}(\phi_2 - \phi_3 + \phi_6 - \phi_7)$$

with the final result

$$\psi_9 = 0.43(\phi_1 - \phi_4 + \phi_5 - \phi_8) + 0.26(\phi_2 - \phi_3 + \phi_6 - \phi_7)$$
$$\psi_{10} = 0.26(\phi_1 - \phi_4 + \phi_5 - \phi_8) - 0.43(\phi_2 - \phi_3 + \phi_6 - \phi_7)$$

Thus the original 10×10 secular determinant has been factored into two 3×3's and two 2×2's by the use of symmetry orbitals which could be set up by inspection. This procedure is always possible and usually reduces the labor tremendously.

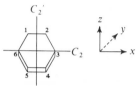

Fig. 10.6   Orientation and symmetry axes of benzene

Although this symmetry reduction was performed in a naïve, straight-forward, logical fashion, it may also be achieved more systematically by using the projection operator techniques outlined in Chapter 6. As an illustrative example we shall generate the MO's of benzene. Instead of point group D_{6h}, to which benzene belongs, we can use point group D_6. The character table for D_6 and the symmetry operations are given in Table A.218 and the coordinate system that we shall use is shown in Fig. 10.6. We now transform each pπ orbital in turn under the symmetry operations of D_6 to generate the reducible representation:

Orbital	I	C_6^z	C_3^z	C_2^z	C_2	C_2'
1	1	0	0	0	0	0
2	1	0	0	0	0	0
3	1	0	0	0	−1	0
4	1	0	0	0	0	0
5	1	0	0	0	0	0
6	1	0	0	0	−1	0
Σ	6	0	0	0	−2	0

The reducible representation corresponds to $A_2 + B_2 + E_1 + E_2$. We now generate the first trial wave function for species A_2 by transforming pπ_1 under *all* operations of D_6 and multiplying the result by the characters of A_2:

	I	$C_6\ C_6'$	$C_3\ C_3'$	C_2^z	$C_2^{1,4}$	$C_2^{2,5}$	$C_2^{3,6}$	$C_2'^{(1,2)}$	$C_2'^{(2,3)}$	$C_2'^{(3,4)}$
pπ_1	1	2 6	3 5	4	−1	−3	−5	−2	−4	−6
A_2	+1	+1	+1	+1		−1				−1

$p\pi_1 \times A_2 = 1(1) + 1(2) + 1(6) + 1(3) + 1(5) + 1(4) + 1(1) + 1(3) +$
$\qquad\qquad\quad 1(5) + 1(2) + 1(4) + 1(6)$
$\qquad\quad = 2(1) + 2(2) + 2(3) + 2(4) + 2(5) + 2(6)$

From this result we have, after normalization:

$$(a_2)\psi_1 = \frac{1}{\sqrt{6}}(\phi_1 + \phi_2 + \phi_3 + \phi_4 + \phi_5 + \phi_6)$$

To solve for the energy of this orbital we evaluate the integral

$E = \int \psi_1 H \psi_1 \, d\tau$
$\quad = \frac{1}{6}\int [(\phi_1 + \phi_2 + \phi_3 + \phi_4 + \phi_5 + \phi_6)H(\phi_1 + \phi_2 + \phi_3 + \phi_4 +$
$\phi_5 + \phi_6) \, d\tau]$
$\quad = \frac{1}{6}(\alpha + \alpha + \alpha + \alpha + \alpha + \alpha + \beta + \beta + \beta + \beta + \beta + \beta + \beta + \beta +$
$\beta + \beta + \beta + \beta)$
$\quad = \alpha + 2\beta = E$

For the orbital of species b_2, we multiply the orbitals to which $p\pi_1$ transforms under all the symmetry operations by the character of species B_2:

$p\pi_1 \times B_2 = 1(1) - 1(2) - 1(6) + 1(3) + 1(5) - 1(4) + 1(1) + 1(3) +$
$\qquad\qquad\quad 1(5) - 1(2) - 1(4) - 1(6)$
$\qquad\quad = 2(1) - 2(2) + 2(3) - 2(4) + 2(5) - 2(6)$

$(b_2)\psi_6 \quad = \dfrac{1}{\sqrt{6}}(\phi_1 - \phi_2 + \phi_3 - \phi_4 + \phi_5 - \phi_6)$

Solving for the energy as above, we obtain

$$E = \alpha - 2\beta$$

Now we turn to species E_1 and obtain the first trial function by multiplying the transformed $p\pi_1$ by the characters of E_1. Since we will need an additional trial wave function, we go through the same procedure for $p\pi_2$:

	I	$C_6 \, C_6'$	$C_3 \, C_3'$	C_2^z	$C_2^{1,4}$	$C_2^{2,5}$	$C_2^{3,6}$	$C_2'^{(1,2)}$	$C_2'^{(2,3)}$	$C_2'^{(3,4)}$
$p\pi_1$	1	2 6	3 5	4	-1	-3	-5	-2	-4	-6
$p\pi_2$	2	3 1	4 6	5	-6	-2	-4	-1	-3	-5
E_1	$+2$	$+1$	-1	-2		0			0	

$p\pi_1 \times E_1 = 2(1) + 1(2) + 1(6) - 1(3) - 1(5) - 2(4)$

$$\psi_3' = \frac{1}{\sqrt{12}}(2\phi_1 + \phi_2 + \phi_6 - \phi_3 - 2\phi_4 - \phi_5)$$

$p\pi_2 \times E_1 = 2(2) + 1(3) + 1(1) - 1(4) - 1(6) - 2(5)$

$$\psi_2' = \frac{1}{\sqrt{12}}(\phi_1 + 2\phi_2 + \phi_3 - \phi_4 - 2\phi_5 - \phi_6)$$

Since these two trial functions are not orthogonal, we use the Gram-Schmidt method to orthogonalize ψ_2' and ψ_3' by taking the linear combination, $\psi_3'' = a\psi_2' + b\psi_3'$, such that $\int \psi_2'\psi_3'' \, d\tau = 0$, and solve as outlined in Chapter 6.

$$(e_1)\psi_3'' = \tfrac{1}{2}(\phi_2 + \phi_3 - \phi_5 - \phi_6)$$

To solve for the energies of this degenerate pair we evaluate either of the integrals $E = \int \psi_3'' H\psi_3'' \, d\tau = \int \psi_2' H\psi_2' \, d\tau$; since ψ_3'' and ψ_2' are degenerate and orthogonal, the two integrals must be equal.

$$E = \int \psi_3 H\psi_3 \, d\tau = \tfrac{1}{12}[4\alpha + \alpha + \alpha + \alpha + 4\alpha + \alpha + 2\beta + 2\beta + 2\beta -$$
$$\beta + 2\beta - \beta + 2\beta + 2\beta + 2\beta - \beta + 2\beta - \beta) = \alpha + \beta$$

Another procedure for generating the two orthogonal wave functions belonging to species e_1 is to take linear combinations (addition and subtraction) of ψ_2 and ψ_3 whereby two new orthonormal functions, ψ_2 and ψ_3 are formed. These are actually the functions listed below; their choice rather than ψ_2 and ψ_3 is simply a matter of convention.

If we repeat the procedure for the calculation of the energy of the e_2 species, we obtain an energy of $\alpha - \beta$. Accordingly, the six MO's of benzene and their energies and symmetry species in D_{6h} are:

$$\psi_1(a_{2u}) = \frac{1}{\sqrt{6}}(\phi_1 + \phi_2 + \phi_3 + \phi_4 + \phi_5 + \phi_6) \qquad \alpha + 2\beta$$

$$\psi_2(e_{1g}) = \tfrac{1}{2}(\phi_1 + \phi_2 - \phi_4 - \phi_5) \qquad \alpha + \beta$$

$$\psi_3(e_{1g}) = \frac{1}{\sqrt{12}}(\phi_1 - \phi_2 - 2\phi_3 - \phi_4 + \phi_5 + 2\phi_6) \quad \alpha + \beta$$

$$\psi_4(e_{2u}) = \frac{1}{\sqrt{12}}(\phi_1 + \phi_2 - 2\phi_3 + \phi_4 + \phi_5 - 2\phi_6) \quad \alpha - \beta$$

$$\psi_5(e_{2u}) = \tfrac{1}{2}(\phi_1 - \phi_2 + \phi_4 - \phi_5) \qquad \alpha - \beta$$

$$\psi_6(b_{2g}) = \frac{1}{\sqrt{6}}(\phi_1 - \phi_2 + \phi_3 - \phi_4 + \phi_5 - \phi_6) \qquad \alpha - 2\beta$$

It is interesting to note that the benzene case just treated was particularly simple in that each MO (or pair of degenerate MO's) belongs to a different symmetry species from all others. Consequently, the secular determinants we might have set up are 1×1 determinants, and we can evaluate the energy and the c's directly.

10.11   The Nonbonding Molecular Orbital in Odd Alternant Hydrocarbons

An *alternant hydrocarbon* (AH) was defined in Section 10.7 as a hydrocarbon in which the π electron system consists only of straight or branched chains and rings with an even number of atoms. This definition results from a more basic one which requires that all atoms in the molecule can be divided into two classes, one starred and one unstarred, such that no two atoms of the same set are bonded together. A little exercise with pencil and paper will readily convince the reader that this division cannot be performed on any ring with an odd number of atoms, but can always be done if no such ring is present. By convention, whenever the two sets contain different numbers of atoms, we choose as the starred set the one which has the larger number of atoms. Whenever the numbers are different, the AH has (at least) as many nonbonding molecular orbitals (NBMO's) as the difference between the number of atoms in the two sets. A particularly important case is the one in which this difference is 1. Such molecules are called *odd alternant hydrocarbons* (OAH). The simplest OAH is the allyl radical, $C_1^*\!-\!C_2\!-\!C_3^*$.

There is a very simple method, developed by Longuet-Higgins, for determining the coefficients of the atomic orbitals in the NBMO without solving the secular determinant. This method depends on the fact that in the NBMO the coefficients of each unstarred atom must be zero (there is a node through these atoms). Such a node requires that the coefficients of the starred atoms surrounding each unstarred atom sum to zero. Thus, in the allyl system, if the coefficient of C_1 is arbitrarily assigned the value of a, in order for the sum of the coefficients of the atoms surrounding C_2 to be zero, the coefficient of the AO on C_3 must be $-a$. According to the normalization condition, the sum of the squares of the coefficients c_{jr} of the AO's making up the ψ_j molecular orbitals must be unity, that is,

$$\sum_r c_{jr}^2 = 1.$$

In the allyl system NBMO this means that

$$(a^2) + 0^2 + (-a^2) = 1 \quad \text{and} \quad a = \frac{1}{\sqrt{2}}$$

whence

$$\psi_{\text{NBMO}} = \frac{1}{\sqrt{2}}\phi_1 - \frac{1}{\sqrt{2}}\phi_3$$

The same procedure may be applied to the benzyl system (see Fig. 10.7a). If the coefficient on C_5 is called $+a$, the coefficients C_3 and C_7 are $-a$ because the sum of the coefficients on C_3 and C_5 and on C_5 and C_7 must vanish. The coefficient at C_1 must be $2a$ since then the sum of the starred coefficients around the unstarred atom C_2 is $2a - a - a = 0$. Using the normalization

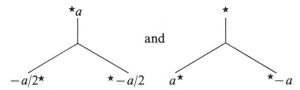

Fig. 10.7 The benzyl system (a) and the nonbonding molecular orbital (b).

condition, we obtain the coefficients:

$$(2a)^2 + 2(-a)^2 + (a)^2 = 1$$
$$7a^2 = 1$$
$$a = \frac{1}{\sqrt{7}}$$

and thus the complete wave function is:

$$\psi_{nb} = \frac{1}{\sqrt{7}}(2\phi_1 - \phi_3 + \phi_5 - \phi_7)$$

which can be drawn as in Fig. 10.7b. Benzyl belongs to point group C_{2v}, and application of the character table to the NBMO pictured shows it to transform as symmetry species b_1.

If the difference between the number of starred and unstarred atoms is greater than 1, more nonbonding MO's occur. Thus, in trimethylenemethane (cf. Section 10.9), $C(CH_2)_3$, we have seen that there exist two orbitals with $x = 0$, $E = \alpha$, that is, NBMO's. It is interesting to note that, for such molecules, one cannot write a normal resonance structure except as a diradical. Application of the Longuet-Higgins method provides two sets of answers for the coefficients:

these, upon normalization, give us exactly the MO's found above. If the coefficients so determined should give nonorthogonal orbitals, a Gram-Schmidt orthogonalization provides suitable answers.

10.12   The Hückel Molecular Orbital Method for Heteroatomic Compounds

The HMO method, which we have developed in this chapter strictly for conjugated systems containing only carbon atoms, has been extended almost from the beginning to heteroatomic systems. Many problems arise in principle, but in practice they all boil down to the task of finding proper numerical values for the various integrals involved. Particularly important is the choice of the Coulomb integrals, α, for the non-carbon atoms. In the first section of this chapter we outlined one way of choosing these integrals, usually called *parameters* because we never carry out an integration, but frequently adjust them empirically to fit one or many molecules. The systematic set of choices mentioned in Section 10.4 does not work at all well when the simple HMO method is applied to heteroatomic compounds, but we shall return to this approximation in the last chapter in connection with the extended Hückel method (EHMO).

Aside from the Coulomb integrals, we frequently must choose resonance integrals β for bonds other than C-C bonds. Again the method proposed in Section 10.1 is not very successful.

We can specify some qualitative expectation about proper choices for Coulomb integrals. Since they represent the energy with which an electron is bound to a π orbital of the atom (i.e., the negative of the ionization potential), we expect them to follow more or less the electronegativity of the atoms. In accord with this expectation we find that we must use $\alpha_C > \alpha_N > \alpha_O > \alpha_F$ (actually the way they are usually written means that α_N is a larger negative number than α_C). Furthermore, some heteroatoms, depending on the structure of the molecule, are capable of contributing one electron to the conjugated system in some types of compounds, and two electrons in others. Thus, in

O contributes   1   2   2   electrons

and in

N contributes   2   1   2   electrons

In these cases we need to choose values for α_O and α_N, actually different values $\alpha_O(1)$ and $\alpha_O(2)$, $\alpha_N(1)$ and $\alpha_N(2)$, and in addition β_{C-O} and β_{C-N}.

These values also should be allowed to vary from compound to compound, with bond length and many other variables.

This uncertainty in parameter values makes all work with heteroatomic systems tenuous. However, with some very simple choices many results can be obtained which serve valuable comparison purposes, even if absolute numbers obtained must be viewed with caution.

One of the early systems of expressing the α's and β's of heteroatoms and heterobonds is to use their values relative to α_C and β_{C-C}, by adding to α_C as an increment a certain energy, usually expressed as a multiple of β_{C-C}, since the latter serves as a unit of energy. Wheland and Pauling, among the earliest workers to be concerned with this problem, suggested:

$$\alpha_N(1) = \alpha_C + \tfrac{1}{2}\beta$$
$$\alpha_N(2) = \alpha_C + \tfrac{3}{2}\beta$$
$$\alpha_O(1) = \alpha_C + \tfrac{3}{2}\beta$$
$$\alpha_O(2) = \alpha_C + \tfrac{5}{2}\beta$$

Resonance integrals are most commonly expressed as multiples of β_{C-C}:

$$\beta_{C-x} = k\beta_{C-C}$$

where the constants k usually do not differ too much from unity (maybe 0.6 to 1.5 in the most common cases). Many modern books on MO theory list whole tables of α's and β's.

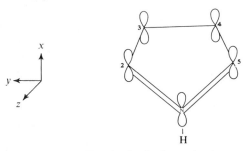

Fig. 10.8   Pyrrole, C_{2v} (in the yz-plane).

Let us examine, as an example, the treatment of pyrrole. We shall use $\alpha_N(2) = \alpha + \tfrac{3}{2}\beta$ and $\beta_{C-N} = \beta_{C-C} = \beta$ (Fig. 10.8). Using the plane of symmetry through the N atom and perpendicular to the molecule, σ_{xz}, we have in the species B_1 and A_2 of the point group C_{2v} of the molecule

B_1	A_2
$c_2 = c_5$	$c_2 = -c_5$
$c_3 = c_4$	$c_3 = -c_4$
c_1	$c_1 = 0$
3 constants, $\therefore$ 3ψ	2 constants, $\therefore$ 2ψ

The secular determinants are:

$$\begin{vmatrix} \alpha + \frac{3}{2}\beta - E & \beta & 0 \\ \beta & \alpha - E & \beta \\ 0 & \beta & \alpha - E - \beta \end{vmatrix} = 0 \qquad \begin{vmatrix} \alpha - E & 1 \\ 1 & \alpha - E - \beta \end{vmatrix} = 0$$

Dividing through by β and setting $x = (\alpha - E)/\beta$ gives:

$$\begin{vmatrix} x + 3/2 & 1 & 0 \\ 1 & x & 1 \\ 0 & 1 & x + 1 \end{vmatrix} = 0 \qquad \begin{vmatrix} x & 1 \\ 1 & x - 1 \end{vmatrix} = 0$$

leading to

$$(x + 1)\left[x\left(x + \frac{3}{2} \right) - 1 \right] - \left(x + \frac{3}{2} \right) = 0 \quad x(x - 1) - 1 = x^2 - x - 1 = 0$$

$$x^3 + \frac{5}{2}x^2 - \tfrac{1}{2}x - \frac{5}{2} = 0$$

The solutions for these equations are:

$x_1 = -2.55$ $\qquad\qquad\qquad$ $x_2 = -1.15$

$x_3 = -0.618$ $\qquad\qquad\quad$ $x_4 = 1.20$

$x_5 = 1.62$

Substituting back into the secular equations based on the symmetry orbitals

$$\chi_1 = \phi_1 \qquad\qquad\qquad \chi_4 = \frac{1}{\sqrt{2}}(\phi_2 - \phi_5)$$

$$\chi_2 = \frac{1}{\sqrt{2}}(\phi_2 + \phi_5) \qquad\qquad \chi_5 = \frac{1}{\sqrt{2}}(\phi_3 - \phi_4)$$

$$\chi_3 = \frac{1}{\sqrt{2}}(\phi_3 + \phi_4)$$

we obtain the MO's:

$$\psi_1 = 0.749\phi_1 + 0.393(\phi_2 + \phi_5) + 0.254(\phi_3 + \phi_4)$$
$$E = \alpha + 2.55\beta$$

$$\psi_2 = 0.503\phi_1 - 0.089(\phi_2 + \phi_5) - 0.605(\phi_3 + \phi_4)$$
$$E = \alpha + 1.15\beta$$

$$\psi_3 = 0.601(\phi_2 - \phi_5) + 0.372(\phi_3 - \phi_4)$$
$$E = \alpha + 0.618\beta$$

$$\psi_4 = 0.431\phi_1 - 0.581(\phi_2 + \phi_5) + 0.265(\phi_3 + \phi_4)$$
$$E = \alpha - 1.20\beta$$

$$\psi_5 = 0.372(\phi_2 - \phi_5) - 0.601(\phi_3 - \phi_4)$$
$$E = \alpha - 1.62\beta$$

Since the molecule has 6π electrons, the total π-electron energy is $6\alpha + (2 \times 2.55 + 2 \times 1.15 + 2 \times 0.62)\beta = 6\alpha + 8.64\beta$. Since the localized structure has two C-C double bonds and two N electrons, the localized energy is $2(2\alpha + 2\beta) + 2(\alpha + 3/2\beta) = 6\alpha + 7\beta$. Then the DE is 1.64β.

PROBLEMS

10.1   Derive the expression for the charge density, q_r, given that $\psi = \sum_s c_s\phi_s$, that is,
$$q_r = \sum_i n_i c_{ir}^2,$$
where n_i is the orbital occupation number, 0, 1, 2.

10.2   Using the HMO treatment, set up the secular determinant for the compounds shown below and then solve for the roots of the secular determinant. Calculate also the π delocalization energy, DE_π, for these compounds. (*Hint:* In handling the calculations for compound (b) the secular determinant leads to a 6th order polynomial in x, and substituting $y = x^2$ leads to a 3rd order polynomial that is readily solved graphically.)

(a) $\square$ (b) $\boxed{| \ \ | \ \ |}$

10.3   Repeat problem 10.2 for the heterocyclic compounds shown and also calculate the π delocalization energy, DE_π.

(a)

(b)

$\alpha_N = \alpha_C + h_N\beta_{C-C}$ $\alpha_O = \alpha_C + h_O\beta_{C-C}$

$\beta_{C-N} = k_{CN}\beta_{C-C}$ · $\beta_{C-O} = k_{CO}\beta_{C-C}$

where $h_N = 0.5$, $k_{CN} = 0.8$.   where $h_O = 2.0$, $k_{CO} = 1.0$.

10.4   Using the HMO treatment determine the orbital coefficients of the four π orbitals of trimethylenemethane.

10.5   Using the HMO treatment set up the appropriate secular determinant for 1,3,5-hexatriene and solve the sixth order equation graphically. Determine the coefficients for all MO's and sketch these as amplitudes along a box of the appropriate length, as in the FEM model.

10.6   Using the symmetry simplification procedures for the HMO method determine the roots and orbital coefficients for the cross conjugated compound:

10.7   Interestingly, the n → π^* transition for o-benzoquinone occurs at 610 nm while that for p-benzoquinone occurs at 450 nm. One possible explanation for this is that the lowest vacant π acceptor orbital of the p-isomer is higher in energy than that of the o-isomer. By means of the symmetry simplifications of the HMO method determine the energies and all the wave functions for p-benzoquinone (assume that $\alpha_O = \alpha_C + 2\beta$) and sketch the molecular orbitals. (See the discussion of dicarbonyl compounds in J. N. Murrell, *The Electronic Spectra of Organic Compounds*, John Wiley and Sons, New York, 1963, p. 168.)

10.8 Generate the six pπ orbitals of benzene by the symmetry simplication method, described in this chapter, but use the point group C_6 instead of D_6 for benzene.

10.9 Determine the relative energies for the ligand symmetry orbitals in O_h in Fig. 6.6. (*Hint :* Define the orbital interactions in terms of α and β.)

GENERAL REFERENCES

1. A Streitwieser, Jr., *Molecular Orbital Theory for Organic Chemists*, John Wiley and Sons, New York, 1961.
2. J. D. Roberts, *Notes on Molecular Orbital Calculations*, W. A. Benjamin, New York, 1962.
3. C. J. Ballhausen and H. B. Gray, *Molecular Orbital Theory*, W. A. Benjamin, New York, 1965.
4. C. Sandorfy, *Electronic Spectra and Quantum Chemistry*, Prentice-Hall, Englewood Cliffs, N.J., 1964.

11 Excited states, photochemistry, and conservation of orbital symmetry (Woodward-Hoffmann Rules)

11.1 The Fate of the Energy Absorbed by Excited Molecules; Jablonski Diagrams

In our discussions of electronic absorption spectra we have concentrated almost exclusively on the absorption process itself, consisting of the interaction of photons of various energies with the molecule in its ground state. We would now like to consider what happens to the excess energy in the excited molecule which has absorbed a quantum of light.

It is of interest to calculate the approximate time required for the absorption process. A photon travels at the velocity of light, 3×10^{10} cm (or 3×10^{18} A)/sec. If the diameter of the molecule with which it interacts is about 3 A, the photon will remain in the vicinity of the molecule for $3/3 \times 10^{18} = 10^{-18}$ sec. The absorption process accordingly has to occur in about 10^{-18} sec or less. In this time interval, atoms undergo virtually no motion, and this fact forms the basis of the Franck-Condon principle (cf. Section 8.6), according to which most molecules, upon electronic excitation, enter a vibrationally excited level of the excited state.

Virtually all organic molecules, except free radicals, have singlet ground states; these are usually denoted by photochemists as S_0, which is equivalent to the Mulliken notation N used in previous chapters to denote the ground state. Since photochemists are usually not interested in the electronic nature of excited states other than their multiplicity, they refer to excited singlets as S_1, S_2, etc., in the order of increasing energies, and similarly to triplets as T_1, T_2, etc. Excitation normally leads to singlet excited states rather than to triplet states because the $T_1 \leftarrow S_0$ transition involves a change in spin multiplicity and is highly forbidden (cf. Section 8.3).

The excitation process takes the ground-state molecule from the lowest vibrational level, in which it usually exists, to an excited singlet state. There are several possible excited singlet states, S_1, S_2, etc., and they, together with the singlet ground state, are called the *singlet manifold*. What happens to the

molecule in the excited singlet states is perhaps best illustrated by diagrams such as Fig. 11.1, which are known as *Jablonski diagrams*. In these
diagrams solid lines represent radiative processes, and wavy lines nonradiative processes. We will now discuss, individually, a number of the

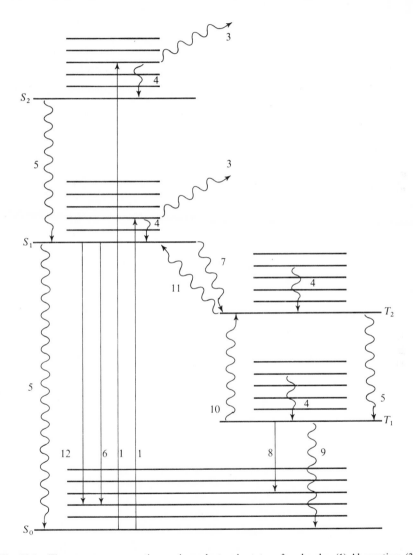

Fig. 11.1   The processes connecting various electronic states of molecules. (1) Absorption, (3)
dissociation and predissociation, (4) vibrational relaxation, (5) internal and external conversion,
(6) fluorescence, (7) intersystem crossing, (8) phosphorescence, (9) quenching, (10) vibrational
reactivation, (11) reverse intersystem crossing, and (12) delayed fluorescence, preceded by (11).

possible events which lead to deactivation or involve other means of dissipating the excess energy. The events or processes will be discussed in the order of the increasing time which they require, the first being the fastest.

11.1a   Vibrational Redistribution

After excitation (process 1 in Fig. 11.1), probably the most rapid process to occur in the now electronically and vibrationally excited molecule is a redistribution of the excess vibrational energy among the different vibrational modes (process 2, not shown in Fig. 11.1). Although nonlinear molecules have $3n - 6$ vibrational modes, the vibrational excitation generally involves only one or at most a few of these modes. These vibrational frequencies are the ones which, because of the Franck-Condon principle, characterize the differences between ground- and excited-state geometry (cf. Section 8.6). Since complex molecules have a large number of degrees of freedom, there is a high probability that the vibrational energy moves between modes toward an equilibrium distribution (which classically would be equipartition). Unless this redistribution occurs in a time shorter than one collision (i.e., $\sim 10^{-13}$ sec), it is unlikely to be of importance. Nevertheless, it is quite likely to be of major significance in the dissociation and predissociation of large molecules. It involves no loss of energy by the molecule. This process is not represented in Fig. 11.1 because the different vibrational modes are not distinguished in the figure.

11.1b   Dissociation

Dissociation (Fig. 11.1, process 3) is almost certain to occur, if at all, within the period of time of a very few vibrations. For a vibration of an energy between 300 and 3000 cm^{-1}, for example, the vibrational frequency is between about 10^{13} and 10^{14} sec^{-1}. This means that the time required for one vibration is, classically, about 10^{-14} to 10^{-13} sec. Given sufficient energy, a diatomic molecule, such as H_2, *must* dissociate in the time of a single vibration, barring the intervention of some other process (e.g., a collision).

The dissociation process may be better understood by reference to the Morse curves of Fig. 11.2. The total energy as a function of the distance between the two H atoms for the ground state molecule (S_0) is represented by curve A. Our earlier discussions (cf. Section 3.3) showed that the occupation of an antibonding orbital resulting from promotion of one electron from the bonding to the antibonding MO slightly more than cancels the bonding contribution of the electron remaining in the bonding MO. The result is shown in Fig. 11.2 in the potential energy curve, B, which is called a *repulsive curve*. What happens when we excite one electron from σ_g to σ_u^*? In the ground

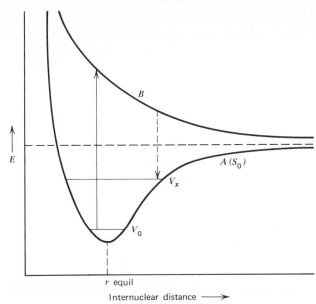

Fig. 11.2   The morse curve, showing photodissociation.

state the molecule is vibrating, that is, the two H atoms are moving periodic-
ally together and apart along the short horizontal line, V_0, in curve A; in
other words the bond distance is oscillating around its equilibrium value.
The excitation raises the molecule along the solid arrow to curve B, but the
relative motion of the atoms persists. As soon as the atoms move apart, there
is no potential to reverse the motion, as in the ground-state situation represented by curve A, and the atoms dissociate. This type of dissociation is called
photodissociation, and it is the simplest photochemical process. It occurs
in H_2 and may occur in any diatomic molecule which is held together by a
single bond, and in many polyatomic molecules as well.

There is, however, one restriction. If the excitation energy is lost by
dropping to the excited vibrational level V_x before the atoms are very far
apart, the molecule is "saved"; it fails to dissociate. The time scale here is
quite interesting. The time required for a molecule to execute a vibration
is of the order of 10^{-14} to 10^{-13} sec. Thus, the molecule is saved only if the
energy loss follows the excitation within an extremely short time period.
Although photodissociation is not uncommon in small molecules, it occurs
in polyatomic molecules only when the excess vibrational energy is concentrated in appropriate vibrational modes; and thus, even though the excited
molecule possesses sufficient energy to dissociate, it does not necessarily do so.

11.1c   Predissociation

In polyatomic molecules, there are usually additional states between the attractive and repulsive states shown in Fig. 11.2 because many electrons can be promoted out of any one of several high-lying occupied orbitals into any one of many low-lying *virtual* (i.e., unoccupied) *orbitals.* An excited state which is not a repulsive state is shown as curve *C* in Fig. 11.3. Excitation to a low vibrational level of curve *C* (arrow *a*) behaves like a normal electronic transition. But excitation to a high vibrational level of curve *C* (arrow *b*) places the molecule energetically above the intersection of curves *B* and *C*. Its vibrational motion on curve *C*, from points *x* to *y* and back, passes this intersection point, and there is a fair probability that the molecule will "lose its way" and continue on curve *B*, which again results in the molecule dissociating. This process, called *predissociation,* is very common in heavy diatomic and polyatomic molecules.

When the molecules are in the higher vibrational excited state, it is also possible that they can undergo rearrangements involving bond formation,

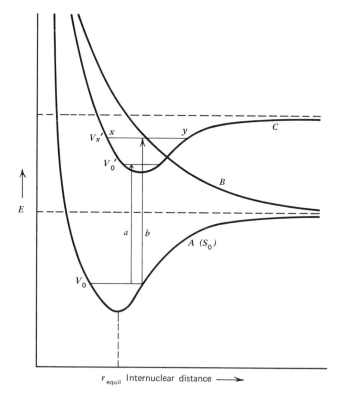

Fig. 11.3   The Morse curve, showing predissociation.

bond breaking, or both. However, it must be remembered that the lifetime in excited singlet states higher than S_1 is very, very short, and even in the lowest vibrational level of S_1 the lifetime, although somewhat longer, is still relatively short. Although photochemical reactions frequently occur in S_1, most organic photochemistry in solution involves molecules in the T_1 state, where, as we shall see, they have much greater lifetimes.

11.1d   Vibrational Relaxation

In terms of a time scale, the next process (process 4, Fig. 11.1) is vibrational relaxation or the dropping to the lowest vibrational level that does not involve a change in *electronic* energy. The mechanism of this process undoubtedly requires either collision (in the gas phase) or a collision-like interaction with the environment (in the liquid phase). Gas-phase collisions require, depending on pressure, about 10^{-13} to 10^{-12} sec or longer, while the liquid-phase collision-like process occurs in about the same time as a vibration (10^{-14} to 10^{-13} sec).

11.1e   Internal and External Conversion

These processes (shown as numbers 5 in Fig. 11.1) are slower than those previously discussed and result in the relaxation of the molecule into a state of lower energy in a given multiplicity manifold.

As originally defined on the basis of observations of the luminescence of gases, *internal conversion* referred to the deactivation (quenching) of excited molecules by collision with other molecules of the same species, while *external conversion* referred to deactivation by collision with dissimilar molecules. In solution this distinction becomes meaningless, and we speak only of internal conversion, referring thereby to a deactivation process in which the electronic excitation energy is dissipated as thermal energy by collision with any other (usually solvent) molecule. It will be noted from Fig. 11.1 that this nonradiative conversion occurs between different electronic states. These processes are shown more clearly in Fig. 11.4, where the "jump" between states is from the lowest vibrational level of S_1 to a high vibrational level of S_0 and is then followed again by vibrational relaxation to the lowest vibrational level of S_0.

The condition for the conservation of momentum makes the jump from the lowest level of S_1 to one of the higher levels of S_0 the more probable, the lower is the quantum number, v, of this vibrational state of S_0. This fact has a direct application to the deuterium isotope effect on the probability of internal conversion and therefore on the lifetime of excited singlet states. Since the C—D stretching frequency is significantly smaller than the C—H stretching frequency (cf. Section 9.3), the spacing of vibrational levels is closer in C—D; cf. Fig. 11.4*b*. As a consequence the lifetime of excited singlets of deuterio aromatic hydrocarbons is longer than that of the corresponding protio compounds.

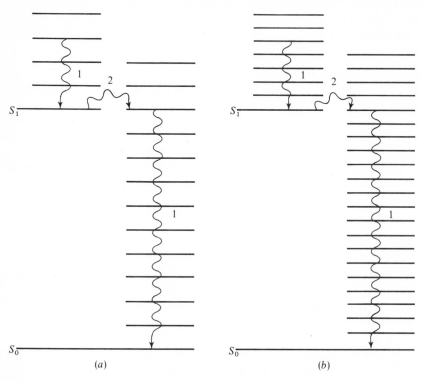

Fig. 11.4   Mechanism of conversion to ground state. (1) Vibrational relaxation, (2) transition from lowest vibrational level of excited state to a high vibrational level of ground state. If (*a*) represents a C—H vibrational spacing, the corresponding C—D is as in (*b*).

11.1*f*   Fluorescence

All the processes discussed to this point are relatively very fast, and only in rare cases are they separately observable. As a result, the detailed knowledge we have about these processes is sketchy and largely inferential.

The emission of radiation (process 6, Fig. 11.1), which results in a transition of the molecule from an excited state to the ground state, without a change in multiplicity, is called *fluorescence* and occurs typically with a half-life of about 10^{-9} to 10^{-8} sec. Therefore, fluorescence practically always occurs from the lowest excited state of the singlet manifold, since this is the only state in the manifold with a half-life longer than the time required for the various collision-dependent relaxation and conversion processes.

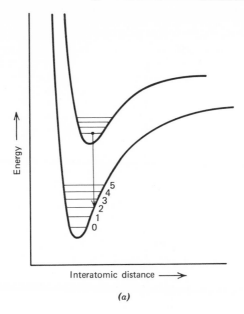

(a)

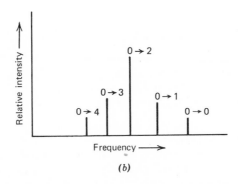

(b)

Fig. 11.5 (*a*) Typical Franck-Condon emission from an excited state with an equilibrium separation slightly greater in the excited state than in the ground state. (*b*) Typical emission spectrum for a diatomic molecule with Franck-Condon curves disposed as shown in Fig. 11.5(*a*).

Since the fluorescence usually occurs from the lowest vibrational state of S_1, and emission, like absorption, is always *vertical*, the molecule descends to an excited vibrational level of the ground state, Fig. 11.5. This is just the reverse of the usual case in absorption, in which promotion occurs from the $v = 0$ level in S_0 and the molecule ends up in higher vibrational levels,

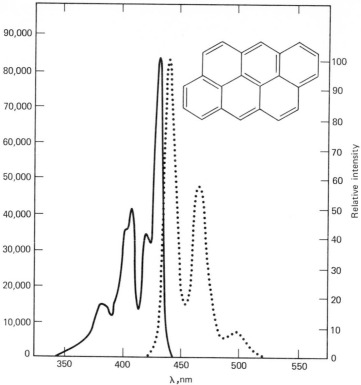

Fig. 11.6 Emission (right) and absorption (left) spectra of anthanthrene, 5×10^{-5} M, in benzene at 298°K. The intensity of the fluorescence emission spectrum is in relative units. Note the approximate "mirror-image" relationship between the emission and absorption spectra. [Courtesy of Dr. David Morgan.]

$v > 0$, in S_1. Accordingly, we expect[a] a "mirror-image" relationship between absorption and fluorescence spectra, with the fluorescence spectrum appearing at longer wavelength (or lower frequency) than the absorption spectrum.[b] A typical example is shown in Fig. 11.6. At sufficiently low temperatures, where virtually all molecules exist in the lowest vibrational state, the $0 \to 0$ band should always be the longest-wavelength band in the absorption

---

[a] This mirror-image relation can be only approximate, since the shape of the absorption band is determined by the vibrational structure of the excited state, and the shape of the emission band by the vibrational structure of the ground state (cf. Section 8.6). Consequently, the mirror-image relations observed are usually only approximate.

[b] The emission spectrum of a compound always appears at longer wavelength than the absorption spectrum; the red shift is called the *Stokes shift*. In unusual cases absorption may occur from a vibrational state in S_0 higher than $v = 0$, whence the $0 \to 0$ band in the fluorescence spectrum will appear to be at shorter wavelength than the exciting radiation, giving rise to an *anti-Stokes shift*.

spectrum and the $0 \to 0$ band should be at the shortest wavelength in the fluorescence spectrum; also, these transitions should be of equal energy. Theoretically, then, the energy of the $0 \to 0$ band should correspond to the frequency at the intersection of the excitation and fluoresecnce spectra. This usually occurs in vapor spectra but is much less common in solution spectra.

The *fluorescence quantum yield*, Φ_f, is a measure of the efficiency of the fluorescence process and is equal to the ratio of the number of photons emitted to the number absorbed. Φ_f can have values between 0 and 1, and in some cases, for instance, 9,10-diphenylanthracene, Φ_f is close to the upper limit. Of course in all non-fluorescing compounds, $\Phi_f = 0$.

The fluorescence spectrum of a compound is determined with an instrument called a *spectrophotofluorimeter*. This instrument is also used to obtain an *excitation spectrum*. For determining the emission spectrum, the exciting light is monochromatic and the emission spectrum is scanned. When the excitation spectrum is desired, the emission monochromator is set at a fixed, convenient wavelength where fluorescence is intense, and the excitation spectrum is scanned. The intensity of the fluorescence, I_f, at this set wavelength as a function of the variable exciting wavelengths, the excitation spectrum, is then determined. This intensity depends on the absorbance according to the equation:

$$I_f = I_0(1 - 10^{-A})\Phi_f$$

where I_0 is the fixed intensity of the monochromatic light, and A is the absorbance, equal as usual to $\epsilon l c$, where ϵ is the extinction coefficient, l the length of the cell, and c the concentration of the absorbing species in moles per liter. The value of I_f is thus low where A is low, and the resulting excitation spectrum is essentially identical to the absorption spectrum obtained with the conventional ultraviolet spectrophotometer. One advantage of the excitation spectrum is that it may be obtained with very dilute solutions, for example, $10^{-7}M$, by using light from a high-intensity source passing through a monochromator; frequently the absorption spectrum cannot be obtained in solutions more dilute than $10^{-4}M$. Usually, however, the absorption spectrum is obtained on other instruments, whose features are designed to give optimum results.

Next, let us focus on the lifetime of the excited state. The time that the excited state takes before it emits light as fluorescence is measured by the natural or inherent radiative lifetime, τ^0, the time required for the concentration of excited molecules to decrease to a value of $1/e$ of its original value if fluorescence were the only means of decay. To consider this problem we return to the Einstein coefficients, A_{if} and B_{if}, discussed in Section 8.2. The radiative lifetime is given by

$$\tau^0 = \frac{1}{A_{if}}$$

But, using the relation between A_{if} and B_{if},

$$\tau^0 = \frac{c^3}{8\pi h v^3 B_{if}}$$

Again using relations given in Section 8.2, we can express B_{if} in terms of the measurable integrated absorption intensity for the same band, $\int \epsilon \, dv$, and substitution gives

$$\tau^0 = \frac{Nc^3}{0.102 \times 8\pi v^2 \int \epsilon \, dv}$$

$$= \frac{3.5 \times 10^8}{\tilde{v}^2 \int \epsilon \, d\tilde{v}} \tag{11.1}$$

Here $\tilde{v}$ is the mean frequency (in cm^{-1}) of the $0 \to 0$ absorption band, and $\int \epsilon \, d\tilde{v}$ may be approximated by $\epsilon \, \Delta\tilde{v}_{1/2}$, where $\Delta\tilde{v}_{1/2}$ is the width of the band at half its maximum intensity, the *half-width*, if the band is fairly symmetrical. Thus equation 11.1 becomes:

$$\tau^0 = \frac{3.5 \times 10^8}{\tilde{v}^2 \epsilon \, \Delta\tilde{v}_{1/2}} \tag{11.2}$$

where now $\tilde{v}$ is the frequency at the maximum intensity of the $0 \to 0$ band. For the low-intensity $n \to \pi^*$ state, say of benzophenone, the radiative lifetime is

$$\tau^0 \sim \frac{3.5 \times 10^8}{(27,000)^2 \, (65) \, (1000)} = 7 \times 10^{-6} \, \text{sec}$$

Actually, instead of considering only the $0 \to 0$ transition, a more accurate estimate is made by integrating over the entire absorption band, including all vibrational transitions. Equations 11.1 and 11.2 show that the emission lifetime of the S_1 state is relatively short when absorption to that state is intense; correspondingly, the less intense the absorption band, the longer is the emission lifetime.

Although we will consider several photochemical reactions in a later section, it is appropriate to point out here that such reactions compete with fluorescence.

11.1g   Intersystem Crossing and Phosphorescence

Another competitive process determining the fate of the excited molecule is the *forbidden* one called *intersystem crossing*, in which there is a change of of spin (Fig. 11.1, process 7). This occurs through spin-orbit coupling, in which states with different spin angular momenta and orbital angular momenta mix slightly because they have the *same* total angular momentum.

Once a different multiplicity manifold is attained by intersystem crossing (e.g., singlet → triplet), the same relaxation processes are important as were relevant in the original manifold. Following the same reasoning, the molecule will relax to the lowest vibrational level of the lowest electronic state of the new manifold. Once this point is reached, however, the final decay to the ground state is much slower and gives rise to a great variety of phenomena.

Intersystem crossing from the lowest excited singlet to the lowest triplet is one of the most important photochemical processes because of the long lifetime of the lowest triplet. The loss of energy from the lowest triplet to the ground state may occur by a radiative process called *phosphorescence* (process 8, Fig. 11.1) or by a relaxation process consisting of a reverse intersystem crossing involving internal or external conversion. The lowest triplet is difficult to populate by direct absorption (singlet → triplet absorption) and equally difficult to depopulate by emission of radiation. If the nonradiative paths for deactivation are eliminated (very low temperature), the natural radiative lifetime of the triplet can be calculated from equations

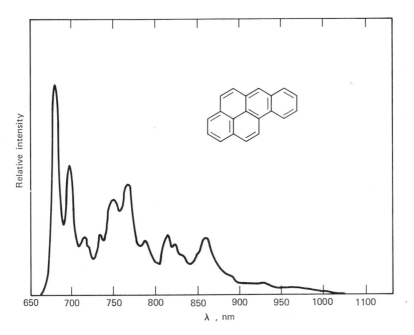

Fig. 11.7   Phosphorescence spectrum of benzo[a]pyrene, 1×10^{-3} M, in ether-ethanol, 2:1 by volume, at 77°K. [Courtesy of Dr. David Morgan.]

11.1 and 11.2. The intensity which appears in the denominator of equation 11.2 is very difficult to measure because it is so low. In benzophenone, ϵ_{max} for the singlet $\rightarrow$ triplet absorption is estimated to be about 10^{-3}; and, assuming $\tilde{v}$ to be the same as for the singlet, we calculate

$$\tau^0 = \frac{(3)(3.5 \times 10^8)}{(27,000)^2 (10^{-3})(1000)} = 1.4 \text{ sec}$$

The factor of 3 in the numerator (i.e., the factor G_f of the Einstein coefficient; cf. Section 8.2) is required in the calculation because of the three-fold degeneracy of the triplet. As can be seen from the above examples with benzophenone, the radiative lifetime of the T_1 state is about 200,000 times that of the S_1 state.

Phosphorescence spectra are usually determined in solutions that are frozen to glasses at about 70–80°K. The phosphorescence spectra occur at longer wavelength (shorter frequency) than the fluorescence spectra. A typical phosphorescence spectrum is shown in Fig. 11.7.

11.1*h*   Delayed Fluorescence and Excimer Fluorescence

When the lowest triplet, T_1, lies close to the lowest singlet, it is sometimes possible to observe *delayed fluorescence*. This fluorescence (*E-type*) is an emission from S_1, but in this case S_1 is populated by a vibrational reactivation of T_1 (usually thermally), followed by a reverse intersystem crossing. The resulting emission spectrum is identical to the fluorescence spectrum. However, this type of delayed fluorescence is recognized by the lifetime typical of phosphorescence and, since thermal activation is involved, by an increase of intensity with temperature.

Another type of delayed fluorescence (*P-type*) occurs when S_1 is formed from T_1 upon the collision of two T_1 by a process called *triplet-triplet annihilation*:

$$T_1 + T_1 \rightarrow S_1 + S_0$$

Here, again, a typical fluorescence spectrum is observed with a lifetime characteristic of phosphorescence. However, since the concentration of S_1 depends on a bimolecular process in T_1 and hence on the square of the concentration of T_1, the intensity is proportional to the square of this concentration and hence to the square of the intensity of the exciting light. This behavior contrasts with most other luminescence intensities, which are proportional to the first power of the intensity of exciting light.

In some cases, triplet-triplet annihilation leads to a complex made up of one molecule in S_1 and one in S_0, called an *excimer* (*excit*ed di*mer*). Generally, excimers have two states, one just above, and the other just below,

S_1. It is the latter one that gives rise to the excimer fluorescence, which consequently is at slightly longer wavelength (slightly lower energy) than normal fluorescence.

11.1*i*   Fluorescence and Phosphorescence Quenching

The most readily observed and most studied of the processes described are the radiative ones, fluorescence and phosphorescence. When they are not observed or, particularly, when they disappear under certain circumstances, they are said to be *quenched*. What this means is that, in the competition of the various processes, the radiative one has lost out because some alternative process was more rapid. Thus, fluorescence competes with internal and external conversion, with intersystem crossing, and possibly with some photochemical processes. Only when the fluorescence is more rapid than, or at least as fast as any of the others, can it be observed. Similarly, phosphorescence competes with the following: relaxation (by reverse intersystem crossing) to the ground state, delayed fluorescence, photochemical reactions, and other processes such as triplet-triplet annihilation and excimer fluorescence.

We have discussed essentially all the processes that an excited molecule can undergo, from those occurring in very, very short times to those which take place over rather long time intervals. We will now consider other aspects of interest in understanding excitation spectra and photochemistry.

11.2   Molecular Orbital Energy Diagrams; Singlet and Triplet States

In Chapter 8 we discussed the relationship between one-electron molecular orbital energy diagrams (MOED's) and state (or term level) diagrams, particularly with respect to the formaldehyde molecule. In Chapter 12 we will deal with the difference between singlet and triplet states. This difference becomes important in excited-state chemistry because of the difference in lifetimes of the two types of states, and also because of the difference in energy between them. The existence of the two types of states arises, of course, out of the fact that, in the ground state, S_0, all electrons are paired, two in each orbital, whereas, upon excitation, one electron is promoted to an orbital of higher energy. Either the spins of the two electrons in the excited state can be the same (i.e., both $\frac{1}{2}$ or both $-\frac{1}{2}$), or the two electrons can have opposed spin (i.e., one $+\frac{1}{2}$ and the other $-\frac{1}{2}$). The multiplicity of a state is equal to $2|S| + 1$, where S is the sum of the spin numbers of either $\pm\frac{1}{2}$. When both electrons have the same spin, $|S| = 1$, and $2|S| + 1 = 3$, we have a triplet state; when the electrons have opposed spin, $S = 0$, and $2|S| + 1 = 1$, a singlet state results.

One of the important limitations of using MOED's resides in the complete neglect of the difference between singlet and triplet states. The equality of the energy of singlets and triplets arises out of the implied assumption that a definite energy is associated with each MO, independently of electron occupation and neglecting all electron repulsions.

The neglect of electron-electron repulsions leads not only to the incorrect conclusion that states of different multiplicity are energetically indistinguishable, but also to predictions about spectra which are contrary to facts. We would like to discuss some systems in which such contradictions arise and to see how they are resolved.

A very simple example is pyridine. The molecule has two ionization potentials, at 9.26 and 10.53 eV. A careful study of the vacuum ultraviolet spectrum has been used to determine that the lowest ionization potential corresponds to the removal of a π electron. Since, according to *Koopman's theorem*, the ionization potential for an electron in an orbital μ is equal to the negative of the orbital energy, ϵ_{μ}, the ionization potential indicates that the highest occupied orbital in pyridine must be a π orbital, and that the n orbital lies slightly below the π. This information permits us to construct the MOED shown in Fig. 11.8, from which we would then predict that the longest-wavelength absorption is a $\pi \rightarrow \pi^*$ transition, with the n $\rightarrow \pi^*$ transition at slightly shorter wavelength. However, these two transitions actually occur at 252 and ~ 287 nm, respectively, in contradiction to the expectation. Why?

To answer this question, we must first examine the excited states, both singlets and triplets, more closely. The excited states that interest us are those in which one electron has been promoted from orbital ψ_{μ} to ψ_{ν}. The total wave function of these states consists of a space and a spin part. At the

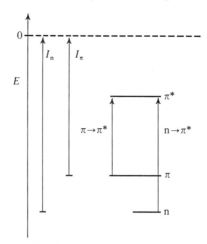

Fig. 11.8 The molecular orbital energy diagram of pyridine; I_{π} and I_n are the ionization potentials of a π and an n electron, respectively.

beginning of this section we described the triplet spin qualitatively; we now proceed to develop the spin wave functions in detail.

In the common convention, the two possible spins of an electron, $+\frac{1}{2}$ and $-\frac{1}{2}$, are denoted as α and β, respectively, and the two electrons are designated as electrons 1 and 2, although of course they cannot be distinguished. The possible spin combinations of the two electrons in different orbitals are:

$$
\begin{array}{ll}
(a) & \alpha(1)\alpha(2) \\
(b) & \beta(1)\beta(2)
\end{array} \left.\right\} \quad \text{components of triplet}
$$

$$
\begin{array}{ll}
(c) & \alpha(1)\beta(2) \\
(d) & \beta(1)\alpha(2)
\end{array}
$$

Thus, in (a) both electrons have a spin of $+\frac{1}{2}$; in (b), both $-\frac{1}{2}$. In either (a) or (b) it makes no difference which of the two electrons is 1 and which is 2, since, in either case, both electrons have the same spin. Thus, in (a), $\alpha(2)\alpha(1)$ is identical with $\alpha(1)\alpha(2)$. Both (a) and (b) are therefore symmetric with respect to exchange of electrons. However, in (c) and (d) we have a different situation. If we exchanged electrons in (c), we would get $\alpha(2)\beta(1)$, which is neither the same as, nor minus the same as (i.e., neither symmetric nor antisymmetric with respect to), the original (c), but is, in fact, something different, namely, (d). The same is true of (d); it is neither symmetric nor antisymmetric with respect to exchange of electrons. As usual, in such cases we take linear combinations of the two functions:

$$(c + d): \quad \frac{1}{\sqrt{2}}\left[\alpha(1)\beta(2) + \alpha(2)\beta(1)\right] \quad \text{(symmetric, triplet)}$$

$$(c - d): \quad \frac{1}{\sqrt{2}}\left[\alpha(1)\beta(2) - \alpha(2)\beta(1)\right] \quad \text{(antisymmetric, singlet)}$$

If we exchange electrons 1 and 2 in the first combination $(c + d)$, we get no change in sign and hence this function is symmetric. But exchanging electrons 1 and 2 in $(c - d)$ gives $\alpha(2)\beta(1) - \alpha(1)\beta(2)$, which is precisely the result obtained by multiplying $(c - d)$ by -1; hence, $(c - d)$ is antisymmetric with respect to exchange of electrons. The antisymmetric spin function characterizes the singlet state (the total wave function must be antisymmetric, and in the singlet state the orbital part of the wave function is symmetric and the spin part of the wave function is antisymmetric), and the symmetric spin function characterizes the triplet state. As a point of fact, the energies of the total wave function, of which the spin functions (a), (b), and $(c + d)$ are a part, are all equal (degenerate) (in the absence of a magnetic field) and together constitute the triplet state. Accordingly, the triplet state is defined as the state which has a spin function symmetric with respect to exchange of electrons.

We must now examine further the orbital or space part of the singlet and triplet wave functions. We are considering the excitation of an electron from the configuration $\psi_\mu^2 \rightarrow \psi_\mu\psi_\nu$. The ground configuration, $\psi_\mu^2 = \psi_\mu(1)\psi_\mu(2)$, is symmetric with respect to exchange of electrons 1 and 2, and should be multiplied by the antisymmetric spin function obtained in the preceding paragraph in order that the total wave function be antisymmetric, as required for all total wave functions; consequently the ground state is a singlet. But the excited configuration corresponds to two functions, $\psi_\mu(1)\psi_\nu(2)$ and $\psi_\mu(2)\psi_\nu(1)$, neither of which is symmetric or antisymmetric. Forming linear combinations of these, as usual, we obtain

$$\Psi_+ = \frac{1}{\sqrt{2}}\left[\psi_\mu(1)\psi_\nu(2) + \psi_\mu(2)\psi_\nu(1)\right] \quad \text{(symmetric, singlet)}$$

$$\Psi_- = \frac{1}{\sqrt{2}}\left[\psi_\mu(1)\psi_\nu(2) - \psi_\mu(2)\psi_\nu(1)\right] \quad \text{(antisymmetric, triplet)}$$

Ψ_+, the singlet, must then be multiplied by an antisymmetric spin function; Ψ_-, the triplet, by a symmetric one.

Thus, we obtain the total wave functions:

$$^1\Psi = \tfrac{1}{2}\left[\Psi_\mu(1)\psi_\nu(2) + \psi_\mu(2)\psi_\nu(1)\right]\left[\alpha(1)\beta(2) - \alpha(2)\beta(1)\right]$$

$$^3\Psi = \frac{1}{\sqrt{2}}\left[\psi_\mu(1)\psi_\nu(2) - \psi_\mu(2)\psi_\nu(1)\right]\begin{cases}\alpha(1)\alpha(2) \\ \dfrac{1}{\sqrt{2}}\left[\alpha(1)\beta(2) + \alpha(2)\beta(1)\right] \\ \beta(1)\beta(2)\end{cases}$$

Now let us consider the energies of the two states, or rather the energy difference between either of them and the ground state. In a strictly orbital sense, the orbital energies corresponding to ψ_μ and ψ_ν are ϵ_μ and ϵ_ν, respectively, and the excitation energy for the process $\psi_\mu \rightarrow \psi_\nu$ is $\epsilon_\nu - \epsilon_\mu$, independently of the behavior of the spins. This is the information contained in the MOED. These orbital energies contain an averaged allowance for electron repulsions, which is introduced through empirical parameters, as in the Hückel theory, or explicitly in more elaborate theories. However, these ϵ's are calculated for the molecule in the ground state; and when electron repulsions are treated explicitly, the center of gravity of the singlet and the triplet is lowered by an amount $J_{\mu\nu} - K_{\mu\nu}$. The terms $J_{\mu\nu}$ and $K_{\mu\nu}$ are the Coulomb and exchange integrals, defined, respectively, in equation 12.16 and 12.17, and represent, respectively, the repulsions of charge distributions ψ_μ^2 and ψ_ν^2 and of two charge distributions $\psi_\mu\psi_\nu$. This lowering of the center of gravity is shown schematically in Fig. 11.9b. However, the singlet and the triplet energies are found to differ by an amount $2K_{\mu\nu}$, the singlet lying $K_{\mu\nu}$

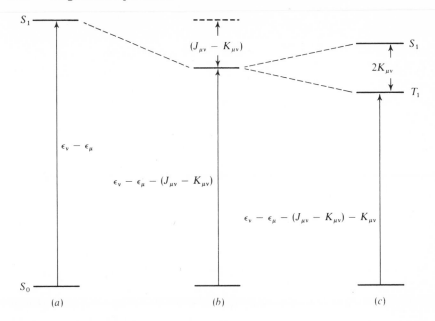

Fig. 11.9   The splitting of singlets and triplets. (*a*) Energy according to the molecular orbital energy diagram. (*b*) The center of gravity including electron repulsion. (*c*) The final singlet and triplet.

above the center of gravity and the triplet the same amount below it; cf. Fig. 11.9.

With this information about singlet and triplet energies, we may return to the pyridine case and the apparent contradiction that ionization potentials show the π orbital as the highest occupied orbital, whereas the absorption spectrum, interpreted in terms of the MOED, suggests that the n orbital is highest. But the MOED, Fig. 11.8, with the implied neglect of electron repulsions, provides inference only about the *center of gravity* or weighted average of the energies of the singlet and triplet states. To examine the singlets and triplets separately we must proceed to an examination of state (term level) diagrams. We should recall (cf. Chapter 8) that a MOED gives the "energy" of each separate orbital, while the term level diagram gives the total energy of the various states of the molecule. The term level diagram for pyridine is given in Fig. 11.10; the center of gravity of singlets and triplets has been added in the center of the Figure for reference. We have seen in Fig. 11.9 that the singlet state lies above the center of gravity by an amount equal to $K_{\mu\nu}$. But it can readily be shown that $K_{n\pi^*} \ll K_{\pi\pi^*}$. Since the center of gravity of the $\pi \to \pi^*$ state lies only slightly below that of the $n \to \pi^*$ state, the $K_{\mu\nu}$ pushes the $\pi \to \pi^*$ singlet well above the $n \to \pi^*$ singlet, and

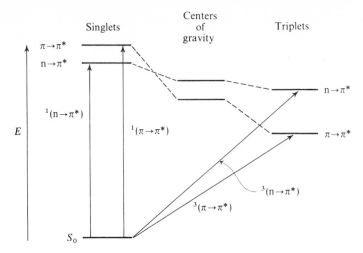

Fig. 11.10 The term level diagram of pyridine, including the centers of gravity of singlet and triplet $\pi \rightarrow \pi^*$ and $n \rightarrow \pi^*$ states.

consequently the $n \rightarrow \pi^*$ transition occurs at longest wavelength. Furthermore, as a consequence of this argument, we can predict that the lowest triplet in pyridine is a $\pi \rightarrow \pi^*$ state, and this is actually found to be so.

We have thus seen that, at least in some cases, the MOED cannot be used to predict the spectra correctly. This happens particularly when we have close-lying states with very different singlet-triplet separation ($2K_{\mu\nu}$). As a corollary of this conclusion, we cannot confidently construct a MOED from spectroscopic information in such cases, at least not without considering both singlet-singlet and singlet-triplet transitions. On the other hand, because of the wide applicability of Koopman's theorem, we can use MOED's with some confidence in the prediction and assignment of ionization potentials.

Another similar and important case of contradiction between information obtained by a MOED and that due to experiment occurs in carbonyl compounds. First let us consider the MOED of the simple carbonyl group shown in Fig. 11.11. From this diagram we predict correctly that there are two transitions and that the longer-wavelength one is the $n \rightarrow \pi^*$ transition. Actually this transition is the only one observed for simple carbonyl compounds; it usually occurs at about 280 nm, corresponding to a transition energy of roughly 100 kcal/mole. Although we know that the MOED of Fig. 11.11 gives the center of gravity of the singlet and triplet states, the observed spectrum and the MOED are not necessarily in conflict.

However, let us now consider the example of a carbonyl group attached to an atom possessing a lone pair of electrons. For such an example, we will

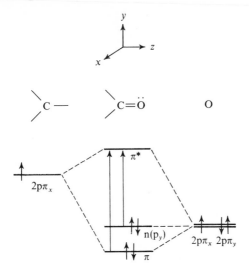

Fig. 11.11   Molecular orbital energy diagram for the $\diagdown$C=O system.

replace one of the R groups in the ketone RCOR by an hydroxyl group, so that we are examining the carboxylic acid, RCO_2H. The lone pair on the oxygen of the hydroxyl is in an orbital approximately degenerate with the 2p orbital of the carbonyl oxygen (the two lone pairs are exactly degenerate in the anion) and becomes part of the π system of the carbonyl group:

$$\begin{matrix} H-\ddot{O}: \\ \diagdown \\ C=O: \\ \diagup \\ R \end{matrix} \leftrightarrow \begin{matrix} H-\ddot{O}^+ \\ \diagdown\diagdown \\ C-O:^- \\ \diagup \\ R \end{matrix}$$

The interaction of the hydroxyl oxygen $p\pi$ orbital with the carbonyl group π orbital can be described by the MOED shown in Fig. 11.12. According to to this MOED, the n orbital in the resulting chromophore cannot be the highest orbital. However, the ultraviolet spectrum of the resulting compound shows that the $n \rightarrow \pi^*$, rather than the allowed $\pi' \rightarrow \pi^*$, is the transition of lowest energy. The longest-wavelength absorptions of a series of substituted carbonyl compounds are shown in Table 11.1. From the intensities and solvent shift behavior, it is known for certain that these absorption bands arise from $n \rightarrow \pi^*$ transitions. Hence the n level appears to be the highest occupied orbital, contrary to Fig. 11.12.

In order to have the n level higher, it might be suggested that the interaction of the $n(p_y)$ orbitals on each side of the two oxygens, Fig. 11.13, leads to two

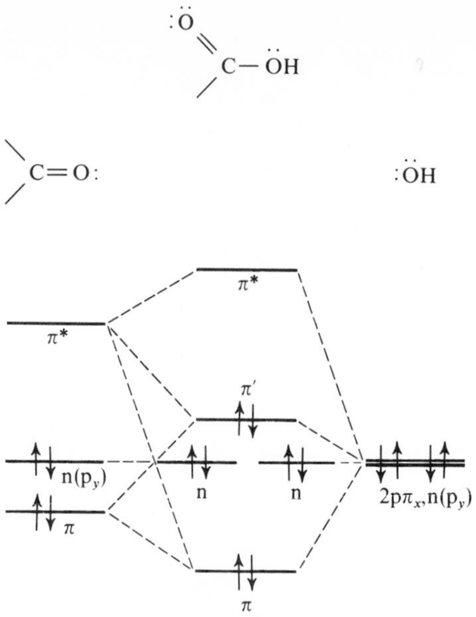

Fig. 11.12 Molecular orbital energy diagram for the carboxyl chromophore.

new levels, n_a and n_s, both occupied, such that the higher n_s level would be above the π' level, thus giving rise to the $n_s \rightarrow \pi^*$ transition at lower energy than the $\pi' \rightarrow \pi^*$, consistent with experimental observation. However, were this the case, the $n_s \rightarrow \pi^*$ transition in such a system would be expected to occur at least with equal, and very likely with lower, energy (longer wavelength) than the $n \rightarrow \pi^*$ in a simple carbonyl, such as acetone or acetaldehyde. However, Table 11.1 shows that, in acetic acid, the $n \rightarrow \pi^*$

TABLE 11.1

$n \rightarrow \pi^*$ **Transitions in Some Substituted Carbonyl Compounds, CH_3COZ^***

Z	Solvent	λ_{max} nm	ϵ^{max}
OH	Ethanol	208	32
OC_2H_5	Ethanol	211	57
NH_2	Water	220	63
Cl	Hexane	220	100
CH_3	Hexane	279	15
H	Hexane	290	17

* J. R. Dryer, *Application of Absorption Spectroscopy of Organic Compounds*, Prentice-Hall, 1956, p. 9.

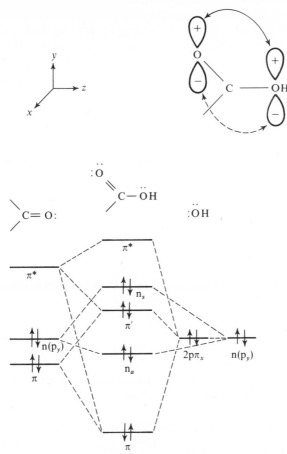

Fig. 11.13 Molecular orbital energy diagram with substantial lone-pair interaction.

transition is considerably *hypsochromic* relative to acetaldehyde (i.e., the absorption band is shifted to shorter wavelengths). Although π^* as well as π' is raised by conjugating C=O with the lone pair on oxygen, π^* would have to be raised unreasonably high to outweigh the raising of the n level by a large n-n interaction. Thus, the n → π^* transitions in carbonyl compounds having a lone pair in conjugation occur at shorter wavelength than in the simple carbonyl compounds, probably because of the raising of the π^* in the conjugated system, rather than because of any substantial n-n interaction. This contradiction, arising out of a literal but misleading use of the MOED's, is readily explained by application of the singlet-triplet energy differences derived above.

The consequences of these considerations again are illustrated in the term level diagram in Fig. 11.10, which we developed for the pyridine and which can now be used to show the excited states corresponding to the MOED for the carbonyl group conjugated to an atom with a lone pair of electrons in carboxyl. Because the n and π^* orbitals occupy different regions of space, their product (before integration) is small everywhere. The π and π^* orbitals, on the other hand, occupy largely the same space and hence their product is large. This difference between the relative spatial arrangement of n and π^* and π and π^* orbitals leads again to $K_{n\pi^*} \ll K_{\pi\pi^*}$, which results in the n $\to \pi^*$ singlet being lower in energy than the $\pi \to \pi^*$ state in carboxylic acids and esters, in accordance with experiment. Of course this difference applies to simple carbonyl compounds as well and again results in a much smaller singlet-triplet separation for the n $\to \pi^*$ transition than for the $\pi \to \pi^*$ transition; for example, the singlet-triplet separation for the n $\to \pi^*$ transition in formaldehyde is of the order of 3000 cm^{-1}, whereas the singlet-triplet energy separation for the ethylene $\pi \to \pi^*$ transition is 24,000 cm^{-1}, eight times as large.

It now becomes clear that the large singlet-triplet separation in the $\pi \to \pi^*$ transition, as compared to the n $\to \pi^*$ separation, accounts for the fact that the MOED shown in Fig. 11.12 incorrectly predicts the $\pi \to \pi^*$ to be the long-wavelength transition. These arguments also again predict that the lowest triplet state should have $\pi \to \pi^*$ character. The photochemistry of 2-acetylnaphthalene and of 1-naphthalenecarboxaldehyde indicates that, although the n $\to \pi^*$ singlets are lower than the $\pi \to \pi^*$ singlets in these compounds, the triplet energies are reversed. An experimental determination of this nature would be highly desirable in the simpler systems described herein.

11.3   Intermolecular Energy Transfer; Photosensitization

In our discussion of the fate of excited molecules we have dealt only with internal and external conversion processes. These processes involve collision of the excited molecule with solvent molecules, and the excited molecule, whether a triplet or a singlet, ends up in the lowest excited state of the appropriate manifold. We will now discuss the energy transfer between an excited molecule (donor), usually in its lowest excited state, and another (acceptor) molecule in the ground state. The transfer of energy may result in the ground-state acceptor molecule being excited and the excited donor molecule returning to the ground state.

The nonradiative electronic energy transfer between a molecule in an excited singlet state and another molecule in a singlet ground state may take place over a relatively large distance, for example, 50–100 A. Such an energy transfer has been shown to occur between 1-chloroanthracene (donor, D) and perylene (acceptor, A). The evidence for this process involves the demonstration that, in the presence of perylene, the fluorescence intensity of the 1-chloroanthracene is diminished, whereas the fluorescence intensity of perylene in the mixture is greater than that of pure perylene. The reactions may be represented as follows:

$$S_0(D) + h\nu \rightarrow S_1(D) \qquad (1)$$
$$S_1(D) + S_0(A) \rightarrow S_0(D) + S_1(A) \quad (2)$$
$$S_1(A) \rightarrow S_0(A) + h\nu \qquad (3)$$

In this example, a poorly fluorescent compound (1-chloroanthracene) "sensitizes" the fluorescence of a highly fluorescent compound (perylene). In the excitation process (1), light of the appropriate frequency must be used so as to excite the donor but not the acceptor. In order for the transfer step (2) to occur efficiently, the singlet energy of the acceptor must be slightly lower than that of the donor.

The actual mechanism of the relatively long-range energy transfer is not very well understood. Much of the early work on equations describing the rate of such transfer is due to Förster. The transfer, which proceeds through overlapping of the electric dipole fields of the donor and acceptor molecules, is called *dd*, or *resonance-excitation transfer*. The rate of this transfer falls off as $1/R^6$ and hence is very sensitive to the distance, R, separating the centers of interacting molecules. The equation also involves an energy overlap integral in which appear the frequencies of the emission band of the donor and of the absorption band of the acceptor.

The lifetime of the excited singlet is relatively short compared to that of the triplet, and thus, if we are to study photochemical reactions, especially those in solution, we are particularly interested in the ways by which the triplet state of a molecule can be populated. We have already described (Fig. 11.1) the process which involves excitations to S_1, followed by the forbidden intersystem crossing to T_1. The state T_1 can then return to S_0 by forbidden emission (phosphorescence) or by the nonradiative process of reverse intersystem crossing to a high vibrational level of S_0.

We will now discuss the nonradiative transfer of triplet energy to another molecule via a donor-acceptor process:

$$T_1(D) + S_0(A) \rightarrow T_1(A) + S_0(D)$$

The exchange of triplet energy requires close contact of the donor-acceptor pair with overlap of orbitals. In a well documented example of this process, benzophenone acts as the donor and naphthalene as the acceptor. Irradiation of the mixture of compounds at 366 nm results exclusively in excitation of benzophenone. This irradiation excites benzophenone to the first excited singlet (n → π*), S_1, which undergoes efficient intersystem crossing to T_1. The n → π* T_1 state, as we indicated earlier, is estimated to have a radiative

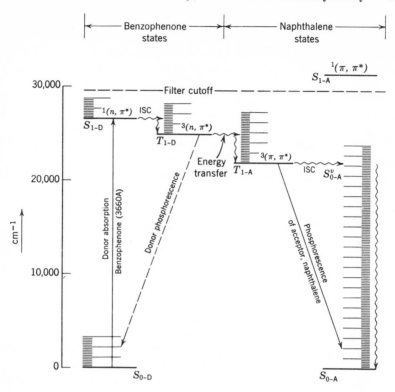

Fig. 11.14 Energy level diagram, showing absorption of radiation by benzophenone and triplet energy transfer to naphthalene. [From J. G. Calvert and J. N. Pitts, *Photochemistry*, John Wiley and Sons, 1966, by permission.]

lifetime of about 1.5 sec, and during this time its energy is transferred to the ground-state naphthalene molecule, thereby raising the latter to its T_1 state and quenching the benzophenone emission. These processes are shown in Fig. 11.14. It will be observed that the triplet energy of the acceptor (naphthalene, 61 kcal/mole) must be slightly lower than the triplet energy of the donor (benzophenone, 69 kcal/mole). It should also be noted that donor phosphorescence may compete with energy transfer, and the observed decreased lifetime of benzophenone phosphorescence provides further evidence for the transfer process.

11.4   Some Photochemical Reactions of the Excited Triplet
11.4a   Photoreduction

In the preceding example of triplet energy transfer, the acceptor molecule returns to its ground state S_0 either through a radiative process (phosphorescence) or by intersystem crossing to a high vibrational level of the ground state (nonradiative). Now we want to consider a few examples of photochemical transformations of acceptor molecules.

One of the oldest known photochemical reactions is the conversion of aromatic ketones to the corresponding pinacol:

$$2(C_6H_5)_2CO + h\nu + (CH_3)_2CHOH \rightarrow [(C_6H_5)_2C-OH]_2 + (CH_3)_2CO$$

The alcoholic solution of benzophenone ($\sim 10^{-2}M$) is irradiated with light at about 350 nm. The reaction sequence consists of the following steps:

$$(C_6H_5)_2C{=}O(S_0) \xrightarrow{h\nu} (C_6H_5)_2C{=}O(S_1) \rightarrow (C_6H_5)_2\dot{C}-\dot{O}(T_1) \qquad (1)$$

$$(C_6H_5)_2\dot{C}-\dot{O}(T_1) + (CH_3)_2CHOH \rightarrow (C_6H_5)_2\dot{C}-OH + (CH_3)_2\dot{C}-OH \qquad (2)$$

$$(CH_3)_2\dot{C}-OH + (C_6H_5)_2C{=}O \rightarrow (CH_3)_2C{=}O + (C_6H_5)_2\dot{C}-OH \quad (3)$$

$$2(C_6H_5)_2\dot{C}-OH \rightarrow (C_6H_5)_2C(OH)C(OH)(C_6H_5)_2 \qquad (4)$$

According to this scheme, 2 moles of benzophenone are converted to product per photon absorbed, or in other words the quantum yield based on the disappearance of benzophenone is:

$$\Phi = \frac{\text{moles reacted}}{\text{photon absorbed}} = 2$$

The quantum yield in the photoreduction depends on the solvent. When isopropyl alcohol is used as solvent, step 3, which involves the interaction of the radical $(CH_3)_2\dot{C}OH$ with ground-state benzophenone to produce acetone and a ketyl radical, $(C_6H_5)_2\dot{C}OH$, proceeds in good yield. However, when a hydrocarbon solvent is used as the hydrogen donor, the abstraction of the second hydrogen atom in step 3 is not a very probable step, and the quantum yield is accordingly quite low. Quantum yields for the benzophenone conversion range from 0.05 in benzene to almost 2 in isopropyl alcohol. It has been clearly established that the chemically active state in the photoreduction is the $(n \rightarrow \pi^*)$ T_1 state. It should be noted that, in step 2, the excitation energy is lost in the abstraction of the H atom and the formation of two reactive radicals.

11.4b   Photochemical *cis* $\rightleftharpoons$ *trans* Isomerization

A second well-established photochemical reaction is the *cis-trans* isomerization of olefins. Although the *trans* isomer is usually more stable and therefore is the predominant isomer at equilibrium, frequently the *cis* isomer predominates in the photostationary state.

Usually *cis* and *trans* isomers differ only slightly in their ground-state energies, but interconversion is difficult to achieve thermally because of the very high energy of activation needed for the process. The *cis* $\rightleftharpoons$ *trans*

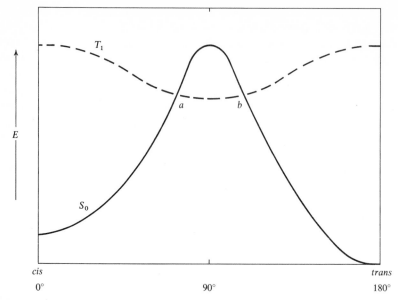

E

cis
$0°$

$90°$

trans
$180°$

Fig. 11.15   Potential energy as a function of angle of twist of an olefin in the ground state, S_0, and the excited triplet state, T_1.

interconversion requires a twist of 180° (assuming a planar conformation for both isomers) around the C-C double bond. A plot of the potential energy as a function of the angle of twist for a typical example is shown as curve S_0 in Fig. 11.15. The highest-energy conformation occurs at 90°, where, with ethylene, for example, the two CH_2 groups are at right angles to each other and thus ethylene goes from D_{2h} symmetry in the planar model through conformations of D_2 symmetry to D_{2d} symmetry in the perpendicular model. In ethylene, the two planar conformations are of equal energy and the maximum occurs at 90°. With substituted ethylenes, the maximum need not occur at exactly 90° and the ground-state energies of the two isomers are different, the less hindered *trans* isomer being lower in energy.

The plot of potential energy as a function of angle of twist for the excited triplet state, shown as curve T_1 in Fig. 11.15, is very different from the ground-state curve, S_0. In the triplet state, where the π electrons occupy different orbitals and have parallel spin, the lowest-energy conformation is the one, for example, in ethylene, with D_{2d} symmetry. In this conformation, the electron repulsion is minimized and the staggered conformation of the H atoms minimizes atom repulsions. It will also be noted that the T_1 curve is much shallower than the S_0 curve because the triplet has none of the restraints of a double bond. It is further predicted from the shapes of these curves that, were an olefin fixed in the 90° conformation, it should exist as a triplet in the ground state; the synthesis of such a molecule should be a challenge to the organic chemist.

Although there is still considerable controversy regarding the exact mechanism of the photochemical *cis* $\rightleftharpoons$ *trans* isomerization, many investigators now believe that a common triplet is involved. In nonsensitized or direct isomerization, either isomer, on irradiation with light of the appropriate wavelength, is promoted to its excited state. The S_1 state then undergoes intersystem crossing to the common T_1. The T_1 state crosses S_0 at point *a* or *b*, Fig. 11.15, depending on whether the *cis* or *trans* isomer is being irradiated. At these points T_1 may cross over to S_0, but this is not a highly probable process because of the required change in multiplicity. Most triplet molecules fall into the potential well of the common triplet. From this well the cross-over to S_0 at point *a* or *b* is again forbidden, but the relatively long lifetime and the shallowness of the well assure a substantial number of cross-overs to high vibrational levels of S_0.

A mechanism involving a common triplet is only one of a number of possibilities suggested. Another involves promotion of either *cis* or *trans* S_0 to separate S_1 states which, by intersystem crossing, lead to separate T_1 states, the energy of T_1 (*trans*) usually being lower than T_1 (*cis*). The molecule can drop from T_1 (*cis*) to T_1 (*trans*) by an exothermic process. State T_1 (*trans*) can then return to a high vibrational level of S_0 (*trans*) by intersystem crossing, but the molecule in the high vibrational level of S_0 will have sufficient energy to rotate around the C-C bond and return to the lower vibrational level as either a *cis* or a *trans* isomer.

The composition of the photostationary state depends on the extinction coefficients of the two isomers at the wavelength of irradiation. For 1,2-diphenylethylene (stilbene), for example, the ratio of extinction coefficients at 313 nm, a wavelength commonly used for excitation since this is one of the mercury lines, is as follows: $\epsilon(trans)/\epsilon(cis) = 7.2$. This means that the *trans* isomer, with its higher extinction coefficient, is 7.2 times more likely to absorb the 313 nm radiation than is the *cis*. Accordingly, the *trans* is continuously pumped over to the *cis* form. From other experiments (sensitized isomerizations) the natural decay ratio of the common triplet state of stilbene has been found to be *cis*/*trans* = 1.5. If the ratio of extinction coefficients is multiplied by the decay rate, the product should be an estimate of the photostationary state ratio of *cis*/*trans* = 7.2 × 1.5 = ∼10.8, or about 92 per cent *cis*. The experimentally determined photostationary state consists of about 93 per cent *cis*, in excellent agreement with the calculated value.

The direct photoisomerization of the stilbenes is complicated by competing reactions, principally the conversion to *trans*-4a,4b-dihydrophenanthrene. This cyclization reaction, which probably proceeds via the S_1 state, can be minimized by avoiding populating the S_1 state, and this can be achieved by the use of appropriate photosensitizers. A sensitizer which has the appropriate triplet energy and which absorbs at a wavelength different from that of either *trans-* or *cis*-stilbene is excited to S_1 and crosses over to T_1, where it then reacts with stilbene S_0 to give stilbene T_1, without the stilbene ever passing through its S_1 state. The energy of the photosensitizer triplet has a very important effect on the ratio of isomers in the photostationary state.

One of the several reasons for the lack of a completely satisfactory mechanism for the isomerization reaction is the difficulty of obtaining singlet → triplet energies directly. One experimental method for obtaining this information, especially for aromatic hydrocarbons, is by means of the *oxygen perturbation spectrum*. In this method, the absorption spectrum of the compound is determined in the presence of 50–100 atmospheres of oxygen. It is assumed that under these conditions a loose complex is formed between the O_2 and the hydrocarbon, imparting triplet character to the ground state of the hydrocarbon. The measurement of the absorption spectrum then provides a singlet → triplet spectrum, since this transition is no longer so strictly forbidden because of the partial triplet character of the ground state. Even when a hydrocarbon is amenable to such a determination, however, it is frequently difficult to identify the 0 → 0 band.

11.4c   Photosensitized Dimerization of Butadiene

The photodimerization of butadiene leads to three main products: *trans*-1,2-divinylcyclobutane, **1**; *cis*-1,2-divinylcyclobutane, **2**; and 3-vinylcyclohexene, **3** (Fig. 11.16). Although *s-trans-* and *s-cis*-butadiene are readily interconvertible in the ground state at room temperature, the excited triplets are not as readily interconvertible because of the larger energy barrier to rotation around the 2,3 bond in the excited state (Section 4.3). We are concerned here only with the triplet because the mixture of butadiene and photosenstizer is irradiated at wavelengths longer than those required for butadiene excitation and therefore no S_1 state of butadiene is involved. The triplet energies of the two conformers differ, and by control of photosensitizers one or the other of the two triplet isomers can be made to predominate. Since they react differently with another mole of butadiene S_0, the composition of the product mixture (which is photostable under the conditions of the experiment) can be controlled by the choice of photosensitizers.

Some results are shown in Fig. 11.16. It has been established that the *s-trans* triplet reacts to give the cyclobutanes, **1** and **2**, and the *s-cis* triplet is the precursor leading to **3**. The higher-energy sensitizers transfer triplet energy to either *s-cis* or *s-trans* indiscriminately, and so the ratio of triplets reflects the ratio of isomers in the ground state. This ratio favors the *s-trans* form; and since this route leads to vinylcyclobutanes, the yield of these products exceeds 95 per cent when sensitizers with triplet energies greater than 60 kcal/mole are used. With sensitizers having triplet energies less than 60 kcal/mole, transfer to *s-trans*-butadiene becomes less efficient and larger quantities of *cis* triplet are formed, leading to increased yields of 4-vinylcyclo-hexene.

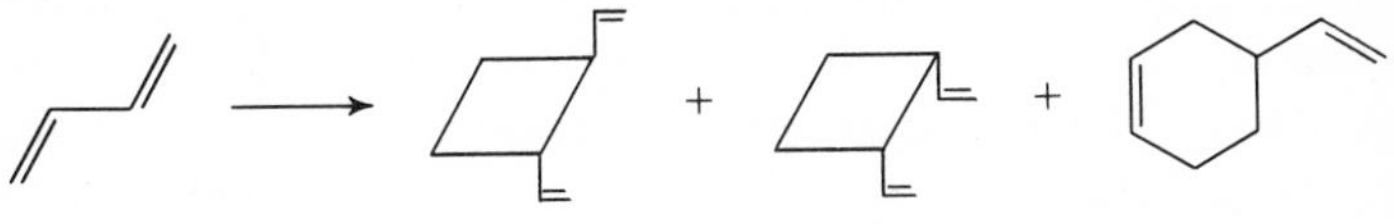

Sensitizers	T_1,(kcal)	Yield of Isomers		
		1	2	3
$C_6H_5COCH_3$	74	78	18	4
$2-C_{10}H_7COC_6H_5$	58	72	16	12
Fluorenone	53	44	13	43

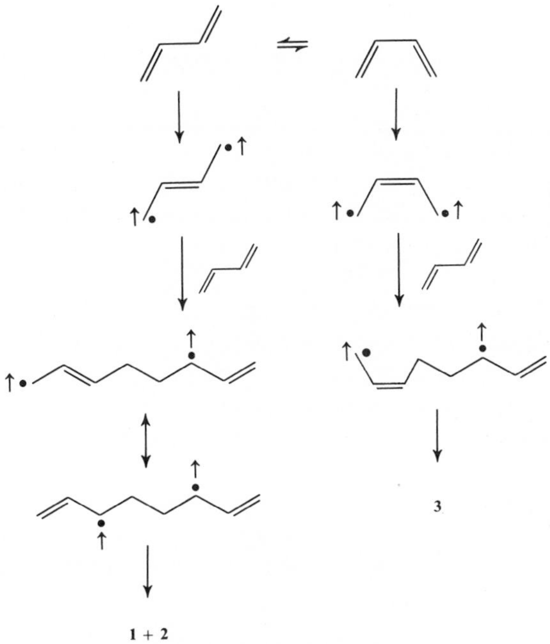

Fig. 11.16   The photosensitized dimerization of butadiene.

11.4d   Photosensitization and Singlet Oxygen

The MOED of oxygen (Section 3.9) explains the triplet character of ground-state $O_2(^3\Sigma_g^-)$, since the highest occupied orbitals are degenerate. The lowest-energy excitation involves an inversion of spin to a singlet state and double occupancy of one of these levels. The transition of next higher energy also involves the inversion of spin, but each degenerate orbital remains singly occupied. In state notation these are both excited singlet states; the lower-energy state, 22 kcal/mole above the ground state, is a $^1\Delta_g$ state, and the higher-energy state, 37 kcal/mole above the $^3\Sigma_g^-$ ground state, is a $^1\Sigma_g^+$ state.

There is increasing evidence that the lowest excited singlet state may play an important role in many reactions. Thus, the 1,4 addition of oxygen to dienes to give peroxides is photosensitized by dyes (Y) with low triplet energies. With 1,3-cyclohexadiene the reaction may be written:

$$Y(S_0) \overset{h\nu}{\Rightarrow} Y(S_1) \rightsquigarrow Y(T_1) \qquad (1)$$

$$Y(T_1) + O_2(T_0) \rightarrow Y(S_0) + O_2(S_1) \qquad (2)$$

$$O_2(S_1) + \text{[cyclohexadiene]} (S_0) \rightarrow \text{[endoperoxide]} \qquad (3)$$

Step 2 involves a triplet-triplet annihilation reaction. It should be appreciated that in such a reaction the spins are preserved, since a triplet with two parallel spins reacts with another triplet, also with two parallel spins. If these four spins align properly, there results a state with all spins paired, that is, a singlet (other spin alignments are possible, leading, e.g. to a quintuplet, with all four parallel, but such cases are of little interest). One fact supporting the correctness of this sequence is that singlet oxygen, which can be generated chemically from sodium hypohalites and hydrogen peroxide, gives the same products as the photoreactions. There is some evidence that singlet oxygen, since it can be populated with such small energy, may be of considerable biochemical importance. It may be that metal catalysts can be found which will facilitate spin inversion and that normal oxygen can be converted by a thermal process into singlet oxygen.

11.5   Woodward-Hoffmann Rules; Electrocyclic Reactions

Many reactions are known in which an unsaturated cyclic compound is converted to a conjugated acyclic diene. Thus, for example, cyclobutene may be converted to butadiene:

The reaction is reversible and can be accomplished either thermally or photochemically. Such reactions have been called *electrocyclic reactions* because they involve ring openings and cyclizations that can be represented formally as occurring by the migration of electrons, as shown above. When substituted cyclobutenes are employed, *cis*- or *trans*-butadienes can be formed (depending on selection rules which will be elaborated below), but the reactions are remarkably stereospecific. Thus heating dimethyl *cis*-1-cyclobutene-3,4-dicarboxylate, **4**, gives exclusively dimethyl *cis, trans*-buta-1,3-diene-1,4-dicarboxylate (Z = CO_2CH_3), **5**.

The stereospecific nature of this reaction and of the reverse cyclization means that the twist or rotation of both Z—C—H groups around the C—C axes must have been in the same direction (both either clockwise or counter-clockwise), since rotation in opposed directions would have resulted in formation of the *trans, trans* isomer, **6**. When the two rotations, for example, those shown as arrows in **5**, conform in direction, the electrocyclic reaction is said to be *conrotatory*.

Actually the *trans, trans* isomer, **6**, can be converted to the *cis*-cyclobutene, **4**, by ultraviolet irradiation. It will be noted that this conversion requires that **6** undergo rotations around the C—C axes in the manner shown above, that is, the directions of rotation at the termini are opposed. When the rotations are in opposite directions, as shown by the arrows in **6**, the electrocyclization is said to be *disrotatory* (see Chapter 4). Since the ring opening and the cyclization are reversible reactions, we have only two possible modes for these reactions; conrotatory or disrotatory. The facts are that the thermal reaction with butadiene is conrotatory and the photochemical reaction is disrotatory.

In the electrocyclic reactions discussed above, the prediction as to whether the reaction will proceed in a conrotatory or disrotatory fashion can be made on the basis of the symmetry of the orbitals involved. In particular, we need focus only on the symmetry of the highest occupied orbital of the acyclic partner, and particularly on the signs of the MO's at the atoms which are to be bonded by the newly formed σ bond in the cyclic partner. In the case of the thermal cyclization of butadiene this requires a knowledge of the sign of the orbitals at carbons 1 and 4 in ψ_2; for the photochemical reaction it is the sign of these atoms in ψ_3 (cf. Fig. 4.4) that must be known.

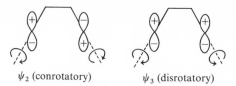

ψ_2 (conrotatory) ψ_3 (disrotatory)

Cyclization in the modes shown in the sketches brings a plus lobe into a plus lobe in each case (or a minus into a minus, which is just as good), resulting in a bonding and attractive interaction that results in the actual formation of the new σ bond between carbons 1 and 4.

In the cyclization of hexatriene, for example, we are concerned with the signs of the orbitals at carbons 1 and 6 in ψ_3 for the thermal reaction and in ψ_4 for the photochemical reaction (cf. Fig. 4.5). In ψ_3, there are two nodes between the ends of the box representing the square well function

TABLE 11.2
Rules for Electrocyclic Reactions

Carbon System	n	k^*	Thermal	Photochemical
$\left\langle\!\!\!= \Box\right.$	1	4	con	dis
$\left\langle\!\!\!= \bigcirc\right.$	1	6	dis	con
$\left\langle\cdot\cdot + \rightleftharpoons \triangle +\right.$	0	2	dis	con
$\left\langle\cdot\cdot - \rightleftharpoons \triangle -\right.$	1	4	con	dis
$\left\langle\cdot\cdot + \rightleftharpoons + \right\rangle$	1	4	con	dis

* The number of $p\pi$ electrons in the acyclic partner. When $k = 4n + 2$ ($n = 0, 1, 2, \ldots$), the reaction is thermally disrotatory; when $k = 4n$, conrotatory. The rules are reversed for the photochemical reactions.

(see the FEM method of Chapter 4). Hence the amplitudes at C_1 and C_6 are equal and of equal sign (or $c_1 = c_6$), and this tells us that the cyclization between atoms 1 and 6 will be disrotatory. In ψ_4, however, $c_1 = -c_6$ and the opposite signs require a conrotatory mode for the photochemical cyclization at carbons 1 and 6. In general, if there are k π electrons in the acyclic system, electrocyclic reactions are thermally disrotatory for $k = 4n + 2$ and conrotatory for $k = 4n$, where n is any integer. In the first excited state (photochemical) this relationship is reversed. Some examples are shown in Table 11.2.

Although we can develop the rules for electrocyclic reactions by considering the symmetry of the highest occupied orbital, the conservation of orbital symmetry during the transformations can be appreciated only by an analysis of all the orbitals in the starting and final products.

As the 1,4 σ bond of cyclobutene is broken, the hydrogen atoms on C_1 and C_4 rotate around the 1,2- and 3,4-bond axes. Figure 11.17 shows that, during the ring opening (or the cyclization) by the disrotatory mode, a plane of symmetry is preserved as the molecule progresses along the reaction coordinate. On the other hand, the conrotatory ring opening (or closing) proceeds with the preservation of a C_2 axis, which is the axis in the plane of the molecule shown in Fig. 11.17.

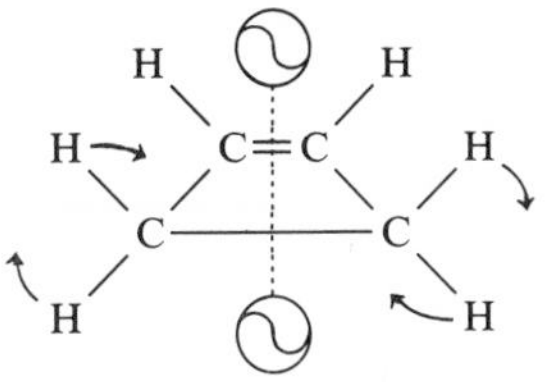

(a) 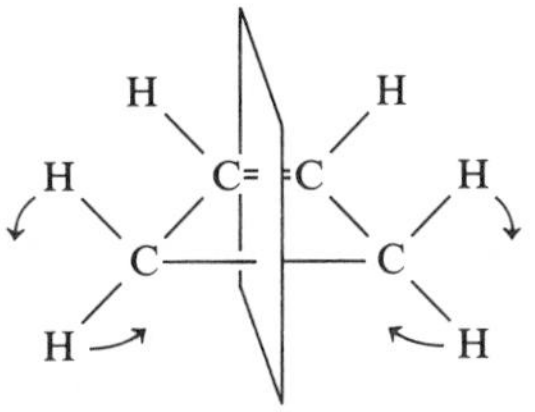   (b)

Fig. 11.17   Alternative modes for the conversion of cyclobutene to *s-cis*-butadiene. (*a*) Conrotatory mode; (*b*) disrotatory mode.

Let us now focus on the behavior of the pertinent orbitals, in the interconversion of cyclobutene to butadiene. The orbitals of interest are the four MO's of butadiene, all of π character, Fig. 11.18, and the π, π^*, σ, and σ^* orbitals of cyclobutene, Fig. 11.19. For the disrotatory mode, all of these orbitals must now be classified with respect to their behavior under reflection on the plane of symmetry; for the conrotatory mode, classification is made on the basis of behavior with respect to the C_2 rotational axis.

These symmetry elements are the appropriate ones to consider because they pass through the bonds being made or broken and represent the symmetry that is preserved during the electrocyclic transformation from reactants to

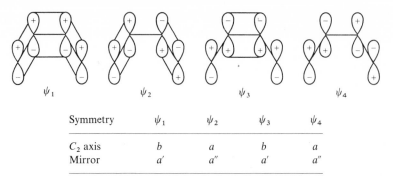

Symmetry	ψ_1	ψ_2	ψ_3	ψ_4
C_2 axis	b	a	b	a
Mirror	a'	a''	a'	a''

Fig. 11.18   The orbitals of *s-cis*-butadiene and their symmetries.

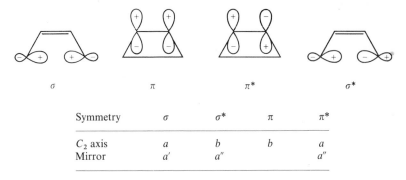

Symmetry	σ	σ^*	π	π^*
C_2 axis	a	b	b	a
Mirror	a'	a''		a''

Fig. 11.19   The orbitals of cyclobutene and their symmetry.

products via the two different modes. A molecule having only a plane of symmetry belongs to point group C_s, and for our purposes we can consider the behavior of the orbitals of the molecule undergoing the disrotatory mode in terms of their behavior with respect to reflection in this plane. In point group C_s symmetric behavior is a', antisymmetric a''. In point group C_2, where the only symmetry element is a C_2 axis, symmetric behavior with respect to the rotation belongs to species a and antisymmetric behavior to species b. It is thus convenient to classify the orbitals of butadiene and cyclobutene as a or b when the molecule is undergoing the conrotatory electrocyclic reaction, and as a' or a'' when it is undergoing the disrotatory one. The classification of the orbitals is shown in Table 11.3.

In order to make correlations it is also necessary to be concerned with the relative energies of these four orbitals, and the relative energy level scheme shown in Fig. 11.20 can be justified quite easily. In cyclobutene, we expect σ orbitals to be the lowest- and σ^* the highest-energy orbitals, and π and π^*

TABLE 11.3
Orbital Correlations for Cyclobutene $\rightleftharpoons$ Butadiene Interconversion

	Cyclobutene Orbitals	Butadiene Orbitals	Symmetry
Conrotatory mode	$\begin{cases} \sigma, \pi^* \\ \pi, \sigma^* \end{cases}$	$\begin{cases} \psi_2, \psi_4 \\ \psi_1, \psi_3 \end{cases}$	$\begin{cases} a \\ b \end{cases}$
Disrotatory mode	$\begin{cases} \sigma, \pi \\ \pi^*, \sigma^* \end{cases}$	$\begin{cases} \psi_1, \psi_3 \\ \psi_2, \psi_4 \end{cases}$	$\begin{cases} a' \\ a'' \end{cases}$

to be of intermediate energy, with π, of course, more stable than π^*. The order of energy of the butadiene orbitals corresponds to the order of the number of nodes in the MO's, with ψ_1 (no nodes) being most stable and ψ_4 (three nodes) least stable.

Let us now carry out the conrotatory ring opening of cyclobutene to butadiene. Our problem is to correlate the four orbitals of the starting compound with the four orbitals of the product, both with respect to symmetry and with respect to energy. We start with the σ orbital on cyclobutene, remembering that, during the conrotatory mode of ring opening, the atoms and the orbitals must be either symmetric or antisymmetric with respect to the C_2 axis. The transformation of the σ orbital to ψ_2 of butadiene is shown in Fig. 11.21. Step 1 shows the rotated σ orbital; step 2, the re-hybridization of the sp^3 to sp^2 and $p\pi$ atomic orbitals; and step 3, the "growing in" of the orbitals on C_2 and C_3 to give the ψ_2 molecular orbital of butadiene, which correlates, that is, has the same symmetry behavior with respect to C_2 as does the σ orbital. It should be noted that ψ_4 of butadiene, as

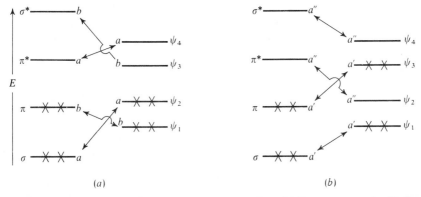

Fig. 11.20   Orbital correlation for cyclobutene $\rightleftharpoons$ butadiene. (a) Conrotatory mode; (b) disrotatory mode. The ground-state cyclobutene correlates with the ground-state butadiene in (a) but with an excited state in (b).

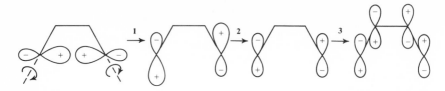

Fig. 11.21 Conrotatory mode and conversion of the σ orbital in cyclobutene to ψ_2 of butadiene.

well as ψ_2, transforms like the σ orbital of cyclobutene under C_2. However, this single σ molecular orbital of cyclobutene cannot correlate with both ψ_2 and ψ_4. On energy grounds, the correlation is between it and ψ_2, the lower-energy one of the two possible orbitals of the product, rather than ψ_4. Thus, the σ and π^* orbitals of cyclobutene mix to give a combination of $\sigma + \pi^*$, which emerges as ψ_2, while π^* is transformed into the combination $\pi^* - \sigma$, which emerges as ψ_4.

The π orbital of cyclobutene is antisymmetric with respect to C_2, and hence it correlates with either ψ_1 or ψ_3 of the product. On the basis of relative energy, the correlation is with ψ_1, which, quantum-mechanically, is equivalent to a mixing of π with σ^*, that is, $\sigma^* + \pi$. This leaves the σ^* orbital of cyclobutene to correlate with ψ_3, both again antisymmetric with respect to C_2, and quantum-mechanically the mixing may be regarded as the combination $\pi - \sigma^*$, emerging into ψ_3. The correlations of cyclobutene and butadiene orbitals for both the conrotatory and disrotatory modes are summarized in Fig. 11.20.

The correlation diagram of Fig. 11.20 is very informative. If, in such diagrams, bonding orbitals in the reactant correlate only with bonding orbitals in the product, the thermal reaction is permitted, Fig. 11.20a. However, if there is a correlation between a bonding orbital in the reactant and an antibonding orbital in the product, Fig. 11.20b, the reaction is not allowed by a thermal process but is permitted by the photochemical route.

The discussion in this section has been restricted to molecules which possess some symmetry with respect to the bonds being made and broken and, in particular, to a consideration of overlap of lobes of orbitals with equal or unequal sign, leading to bonding or antibonding interaction. The overlap of orbitals, however, does not depend on molecular symmetry, and the arguments consequently can be carried out with care for less symmetric molecules. It would appear that the arguments of the correlation diagram would break down when the symmetry restrictions are removed, since the noncrossing rule requires the orbitals to be symmetry "correlated" in order of energies. Thus, for example, in considering the thermal cyclization of 2-methyl-1,3-butadiene, a compound lacking symmetry, one might be tempted to think that the disrotatory mode, which is forbidden in the butadiene case, would be permitted for the methyl derivative.

The argument that this consideration is not valid can be made as follows. At the beginning of this section, we pointed out that we are concerned primarily with the *signs* of the lobes of orbitals forming bonds in the electro-cyclic reactions. We introduced symmetry as a convenient means to determine the relative signs. Introduction of substituents (like the methyl group in 2-methyl-1,3-butadiene) does not greatly affect (perturb) the MO's, and consequently does little to change the conclusions previously reached. The same thing may be stated differently by saying that, although the symmetry of the molecule is destroyed, the essential (or local) symmetry of the π-electron system remains unchanged, or at most is only slightly perturbed.

In the latter part of this section we developed the ideas of the Woodward-Hoffman rules in terms of a more formal treatment, based entirely on sym-metry, and using correlation diagrams between orbitals. The correlation lines between orbitals (cf. Fig. 11.20) were allowed to cross, because they represented orbitals of different symmetry. In the absence of symmetry, such crossing is no longer possible because of the noncrossing rule, and it would appear that orbitals should be correlated only in ascending order of energies. This is correct, but the behavior of the correlation line (the line connecting orbitals) must be examined in detail.

Thus, for example, in Fig. 11.20b, the correlation line between π (of cyclo-butene) and ψ_3 (of butadiene) crosses the correlation line between π^* and ψ_2, and this crossing is allowed because the first set belongs to an a' species and the second to an a'' species. In the 2-methyl-substituted derivative, this distinction disappears and crossing is forbidden. However, as we proceed along the reaction coordinate from the cyclobutene to the butadiene, the π orbital still behaves as if it were going into a ψ_3 orbital, and the π^* as if it were going into ψ_2. Only when the energies of the two become very close do the orbitals interact strongly and repel one another in the fashion shown in Fig. 11.22. Thus, although now π and ψ_2 are correlated, a substantial energy barrier intervenes. Hence, the disrotatory cyclization is no more allowed in the 2-methyl derivative than in butadiene itself. Generally, the correlation diagrams are of little value in the absence of symmetry, unless examined in much greater detail.

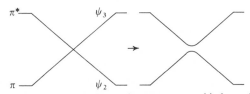

Fig. 11.22   The effect of loss of symmetry on orbital correlation.

11.6   Symmetry State Correlations

In the preceding discussions we have considered electrocyclic transformations in terms of particular MO's and their correlation, stressing the importance of the highest occupied orbital. We now consider the effect of all
the electrons in all the MO's of interest in the reactants and products, that is,
the states of the molecules. The electronic configuration of the ground state
of cyclobutene may be written as $\sigma^2\pi^2$. Instead of writing the configuration
in terms of types of orbitals, we can also write it in terms of the symmetry
species of the orbitals (cf. Section 8.3). Thus, if we are to consider the conrotatory mode of ring opening, where symmetry classification is based on the
behavior of the orbital with respect to the axis of symmetry, $\sigma^2\pi^2$ may be
written as a^2b^2. This configuration obviously corresponds to the totally
symmetric state A since $(+1)^2 \cdot (-1)^2 = 1$. The ground-state configuration
of butadiene, $\psi_1{}^2\psi_2{}^2$, in symmetry notation corresponds to b^2a^2 and is also
an A state. Hence in the conrotatory mode the two ground states, which are
the lowest-energy states of reactant and product, correlate.

Now, let us consider the transformation of the first excited state of cyclobutene to excited butadiene. The first excited state of cyclobutene $(\sigma^2\pi\pi^*)$
has the configuration $a^2ba = (1)^2(-1)(1) = -1$, which is antisymmetric
and designated B; since σ correlates with ψ_2, π with ψ_1, and π^* with ψ_4, the
$\sigma^2\pi\pi^*$ state correlates with $\psi_1\psi_2{}^2\psi_4 = a^2ba = B$ in butadiene. However,
it will be noted that the configuration $\psi_1\psi_2{}^2\psi_4$ is not the lowest excited state
of butadiene, which would be $\psi_1{}^2\psi_2\psi_3$, also $b^2ab = B$. Accordingly, the
probability of cyclobutene in the lowest-energy excited state $(\sigma^2\pi\pi^*)$ being
transformed to butadiene in the correlated excited state $(\psi_1\psi_2{}^2\psi_4)$ is quite
low because of the uphill energy requirement. A higher excited state of cyclo-
butene has the configuration $\sigma\pi^2\sigma^* = B$, corresponding to $\psi_1{}^2\psi_2\psi_3 = B$ of
butadiene, its lowest possible excited state, and these two states are correlated.

We can now draw a new correlation diagram in state notation for the
conrotatory process (see Fig. 11.23a). By exactly the same kind of reasoning,
we would arrive at the correlation diagram shown in Fig. 11.23b for the
disrotatory process, taking $+1$ for a' and -1 for a''. The straight lines (either
broken or full) connect the symmetry states that correlate with each other.
However, states of the same symmetry cannot cross, and the full lines
represent the true correlations. Figure 11.23 shows that the thermal (ground-
state) conrotatory conversion of cyclobutene to butadiene is energetically
more favorable than the disrotatory process because the latter has to over-
come an energy barrier. On the other hand, the conrotatory photochemical
electrocyclization from the first excited state of cyclobutene to the first
excited state of butadiene has an energy barrier, hence a prohibitive activation energy, whereas the corresponding photochemical transformation, proceeding by the disrotatory mode, has no energy barrier, hence is allowed.

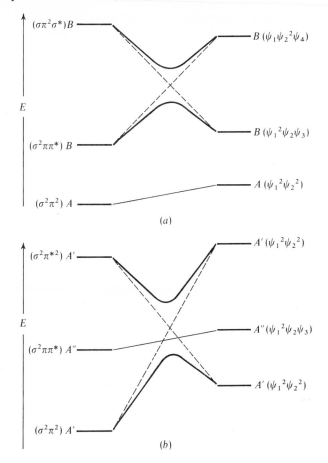

Fig. 11.23 State correlation diagram for cyclobutene ⇌ butadiene. (*a*) Conrotatory mode; (*b*) disrotatory mode.

11.7   Ethylene Dimerization; Cycloaddition Reactions

We wish to examine now the selection rules for a cycloaddition reaction, that is, the concerted reaction between two unsaturated molecules to produce a new ring system. The simplest example is the dimerization of ethylene to cyclobutane:

$$
\begin{array}{c}
\text{C} \\
\| \\
\text{C}
\end{array}
+
\begin{array}{c}
\text{C} \\
\| \\
\text{C}
\end{array}
\rightarrow
\begin{array}{c}
\text{C} - \text{C} \\
| \quad | \\
\text{C} - \text{C}
\end{array}
$$

This cycloaddition reaction requires the conversion of two π bonds to two σ bonds.

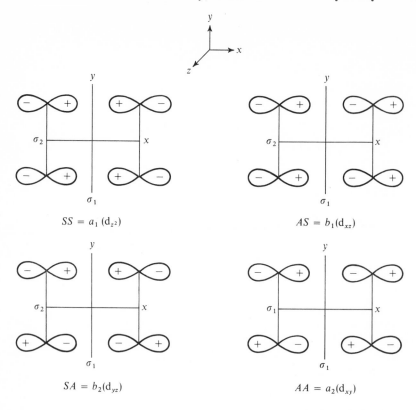

Fig. 11.24   Classification of the combination of π and π^* orbitals of two ethylenes and the σ and σ^* orbitals of cyclobutene. (The designation in parentheses represent atomic d orbitals with corresponding symmetry.)

The π orbitals on the ethylenes can combine in two ways and the two π^* orbitals can also combine in two ways, to give four combinations which must correlate with the combinations of 2σ and $2\sigma^*$ orbitals in the product. If we vizualize the two ethylenes approaching each other sideways with their planes parallel to one another, so that their π systems start to interact, we see that we can classify the four combinations of interest with respect to two planes of symmetry, Fig. 11.24. One plane, $\sigma_1(\sigma_{xy})$, is perpendicular to the plane of the four C atoms, and the other plane, $\sigma_2(\sigma_{yz})$, is the molecular plane. We may now classify the eight orbitals as symmetric or antisymmetric with respect to these planes, that is, S or A with respect to σ_1 and σ_2, respectively. Because two σ's intersecting at right angles require C_2, we may concurrently classify the orbitals according to the symmetry species in point group C_{2v}, which has C_2 and $2\sigma_v$.

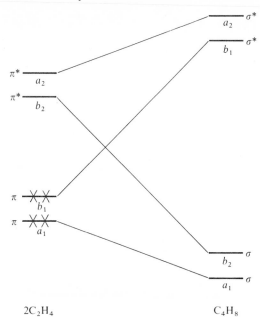

Fig. 11.25 The symmetry species of the four molecular orbitals of two ethylenes and the four σ's of cyclobutane in C_{2v} and their correlation.

We can now draw the correlation diagram of Fig. 11.25. Although π's and π^*'s constitute degenerate sets in two isolated ethylenes, it is clear from Fig. 11.24 that, as the two ethylenes approach each other, of the two σ bonds that are to be formed in cyclobutane, the energy of the incipient one of a_1 symmetry will be lower than that of the one of b_2 symmetry, since, in the former, two π-bonding orbitals are approaching each other and there are no repulsions, whereas π^*'s are being converted to σ's in the latter. Similarly, we expect the σ^* of b_1 symmetry to be lower in energy than the σ^* of a_2 symmetry, because the latter results from the transformation of π^*'s while the former is formed from lower-energy π's. Figure 11.24 shows that the σ^* of a_2 symmetry requires the approach of four lobes, all of different sign, whereas the σ^* of b_1 symmetry involves the approach of only two lobes of different sign. Examination of the correlation diagram of Fig. 11.25 shows the thermal dimerization to be forbidden, since a bonding orbital of the reactant correlates with an antibonding orbital of the product. The state diagram, Fig. 11.26, of course, also shows the thermal process to be unfavorable. However, the photochemical process would involve an electron in b_2 (Fig. 11.25), a π^* orbital, ending up in the $\sigma(b_2)$ of cyclobutane and hence

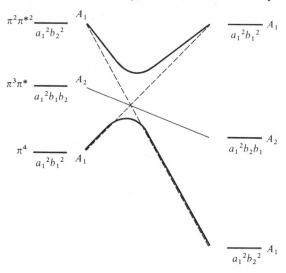

Fig. 11.26   The state correlation diagrams for the ethylene to cyclobutane dimerization.

would be allowed. The state correlation diagram shows the $A_2 \rightarrow A_2$ process as allowed.

There are many examples of concerted cycloaddition reactions. The ethylene dimerization may be considered as a special instance of a four-centered cycloaddition, Fig. 11.27a. There are also six-centered reactions, Fig. 11.27c, the most important examples of which are the Diels-Alder type of reactions. A cycloaddition reaction involving the interconversion of

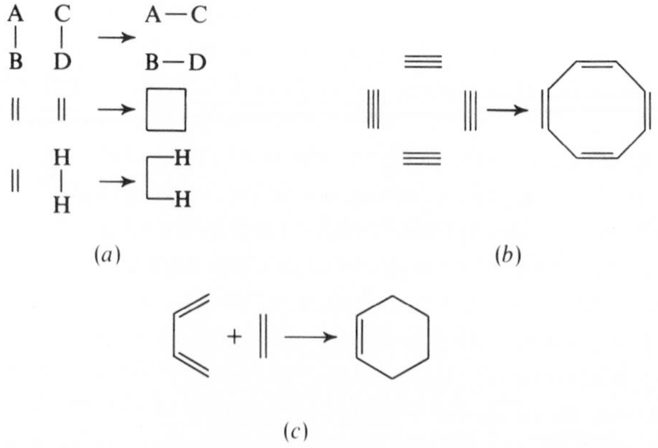

$$
\begin{array}{ccc}
\text{A} & \text{C} & \text{A}-\text{C} \\
| & | & \\
\text{B} & \text{D} & \text{B}-\text{D}
\end{array}
\rightarrow
$$

(a)

(b)

(c)

Fig. 11.27   Cycloaddition reactions. (a) Four-centered cycloadditions; (b) eight-centered cycloadditions.

equal numbers of π and σ bonds is thermally allowed and photochemically forbidden if the number of pπ electrons involved is $4n + 2$, where n is an integer; the reverse is true if the number is $4n$. Eight-centered cycloadditions, like the tetramerization of acetylene, Fig. 11.27b, are also known; this reaction is thermally forbidden but has been achieved catalytically for reasons analogous to those to be discussed below.

The cycloaddition reactions we have been considering are concerted reactions, that is, the bond-making and bond-breaking processes occur simultaneously. The great importance of the orbital symmetry conservation principle is apparent in Hoffman and Woodward's claim that orbital symmetry controls the feasibility and stereochemical consequences of every concerted reaction.

11.8   The Transition-Metal-Catalyzed Dimerization

In one of the most interesting applications of the selection rules of Woodward and Hoffmann, the dimerization of ethylene, forbidden in the ground-state thermal process, is found to be allowed if a transition metal is available to interact with the two ethylenes. In order to understand the catalysis, it is necessary to classify the metal orbitals according to their symmetry species in C_{2v} and to develop the MO level diagram of the $2C_2H_4 \cdot M$ complex. The nine metal orbitals, s, p_x, p_y, p_z, d_{xz}, d_{xy}, d_{yz}, $d_{x^2-y^2}$, and d_{z^2}, belong, respectively, to symmetry species a_1, b_1, b_2, a_1, b_1, a_2, b_2, a_1, and a_1. It will be noted that four of the five d orbitals belong to different species, and, in drawing our MO diagram, we need to show only these four. The MOED is given in Fig. 11.28. We show the d_{xz} and d_{z^2} as the highest d orbitals, raised in energy because of the ligand field of the π systems of the ethylenes. These two d orbitals are shown as empty, and hence they function as acceptor orbitals. The d_{yz} and d_{xy} orbitals, which are shown as filled, can act as the donor orbitals interacting with the π^* orbitals of the olefins in the back-donation process.

Now, the function of the catalyst, if it is to enable the reaction to proceed, is to transfer the electron pair initially in the $b_1(\pi)$ orbital (Fig. 11.29) to the $\sigma(b_2)$ orbital of the cyclobutane. The metal does this by an L $\rightarrow$ M (ligand-to-metal) interaction of its empty acceptor d_{xz} with the filled $b_1(\pi)$ orbital in the process of complexation. At the same time the empty π^* on the olefins of b_2 symmetry interacts with the filled d_{yz} of the metal to give the back-bonding M $\rightarrow$ L interaction. The formation of the complex and its subsequent decomposition transfer the electron pair originally in the d_{yz} to the newly formed σ bonding orbital b_2 of the cyclobutane. This electron-transfer process is shown in Fig. 11.29. The cyclobutane then departs from the metal, Fig. 11.29, and the catalyst is prepared to again complex with two moles of

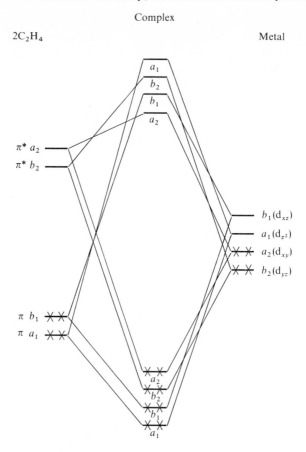

Fig. 11.28 The molecular orbital energy diagram for the transition-metal complex with 2 moles of ethylene.

olefin to again catalyze the dimerization. The liberation of cyclobutane and the metal from the complex, Fig. 11.30, results in the d_{yz} being empty and the d_{xz} filled, whereas in the initial complexing, Fig. 11.28, the d_{xz} is empty and the d_{yz} is shown as filled. This discrepancy should not bother us, however, because the labels on these two orbitals are a matter of convenience. The important feature is that a pair of electrons from the metal flows into the incipient cyclobutane b_2 bonding orbital, while a pair of electrons from the $b_1(\pi)$ olefin orbitals flows into an empty b_1 orbital of the metal.

Naturally the metal is not going to be present in the system as a free metal; it will be present as a soluble complex. The metal-catalyst complex, however, should be such that the b_1 and b_2 orbitals are not involved in bonding with

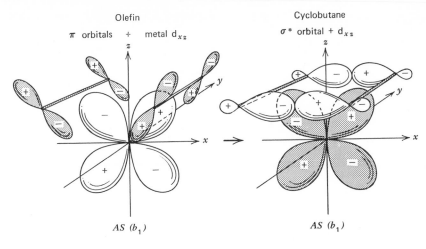

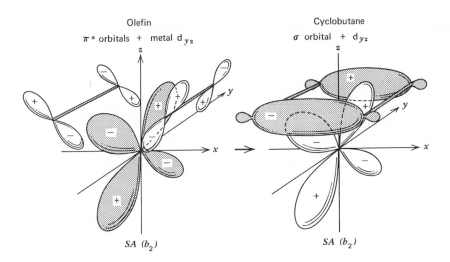

Fig. 11.29   The role of the metal orbitals; the *exchange* of electron pairs. [From F. Mango, *Advances in Catalysis*, Vol. 20, Academic Press, 1970, by permission.]

ligands other than the olefins and are thus available to perform the electron switch from reactant to product. Thus low valency iron, nickel, and cobalt catalysts have been used. Probably the catalyzed olefin disproportionation reaction, with its important industrial potential, also proceeds by the mechanism discussed:

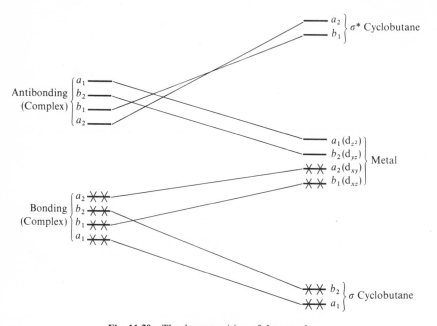

$$CH_3CH_2CH{=}CH_2$$

$$CH_3CH_2CH{=}CH_2$$

$$\rightleftharpoons \begin{bmatrix} CH_3CH_2CH{-}CH_2 \\ \\ CH_3CH_2CH{-}CH_2 \end{bmatrix} \rightleftharpoons$$

$$CH_3CH_2CH{=}CHCH_2CH_3$$

$$+$$

$$CH_2{=}CH_2$$

Fig. 11.30 The decomposition of the complex.

11.9 Sigmatropic Transformations

There are many examples in the literature of the migration of a hydrogen atom from one carbon atom in a conjugated chain to another:

$$\overset{H}{\diagdown}\overset{H}{\diagup}$$
$$C{-}(C{=}C)_n \rightarrow (C{=}C)_n{-}C$$

Such rearrangements have been called *sigmatropic rearrangements* because a σ bond, here a carbon-hydrogen σ bond, is made to undergo change or rearrangement. Such rearrangements are allowed or forbidden in accordance with the Woodward-Hoffmann rules of orbital symmetry. Again, as in electrocyclic transformations, the symmetry of the highest occupied orbital is the controlling factor. Here the orbital of interest is the highest occupied orbital of the radical fragment left by the migrating H atom.

Let us consider the migration of hydrogen to carbon 1 from carbon 5 in *cis*-1,3 pentadiene. This rearrangement is called a $[1,5]$ sigmatropic shift.

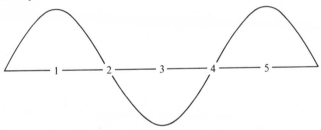

The radical fragment obtained by the removal of the H atom on carbon 5 leads to a pentadienyl radical possessing five $p\pi$ electrons. Accordingly, the highest occupied orbital is ψ_3. If we use our FEM procedures developed in Chapter 4, we can immediately determine that ψ_3 has an amplitude function represented by:

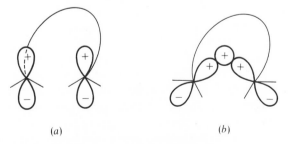

The amplitudes at the terminal C atoms have the same sign, and this means that the orbitals at the termini have the symmetry shown in Fig. 11.31a. The migration of the hydrogen 1s orbital from one of the terminal carbons to the other can be pictured as in Fig. 11.31b. It should be noticed that this transformation involves the disrotatory mode, and that a plane of symmetry is retained during the sigmatropic shift. It will be further noted that the

(a)   (b)

Fig. 11.31   A suprafacial sigma tropic shift

migration from origin to terminus occurs in a continuous path that remains on one face of the $p\pi$ system, and hence this process is designated *suprafacial*; suprafacial processes are allowed, and in this system the sigmatropic shift proceeds thermally.

Now let us consider a $[1,3]$ sigmatropic shift, using propylene as an example. We will focus on ψ_2 of the allyl system, which has the symmetry shown in Fig. 11.32a. The transfer of the 1s hydrogen orbital must then occur

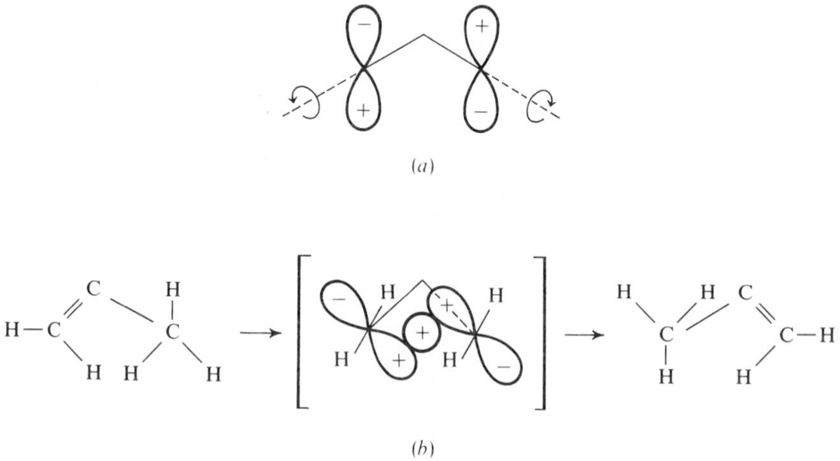

(a)

(b)

Fig. 11.32   An antarafacial sigmatropic shift.

as represented in Fig. 11.32b. It will be noted that the rotation of the termini is conrotatory and that a C_2 axis is preserved. However, here the pathway involves movement from one face to the opposite face of the $p\pi$ system, and accordingly this process is called *antarafacial*. Antarafacial processes are forbidden, and in this three carbon system, the $[1,3]$ migration does not proceed thermally.

The selection rules for sigmatropic rearrangements can be developed by the procedures described above. The $[1,3]$ process is forbidden thermally since it must proceed by an antarafacial process; it is allowed photochemically in molecules where ψ_3, having equal amplitudes at the termini, is the lowest unoccupied orbital in the ground state and the migration proceeds by a suprafacial process. The selection rules are reversed for the $[1,5]$ rearrangement.

The rearrangements known as the Claisen and Cope rearrangements are examples of sigmatropic shifts of order $[3,3]$, in which the σ bond being made and broken is a C-C rather than a C-H bond:

The sigmatropic change of order $[i,j]$ is defined as the migration of a σ bond (here the 3,4 bond), flanked by one or more π-electron systems, to a new position (the 1,6 bond) having termini $i - 1$ and $j - 1$ atoms (here $i = j = 3$) removed from the original bonded loci in an uncatalyzed intramolecular process. The number in brackets defines the position of the new σ bond in relation to the position of the one that was broken. Here the new bond terminates at positions two atoms removed from each of the termini of the old bond.

PROBLEMS

11.1   The natural fluorescence lifetime, $\tau_f{}^0$, of a molecule is defined as the time required for the fluorescence intensity to decrease to $1/e$ of the original intensity, providing of course that fluorescence is the only pathway for decay of the excited singlet state. For most molecules, other processes compete with fluorescence and the actual measured fluorescence lifetime τ_f, the mean lifetime, is always less than $\tau_f{}^0$. The relationship between the two lifetimes is

$$\tau_f{}^0 = \frac{\tau_f}{\Phi_f} \tag{1}$$

where Φ_f is the quantum yield for fluorescence. (*a*) Assuming that the reactions shown below are the only ones occurring, derive expressions for τ_f and $\tau_f{}^0$ in terms of the rate constants given.

Process	Rate
$S_0 + h\nu \longrightarrow S_1$	$I_a \phi_i$
$S_1 \longrightarrow S_0 + h\nu$	$k_1[S_1]$
$S_1 \longrightarrow S_0 + \text{heat}$	$k_2[S_1]$
$S_1 \longrightarrow T_1$	$k_3[S_1]$
$T_1 \longrightarrow S_0 + \text{heat}$	$k_4[T_1]$

where I_a is the intensity of the incident light, and ϕ_i is the quantum yield for absorption. (*b*) Verify equation (1).

11.2   Calculate the fluorescence intensity of a $5 \times 10^{-5} M$ solution of anthracene in ethanol contained in a 1 cm cell when the solution is excited with 366 nm light (Hg vapor), having an intensity, I_0, of 6×10^{16} quanta/cm^3 sec. The molar absorptivity of anthracene at this wavelength is 4600 and ϕ_f is 0.36.

11.3   The quenching of the fluorescence of a substance A by a quencher Q is usually treated experimentally by constructing a Stern-Volmer plot, after collecting certain essential data and assuming that the excited singlet $S_1(A)$ is populated and disappears by the following process:

Process	Rate
$S_0 + hv \longrightarrow S_1(A)$	$I_a \phi_i$
$S_1(A) \longrightarrow S_0(A) + hv$	$k_1[S_1]$
$S_1(A) \longrightarrow S_0(A) + \text{heat}$	$k_2[S_1]$
$S_1(A) \longrightarrow T_1(A)$	$k_3[S_1]$
$S_1(A) + Q \longrightarrow S_0(A) + Q^*$	$k_4[S_1][Q]$

The equation of the Stern-Volmer plot is

$$\frac{\phi_f{}^0}{\phi_f} = 1 + \frac{k_4[Q]}{k_1 + k_2 + k_3}$$

[see F. D. Lewis and J. C. Dalton, *J. Amer. Chem. Soc.*, **91**, 5260 (1969)].
where $\phi_f{}^0$ = fluorescence efficiency of A in the absence of quenching;
      ϕ_f = fluorescence efficiency of A in the presence of quencher Q.
(a) Derive the equation and (b) calculate k_4 in the example of the quenching of phenanthrene fluorescence by hexyl azide where the mean fluorescence lifetime $1/(k_1 + k_2 + k_3)$ in the absence of quencher is 5.5×10^{-9} sec and the following ratios of $\Phi_f{}^0/\Phi_f$ at different molar concentrations $[Q]$ of hexyl azide were found:

$[Q]$	$\phi_f{}^0/\phi_f$
0.01	1.32
0.032	2.02
0.067	3.14

11.4   From the absorption and emission spectrum of anthanthrene shown in Fig. 11.6, determine the natural radiative lifetime of the lowest excited singlet state (a) using the $0 \to 0$ band. (b) What would be the effect on the result of the entire band, rather than if the $0 \to 0$ band were used, in the calculation.

11.5   Although the photoreduction of benzophenones and alkyl phenyl ketones to pinacols by secondary alcohols proceeds in quite good yield, similar reduction by 2-acetylnaphthalene goes in very poor yield. Suggest a possible reason for the difference. (*Hint:* Even though the lowest triplet of benzophenone is 69 kcal/mole and that of 2-acetylnaphthalene is only 59 kcal/mole above their respective ground states, the difference in energy probably does not account for the difference in reactivity.)

11.6   The Diels-Alder reaction between ethylene and butadiene is a common example of a cycloaddition reaction that is thermally allowed by the orbital symmetry rules. Develop the orbital symmetry correlations between reactants and product and then develop the state correlation diagrams.

11.7   It is well known that *cis*-stilbene undergoes a photocyclization to dihydrophenanthrene:

The HMO coefficients for the eight lowest energy MO's of stilbene and dihydrophenanthrene are given in Tables 11.4 and 11.5 opposite. Some authors have attempted to correlate

TABLE 11.4
HMO Coefficients for Stilbene

ψ_i	c_1	c_2	c_3	c_4	c_5	c_6	c_7	c_8	c_9	c_{10}	c_{11}	c_{12}	c_{13}	c_{14}
ψ_1	0.264	0.202	0.183	0.202	0.264	0.381	0.315	0.315	0.381	0.264	0.202	0.183	0.202	0.264
ψ_2	0.291	0.265	0.257	0.265	0.291	0.335	0.109	-0.109	-0.335	-0.291	-0.265	-0.257	-0.265	-0.291
ψ_3	0.049	0.277	0.369	0.277	0.049	-0.204	-0.404	-0.404	-0.204	0.049	0.277	0.369	0.277	0.049
ψ_4	0.135	-0.235	-0.407	-0.235	0.135	0.392	0.182	-0.182	-0.392	-0.135	0.235	0.407	0.235	-0.135
ψ_5	0.354	0.354	0.000	-0.354	-0.354	0.000	0.000	0.000	0.000	0.354	0.354	0.000	-0.354	-0.354
ψ_6	0.354	0.354	0.000	-0.354	-0.354	0.000	0.000	0.000	0.000	-0.354	-0.354	0.000	0.354	0.354
ψ_7	0.274	-0.079	-0.314	-0.079	0.274	0.217	-0.438	-0.438	0.217	0.274	-0.079	-0.314	-0.079	0.274
ψ^*_8	0.274	0.079	-0.314	0.079	0.274	-0.217	-0.438	0.438	0.217	-0.274	-0.079	0.314	-0.079	-0.274

TABLE 11.5
HMO Coefficients for Dihydrophenanthrene

ψ_i	c_1	c_2	c_3	c_4	c_5	c_6	c_7	c_8	c_9	c_{10}	c_{11}	c_{12}
ψ_1	0.094	0.182	0.260	0.323	0.367	0.389	0.389	0.367	0.323	0.260	0.182	0.094
ψ_2	0.182	0.323	0.389	0.367	0.260	0.094	-0.094	-0.260	-0.367	-0.389	-0.323	-0.182
ψ_3	0.260	0.389	0.323	0.094	-0.182	-0.367	-0.367	-0.182	0.094	0.323	0.389	0.260
ψ_4	0.323	0.367	0.094	-0.260	-0.389	-0.182	0.182	0.389	0.260	-0.094	-0.367	-0.323
ψ_5	0.367	0.260	-0.182	-0.389	-0.094	0.323	0.323	-0.094	-0.389	-0.182	-0.260	-0.367
ψ_6	0.389	0.094	-0.367	-0.182	0.323	0.260	-0.260	-0.323	0.182	0.367	-0.094	-0.389
ψ^*_7	0.389	-0.094	-0.367	0.182	0.323	-0.260	-0.260	0.323	0.182	-0.367	-0.094	0.389
ψ^*_8	0.367	-0.260	-0.182	0.389	-0.094	-0.323	0.323	0.094	-0.389	0.182	0.260	-0.367

a high free valence index at particular positions in a molecule in the excited state with ease of photochemical cyclization into these positions. (a) Calculate the free valence index at positions 1 and 14, in stilbene, in the ground and first excited state. (b) The figure shows that photocyclization results in *trans* stereochemistry of the site of ring closure. Rationalize this conclusion. (c) Construct the orbital correlation diagram and the state correlation diagram for the conrotatory and disrotatory cyclization of stilbene to dihydrophenanthrene.

11.8   The ring opening reaction of a cyclopropyl cation to an allyl cation should be thermally allowed and disrotatory according to the Woodward-Hoffmann rules (this is a $4n + 2$ system where $n = 0$). The acetoylsis of the two possible *cis*-1,2-dimethylcyclopropyl-3-tosylates, both of which presumably proceed via the corresponding allyl cation, exhibits enormously different rates:

TsO △	TsO (2) △ CH₃ (with CH₃, 3, 1)	CH₃ △ CH₃ (TsO)

Relative rate: 1.0   4.0   18,000

In the disrotatory ring opening of the dimethyl compounds, the two methyl groups may rotate either toward each other (inward disrotatory) or away from each other (outward disrotatory). From the relative rate data it was concluded that, of the two possible disrotatory modes of ring opening at the 1,2-bond, the process in which this bond is broken by disrotatory outward motion would account for the difference in rate. Explain. [Hint: The breaking of the 1,2-bond by rotation in a direction which places electron density from this bond on the side of the ring opposite that of the leaving group assists the departure of the leaving group and enhances the solvolysis rate. See R. B. Woodward and R. H. Hoffmann, *The Conservation of Orbital Symmetry*, Academic Press, 1970, p. 47; C. H. DePuy, *Accts. Chem. Res.*, **1**, 33 (1968).]

GENERAL REFERENCES

1. J. G. Calvert and J. N. Pitts, Jr., *Photochemistry*, John Wiley and Sons, New York, 1966.
2. N. J. Turro, *Molecular Photochemistry*, W. A. Benjamin, New York, 1965.
3. M. Orchin and H. H. Jaffé, *The Importance of Antibonding Orbitals*, Houghton Mifflin Company, Boston, 1967.
4. F. Mango in *Advances in Catalysis*, Vol. 20, Academic Press, New York, 1970.
5. R. B. Woodward and R. Hoffmann, "The Conservation of Orbital Symmetry," Academic Press (1970).

12 Methods beyond the Hückel Molecular Orbital

In Chapter 10 we have carefully examined the HMO method of obtaining molecular orbitals for π-electron systems. Although this method is simple and straightforward, it is extremely approximate and is in principle restricted to π-electron systems, and consequently to planar (or nearly planar) conjugated and aromatic systems. We must now examine the extensions in various directions. This will require us to look into less approximate methods and into methods applicable to nonplanar systems.

12.1 Improved Hückel Molecular Orbital-Like Methods

As soon as it was recognized that the HMO method could be used for correlating and even predicting many properties of conjugated systems, but that the method was not infallible, or even as general as desired for π-electron systems, investigators searched for improvements. Since the HMO method involves many crude and unjustifiable approximations, they were eliminated one at a time, in the hope of improving the method. None of these modified treatments, however, was very successful.

The first of these "improvements" was not to neglect overlap integrals. The mathematical problem of solving the secular equation in the form of equation 10.9 is quite time-consuming, and is met in the manner indicated in Section 12.5 in connection with the extended Hückel method. However, it was soon realized that no significant improvement is achieved thereby, and this line of investigation was abandoned.

The problem of parametrization, and in particular of making Coulomb and resonance integrals self-consistent, has received more attention than any other aspect of the HMO method. Thus it was recognized early that there must be an interrelation between resonance integrals, bond lengths, and bond orders, and many recipes for making β_{ij} a function of the calculated p_{ij}—usually a proportionality—have been proposed. Similarly, it was recognized that the Coulomb integral for an atomic orbital ϕ_r of atom r, which is believed to depend on the energy required to remove an electron from this orbital, must be a function of the electron density, q_r, on this atom. The most elaborate treatments of this problem are contained in Streitwieser's ω-technique and R. D. Brown's variable electronegativity self-consistent field (VESCF) method. Unfortunately, these techniques require iterative procedures: solution of the secular equation and calculation of q's and p's with an initial set of α's and β's, calculation of a new set of α's and β's from the q's and p's obtained, and repetition of this process until self-consistency is attained, that is, until the q's and p's calculated in one iteration do not differ from those obtained in the previous iteration. Even worse, the iterative process does not always converge; all too frequently cases occur in which no self-consistent solution can be found by the method used.

12.2   Self-Consistent Field (SCF) Methods

In order to proceed to even more detailed methods, we must return to the Schrödinger equation:

$$H\Psi = E\Psi \tag{12.1}$$

and examine the nature of the Hamiltonian operator more closely. We saw in Chapter 10 that this operator represents the kinetic and potential energy terms:

$$H = T + V \tag{12.2}$$

In quantum mechanics, the kinetic energy is given by the sum of all the one-electron differential operators (in atomic units):[a]

$$T = -\tfrac{1}{2}\sum_i \nabla_i^2 \tag{12.3}$$

where the summation extends over all electrons, and ∇^2 represents the Laplacian operator, in Cartesian coordinates, given by:

$$\nabla^2 = \frac{\partial^2}{\partial x^2} + \frac{\partial^2}{\partial y^2} + \frac{\partial^2}{\partial z^2} \tag{12.4}$$

For the present purpose, the potential energy, V, involves only the electrostatic attractions and repulsions of nuclei and electrons:[b]

$$V = \sum_{r<s} \frac{Z_r Z_s}{r_{rs}} - \sum_{r,i} \frac{Z_r}{r_{ri}} + \sum_{i<j} \frac{1}{r_{ij}} \tag{12.5}$$

Here Z_r and Z_s are the nuclear charges on r and s, and r_{xy} is the distance between particles x and y, be they nuclei (r, s) or electrons (i, j). It is the last

[a] In Atomic units, the unit of energy is the hartree: 1 hartree = 27.204 eV; the unit of mass is the mass of an electron, and its charge is the unit of charge.

[b] Throughout the following discussion, we shall use subscripts $i, j, \ldots$ for electrons; $r, s, \ldots$ for atoms (nuclei); $\mu, v, \ldots$ for molecular orbitals; and $\rho, s, \tau, v, \ldots$ for atomic orbitals.

term in V that causes most of the real problems in molecular quantum mechanics. Substituting equation 12.2–12.5 into 12.1, we obtain the following set of coupled differential equations:

$$\left\{ -\frac{1}{2} \sum_i \left[\left(\frac{\partial^2}{\partial x_i^2} + \frac{\partial^2}{\partial y_i^2} + \frac{\partial^2}{\partial z_i^2} \right) + \sum_r \frac{Z_r}{r_{ri}} - \sum_{j>i} \frac{1}{r_{ij}} \right] + \sum_{r<s} \frac{Z_r Z_s}{r_{rs}} - E \right\} \Psi$$

$$= 0 \qquad (12.6)$$

If the last term in the square bracket, the sum over the terms in $1/r_{ij}$, is neglected, Ψ is a product of one-electron functions or orbitals, $\psi_v(i)$,

$$\Psi = \prod_i \psi_v(i) \qquad (12.7)$$

and this set of equations is separable into a set of independent differential equations:

$$\left(\frac{\partial^2}{\partial x_i^2} + \frac{\partial^2}{\partial y_i^2} + \frac{\partial^2}{\partial z_i^2} + \sum_r \frac{Z_r}{r_{ri}} \right) \psi_v(i) = \left(E - \sum_{r<s} \frac{Z_r Z_s}{r_{rs}} \right) \psi_v(i) \quad (12.8)$$

These equations, ignoring the constant summation term on the right-hand side, are, in effect, the equations solved in the HMO method (and the EHMO method; cf. Section 12.5), with the parentheses on the left side identified with H. One hopes that the semiempirical parameters absorb much of the neglect of the $1/r_{ij}$ terms, at least in an averaged fashion. These neglected terms represent the electron-electron repulsions, and their inclusion is crucial.

One more omission has been made to this point. In writing equation 12.7 we have implied two things: (1) that electron spin is included in $\psi_v(i)$, that is, that the ψ_v are *spin orbitals*, and (2) that the ith electron occupies $\psi_v(i)$. The first of these needs to be recognized but produces few problems; the second, however, implies that we can distinguish electrons. Since electrons are, in fact, indistinguishable, it is necessary that the physically observable quantity, that is, the probability function Ψ^2, be unchanged if any pairs of electrons are exchanged. This is so if Ψ either remains unchanged or changes sign only when any two electrons are exchanged. Only the latter, that is, antisymmetric, behavior of electronic wave functions can apply to systems in the physical world we know, and this fact represents the most fundamental statement of the Pauli principle. If, for a two-electron system, we designate an orbital ψ_1 occupied by electron 2 as $\psi_1(2)$, the Pauli principle is not obeyed by either $\Psi = \psi_1(1)\psi_2(2)$ or $\Psi = \psi_1(2)\psi_2(1)$, but the function

$$\psi = \frac{1}{\sqrt{2}} [\psi_1(1)\psi_2(2) - \psi_1(2)\psi_2(1)]$$

does obey the requirement.

This is an expansion of the determinant

$$\psi = \frac{1}{\sqrt{2}} \begin{vmatrix} \psi_1(1) & \psi_2(1) \\ \psi_1(2) & \psi_2(2) \end{vmatrix}$$

Such determinants, which are called *Slater determinants*, always satisfy the Pauli principle. In the rest of this chapter, a many-electron wave function, Ψ, will always refer to a Slater determinant, where the ψ_ν are spin orbitals, that is, products of space and spin functions:

$$\psi = \frac{1}{\sqrt{n!}} \begin{vmatrix} \psi_1(1) & \psi_1(2) & \psi_1(3) & \dots & \psi_1(n) \\ \psi_2(1) & \psi_2(2) & \psi_2(3) & \dots & \psi_2(n) \\ \vdots & \vdots & \vdots & \vdots & \vdots \\ \psi_n(1) & \psi_n(2) & \psi_n(3) & \dots & \psi_n(n) \end{vmatrix}$$

$$= \frac{1}{\sqrt{n!}} \, |\psi_1(1)\psi_2(2) \ \dots \ \psi_n(n)| \tag{12.9}$$

where equation 12.9 is a shorthand form of writing a determinant by placing the leading term of the expansion between bars.

We must now return to equation 12.6 and use it to evaluate the energy. This is done in the usual fashion by premultiplying by Ψ and integrating over all space. In order to simplify the calculations, we shall separate the Hamiltonian into parts and evaluate them separately. Thus,

$$H = \sum_i h^{(i)} + \sum_{i<j} \frac{1}{r_{ij}} + \sum_{r<s} \frac{Z_r Z_s}{r_{rs}} \tag{12.10}$$

The first set of terms, $h^{(i)}$, are the one-electron Hamiltonians, since each depends only on the coordinates of a single electron:

$$h^{(i)} = -\tfrac{1}{2}\nabla_i^2 - \sum_r \frac{Z_r}{r_{ir}} \tag{12.11}$$

and includes the kinetic energy and electron-nuclear attraction terms. The evaluation of

$$\int \Psi \sum_i h^{(i)} \Psi \, d\tau$$

is relatively simple because, when substituting in equation 12.9,

$$\frac{1}{n!} \int |\psi_1(1)\psi_2(2) \dots \psi_n(n)| \sum_i h^{(i)} |\psi_1(1)\psi_2(2) \dots \psi_n(n)| \, d\tau \tag{12.12}$$

we observe the following. First, in this type of expression we can always drop the first determinant, use only its leading term, and multiply by $n!$, since each other term will give exactly the same result $n!$ times. Then equation 12.12 becomes

$$\int \psi_1(1)\psi_2(2) \dots \psi_n(n) \sum_i h^{(i)} |\psi_1(1)\psi_2(2) \dots \psi_n(n)| \, d\tau \qquad (12.12a)$$

We shall evaluate equation 12.12a by breaking it up into many individual terms. First, if we take out one term of the sum, for example, $i = 2$, and the leading term in the remaining determinant, equation 12.9 yields one integral which factors into a product of n separate one-electron integrals:

$$\int \psi_1(1)\psi_2(2) \dots \psi_n(n) h^{(2)} \psi_1(1)\psi_2(2) \dots \psi_n(n) \, d\tau$$
$$= \int \psi_1(1)\psi_1(1) \, d\tau \int \psi_2(2) h^{(2)} \psi_2(2) \, d\tau_2 \dots \int \psi_n(n)\psi_n(n) \, d\tau_n$$

The integrals over $d\tau_i$, $i \neq 2$, are all normalization integrals and hence equal to 1. This leaves

$$\int \psi_2(2) h^{(2)} \psi_2(2) \, d\tau_2 = h_2 \qquad (12.13)$$

Now we need to take the various permuted terms of the determinant. One such term is

$$- \int \psi_1(1)\psi_2(2) \dots \psi_n(n) h^{(2)} \psi_1(2)\psi_2(1) \dots \psi_n(n) \, d\tau$$
$$= - \int \psi_1(1)\psi_2(1) \, d\tau_1 \int \psi_2(2) h^{(2)} \psi_1(2) \, d\tau_2 \dots \int \psi_n(n)\psi_n(n) \, d\tau_n$$

Here we must use a minus sign since one pair of electrons is exchanged; the first factor is an orthogonality integral, and hence the whole term vanishes. This is true for each of these permuted terms, since there is always at least one orthogonality integral. In order to evaluate equation 12.12 completely, we must repeat the process for each i in the central sum, and each will reduce analogously to an integral of the form of 12.13. Thus this part of the integration leads to a term in the final result of

$$\sum_v h_v = \sum_i \int \psi_v(i) h^{(i)} \psi_v(i) \, d\tau_i$$

The last term in the equation 12.10 is independent of the coordinates of *all* electrons (depending only on nuclear coordinates), and hence, on premultiplication by Ψ and integration,

$$\int \Psi \sum_{r<s} \frac{Z_r Z_s}{r_{rs}} \Psi \, d\tau = \sum_{r<s} \frac{Z_r Z_s}{r_{rs}} \int \Psi^2 \, d\tau = \sum_{r<s} \frac{Z_r Z_s}{r_{rs}} \qquad (12.14)$$

leads only to a constant term (representing the nuclear repulsions).

Finally, however, the middle set of terms in equation 12.10 gives the most trouble:

$$\int \psi_1(1)\psi_2(2)\psi_3(3) \dots \psi_n(n) \sum_{i<j} \frac{1}{r_{ij}} |\psi_1(1)\psi_2(2)\psi_3(3) \dots \psi_n(n)| \, d\tau \quad (12.15)$$

Let us again remove one term, $i = 2$, $j = 3$, and the leading term of the determinant:

$$\int \psi_1(1)\psi_2(2)\psi_3(3) \dots \psi_n(n) \frac{1}{r_{23}} \psi_1(1)\psi_2(2)\psi_3(3) \dots \psi_n(n) \, d\tau_n$$

$$= \int \psi_1(1)\psi_1(1) \, d\tau_1 \int \int \psi_2(2)\psi_3(3) \frac{1}{r_{23}} \psi_2(2)\psi_3(3) \, d\tau_2 \, d\tau_3 \dots \int \psi_n(n)\psi_n(n) \, d\tau_n$$

Here again the first and the last integrals are normalization integrals, and consequently equal to 1. The middle integral is called J_{23}:

$$J_{\mu\nu} = \int \int \psi_\mu(i)\psi_\nu(j) \frac{1}{r_{ij}} \psi_\mu(i)\psi_\nu(j) \, d\tau_i \, d\tau_j \qquad (12.16)$$

and will be considered later.

If we take one particular permutation of the leading term of equation 12.15:

$$-\int \psi_1(1)\psi_2(2)\psi_3(3) \dots \psi_n(n) \frac{1}{r_{23}} \psi_1(1)\psi_2(3)\psi_3(2) \dots \psi_n(n) \, d\tau$$

$$= -\int \psi_1{}^2(1) \, d\tau_1 \int \int \psi_2(2)\psi_3(3) \frac{1}{r_{23}} \psi_2(3)\psi_3(2) \, d\tau_2 \, d\tau_3 \dots \int \psi_n{}^2(n) \, d\tau_n$$

again, all integrals but the center are unity, and we have

$$K_{\mu\nu} = \int \int \psi_\mu(i)\psi_\nu(j) \frac{1}{r_{ij}} \psi_\mu(j)\psi_\nu(i) \, d\tau_i \, d\tau_j \qquad (12.17)$$

Next we take all other permutations, for example,

$$-\int \psi_1(1)\psi_2(2)\psi_3(3) \dots \psi_n(n) \frac{1}{r_{23}} \psi_1(3)\psi_2(2)\psi_3(1) \dots \psi_n(n) \, d\tau$$

$$= -\int \psi_1(1)\psi_3(1) \, d\tau_1 \int \int \psi_2(2)\psi_3(3) \frac{1}{r_{23}} \psi_1(3)\psi_2(2) \, d\tau_2 d\tau_3 \dots \int \psi_n{}^2(n) \, d\tau_n$$

Once more, the first integral vanishes since it is an orthogonality integral; it is readily verified that any other permutation also has at least one such integral, and therefore they all vanish. Thus, the middle term of equation 12.10 gives us in the integration

$$\sum_{\mu < \nu} (J_{\mu\nu} - K_{\mu\nu})$$

and the energy becomes

$$E = \sum_\nu h_\nu + \sum_{\mu < \nu} (J_{\mu\nu} - K_{\mu\nu}) + \sum_{r < s} \frac{Z_r Z_s}{r_{rs}} \qquad (12.18)$$

We have now expressed the total energy, E, in terms of three types of *molecular integrals*, h_ν, $J_{\mu\nu}$, and $K_{\mu\nu}$. We must next examine their evaluation in terms of the basis functions (atomic orbitals) we wish to use. The ψ_ν are approximated by a LCAO formulation,

$$\psi_\nu(i) = \left[\sum_\rho c_{\nu\rho}\phi_\rho(i) \right] \sigma_\nu(i)$$

where $\sigma_\nu(i)$ is the spin function (either α or β) since the $\psi_\nu(i)$ represent spin orbitals. In making this substitution, equation 12.13 becomes

$$h_\nu = \int \psi_\nu(i) \left(-\tfrac{1}{2}\nabla_i^2 - \sum_r \frac{Z_r}{r_{ri}} \right) \psi_\nu(i)\, d\tau_i$$

$$= \int \sum_\rho c_{\nu\rho}\phi_\rho(i) \left(-\tfrac{1}{2}\nabla_i^2 - \sum_r \frac{Z_r}{r_{ri}} \right) \sum_\sigma c_{\nu\sigma}\phi_\sigma(i)\, d\tau_i \int \sigma_\nu{}^2(i)\, d\sigma_i \qquad (12.19)$$

The spin integral (the last factor) is unity. Rearranging, we obtain

$$h_\nu = \sum_\rho \sum_\sigma c_{\nu\rho} c_{\nu\sigma} \int \phi_\rho(i) \left(-\tfrac{1}{2}\nabla_i^2 - \sum_r \frac{Z_r}{r_{ri}} \right) \phi_\sigma(i)\, d\tau_i$$

We now shall classify the integrals occurring in this triple sum into two different types:

(1) $$H_{\rho\rho} = \iint \phi_\rho(i) \left(-\tfrac{1}{2}\nabla_i^2 - \frac{Z_r}{r_{ri}} \right) \phi_\rho(i)\, d\tau_i + \int \phi_\rho(i) \left(-\sum_{s \neq r} \frac{Z_s}{r_{si}} \right) \phi_\rho(i)\, d\tau_i$$

Here, the second term is a sum of penetration integrals, $P_{\rho s}$, which represent the attraction of nuclei s for the charge distribution represented by $\phi_\rho{}^2$ (which is assumed centered on nucleus r). These terms are small and are often neglected. The first term in $H_{\rho\rho}$ is an atomic integral, which is usually assumed to have the same value that it has in the free atom:

(2) $$H_{\rho\sigma} = \int \phi_\rho(i) \left(-\tfrac{1}{2}\nabla_i^2 - \sum_r \frac{Z_r}{r_{ri}} \right) \phi_\sigma(i)\, d\tau_i$$

If ρ and σ are centered on the same atom, $H_{\rho\sigma}$ vanishes identically; otherwise, $H_{\rho\sigma}$ is known as a resonance integral and provides the binding force. The evaluation of $J_{\mu\nu}$ (equation 12.16) gives

$$J_{\mu\nu} = \iint \sum_\rho c_{\mu\rho}\phi_\rho(i) \sum_\sigma c_{\nu\sigma}\phi_\sigma(j) \frac{1}{r_{ij}} \sum_\tau c_{\mu\tau}\phi_\tau(i) \sum_\nu c_\mu \phi(j)\, d\tau_i\, d\tau_j \int \sigma_\mu{}^2(i)\sigma_\nu{}^2(j)\, d\sigma$$

$$(12.20)$$

Again the spins integrate to unity. Rearranging, we have

$$J_{\mu\nu} = \sum_\rho \sum_\sigma \sum_\tau \sum_\nu c_{\mu\rho} c_{\nu\sigma} c_{\mu\tau} c_\nu \iint \phi_\rho(i)\phi_\sigma(j) \frac{1}{r_{ij}} \phi_\tau(i)\phi(j)\, d\tau_i\, d\tau_j \quad (12.21)$$

Here, we have several possibilities. If $\rho = \sigma = \tau = \upsilon$, we have a one-center integral called $\Gamma_{\rho\rho}$.

$$\Gamma_{\rho\rho} = \int\int \phi_\rho^2(i)\phi_\rho^2(j)\frac{1}{r_{ij}}\,d\tau_i\,d\tau_j$$

If $\rho = \tau$, $\sigma = \upsilon$, but $\rho \neq \sigma$, we have a Coulomb repulsion integral (not to be confused with the Coulomb integral of HMO), called $\Gamma_{\rho\sigma}$:

$$\Gamma_{\rho\sigma} = \int\int \phi_\rho^2(i)\phi_\sigma^2(j)\,d\tau_i\,d\tau_j$$

which may involve two centers if ρ and σ are centered on different atoms, or one center if they are centered on the same atom. Both these integrals represent physically the repulsion between a charge distribution in ϕ_ρ, on the one hand, and in either ϕ_ρ or ϕ_σ, respectively, on the other hand. Another integral occurs when $\rho = \sigma$, $\tau = \upsilon$, but this is an *exchange integral*:

$$\int\int \phi_\rho(i)\phi_\sigma(j)\frac{1}{r_{ij}}\phi_\rho(j)\phi_\sigma(i)\,d\tau_i\,d\tau_j$$

it also may be a one- or two-center integral.

We shall now introduce a shorthand notation for these integrals: each integral in equation 12.21 in general is written as

$$\langle\rho\tau|\sigma\upsilon\rangle$$

where the left half of the expression contains the subscripts of the ϕ's of electron i, the right half those of electron j. Also, it is readily shown that:

$$\langle\rho\tau|\sigma\upsilon\rangle = \langle\tau\rho|\sigma\upsilon\rangle = \langle\rho\tau|\upsilon\sigma\rangle = \langle\tau\rho|\upsilon\sigma\rangle = \langle\sigma\upsilon|\rho\tau\rangle, \text{ etc.}$$

With this notation we have $\Gamma_{\rho\rho} = \langle\rho\rho|\rho\rho\rangle$, $\Gamma_{\rho\sigma} = \langle\rho\rho|\sigma\sigma\rangle$, and the exchange integral $\langle\rho\sigma|\rho\sigma\rangle$. Continuing to enumerate the integrals that we obtain, we have, in addition to the *two-center* integral $\langle\rho\rho|\rho\sigma\rangle$, two *three-center* integrals $\langle\rho\rho|\sigma\tau\rangle$ and $\langle\rho\sigma|\rho\tau\rangle$ and one type of *four-center* integral $\langle\rho\sigma|\tau\upsilon\rangle$. Although we indicate these integrals as two-, three-, and four-center integrals, the number of centers may be smaller if several of the atomic orbitals are centered on the same atom.

In attempting to evaluate the integral $K_{\mu\nu}$ (equation 12.17), we run into a new complication because of the spin functions; this integral, after the same rearrangement as above, becomes

$$K_{\mu\nu} = \sum_\rho\sum_\sigma\sum_\tau\sum_\upsilon c_{\mu\rho}c_{\nu\sigma}c_{\mu\tau}c_{\nu\upsilon}\int\int \phi_\rho(i)\phi_\sigma(j)\frac{1}{r_{ij}}\phi_\tau(j)\phi_\upsilon(i)\,d\tau_i\,d\tau_j$$

$$\times \int \sigma_\mu(i)\sigma_\nu(i)\,d\sigma_i \times \int \sigma_\mu(j)\sigma_\nu(j)\,d\sigma_j \qquad (12.22)$$

The terms are almost exactly the same as in $J_{\mu\nu}$ (except for the reversal of electrons i and j in the double integral), but the integration over spins now is an integration for each electron, over the product of the spin parts of the two spin orbitals. If the spins of ψ_μ and ψ_ν are equal, both spin integrals integrate to unity and no problem arises. If the spins, on the other hand, are opposite, the spin integrals integrate to zero, and $K_{\mu\nu}$ vanishes! We will often write $K_{\mu\nu}$ only for the space part of equation 12.22, and add $\delta_{\mu\nu}$, which in this connection is 1 if the spins of μ and ν are alike, and is 0 if they are not. The integrals over AO's required for an energy evaluation are summarized in Table 12.1.

TABLE 12.1
Types of Integrals Required

$\int \phi_\rho \phi_\sigma \, d\tau$ Overlap integral

$$\int \phi_\rho(i)\left(-\tfrac{1}{2}\nabla_i - \sum_s \frac{Z_s}{r_{si}}\right)\phi_\rho(i)\, d\tau_i = H_{\rho\rho}$$

$\int \phi_\rho(i) h^{(i)} \phi_\sigma(i)\, d\tau_i = H_{\rho\sigma}$ Resonance integral

$\int \phi_\rho(i) \dfrac{Z_s}{r_{si}} \phi_\rho(i)\, d\tau_i = H_{\rho s}$ Penetration integral

$\langle \rho\rho | \rho\rho \rangle = \Gamma_{\rho\rho}$ One-center Coulomb repulsion integral

$\langle \rho\rho | \sigma\sigma \rangle = \Gamma_{\rho\sigma}$ Two-center Coulomb repulsion integral

$\langle \rho\sigma | \rho\sigma \rangle$ Exchange integral

$\langle \rho\rho | \rho\sigma \rangle$ Hybrid integral

$\langle \rho\sigma | \sigma\tau \rangle$ Three-center integrals

$\langle \rho\sigma | \tau\tau \rangle$

$\langle \rho\sigma | \tau\nu \rangle$ Four-center integral

The energy expression of equation 12.18, with the proper substitutions, represents the energy of the system. As in HMO, we use the variational principle to find the best energy. Along with this energy we determine the set of coefficients $c_{\mu\rho}$ describing the MO's which combine into the Slater determinant best describing the molecule. Unfortunately, however, when we perform the differentiations, we do not obtain a solvable set of equations, since the equations are no longer linear. In other words, to properly evaluate the potential for any one electron, we must have advance knowledge of the distribution of all the other electrons. The procedure then is to start with an initial guess of MO's (HMO frequently provides this guess), calculate all necessary integrals, and perform an energy minimization, obtaining a new set of MO's in the process. This set of steps is repeated until no further changes occur. (Some authors use energy, others use wave functions, as a criterion of convergence.) This method, here outlined crudely, is due to Roothaan and is called the method of the *self-consistent field* (SCF).

12.3　Self-Consistent Field Methods for π Electrons (Pariser-Parr-Pople)

With the formulation of SCF methods, it became immediately desirable to combine their greater meaningfulness with the simplicity of the σ-π separation which has proved so useful in the HMO method. This combination was achieved, in somewhat different ways, by Pariser and Parr and by Pople, and a combination of their methods has become widely known and applied as the PPP or P^3 method. This method starts from the SCF formulation of Section 12.2, but replaces the nuclei in the Hamiltonian by cores, which consist of the complete atom with all of its electrons except the π electrons. The interactions of the core electrons (which include all inner-shell, all lone-pair, and all σ electrons of the atom) with the nuclei and with each other are included in the $H_{\rho\rho}$, $H_{\rho\sigma}$, and penetration integrals without being explicitly considered. Interactions between σ and π electrons are also included in an average manner in $H_{\rho\rho}^{\text{core}}$ and $H_{\rho\sigma}^{\text{core}}$, which are given the superscripts to indicate these facts.

Another, new approximation, analogous to the neglect of overlap integrals in HMO, is introduced under the name of *neglect of differential overlap*. This new approximation neglects the product $\phi_\rho\phi_\sigma$ any time that $\rho \neq \sigma$, not only its integral over all space. In this approximation, then, all integrals involved in the evaluation of $J_{\mu\nu}$ and $K_{\mu\nu}$ are neglected except $\Gamma_{\rho\rho}$ and $\Gamma_{\rho\sigma}$ (this is the reason why these integrals have been given special symbols). Also, values for $H_{\rho\rho}^{\text{core}}$ and $H_{\rho\sigma}^{\text{core}}$, $\Gamma_{\rho\rho}$ and $\Gamma_{\rho\sigma}$ are not determined theoretically, but are obtained from experimental data. Most commonly,

$$H_{\rho\rho}^{\text{core}} = VSIP + A$$
$$\Gamma_{\rho\rho} = VSIP - A$$

where $VSIP$ and A are *valence-state ionization potentials* and *electron affinities*, respectively. For $\Gamma_{\rho\sigma}$, a number of formulations have been tried, with varying success:

$$\Gamma_{\rho\sigma} = \tfrac{1}{2}(\Gamma_{\rho\rho} + \Gamma_{\sigma\sigma}) - ar_{rs} + br_{rs}^2 \qquad \text{(Pariser)}$$

where a and b are a pair of constants for each pair of types of atoms, obtained by fitting this formula to values calculated on the basis of a uniformly charged sphere at $r_{rs} = 2.80$ and 3.70 A, and r_{rs} is the distance between the atoms on which ϕ_ρ and ϕ_σ are centered. In this approximation, the charged sphere model is used for $r_{rs} > 3.70$ A:

$$\Gamma_{\rho\sigma} = \frac{1}{2}\left(\frac{14.399}{\dfrac{1}{\Gamma_{\rho\rho}^0} + r_{rs}} + \frac{14.399}{\dfrac{1}{\Gamma_{\sigma\sigma}^0} + r_{rs}} \right) \qquad \text{(Mataga)}$$

$$\Gamma_{\rho\sigma} = \frac{14.399}{a \cdot \exp\left(-r_{rs}^2/2a^2\right) + r_{rs}} \qquad \text{(Beveridge-Hinze)}$$

with $a = 2 \times 14.399(\Gamma_{\rho\rho} + \Gamma_{\sigma\sigma})$. The $H_{\rho\sigma}^{core}$ have turned out to be the most stubborn, and none of the various general formulations suggested has found wide application, although the empirical values chosen for various bonds are usually taken proportional to overlap integrals. We will not go into further details of the method, since good computer programs are available from the Quantum Chemistry Program Exchange at Indiana University.

The PPP method, as outlined so far, is quite adequate for ground-state calculations. However, application to excited states, and hence to spectra, requires additional work.

12.4 Configuration Interaction

Probably the most severe approximation made in the entire development is the assumption that the total (many-electron) wave function can be factored into a product of molecular orbitals (one-electron wave functions), or rather a determinant of such products. Although such Slater determinants are solutions to the Schrödinger equation when the electron interaction terms are neglected, once these terms are introduced this is no longer true. We shall call these Slater determinants *configurations* or *spin configurations*, depending on whether the orbitals from which they are made up are space orbitals (not containing spin) or spin orbitals, respectively. We need to examine these concepts a little further.

The ground states of most organic molecules have an even number of electrons occupying completely a set of bonding (and possibly some non-bonding) orbitals; such states are called *closed shells* and are readily described as a single configuration or a single spin configuration. Thus, the ground state of benzene with six electrons is described as:

$$\psi_1{}^2\psi_2{}^2\psi_3{}^2: \quad \text{a configuration}$$
$$a_{2u}^2 e_g{}^4: \quad \text{a configuration in symmetry notation}$$
$$|\psi_1(1)\bar{\psi}_1(2)\psi_2(3)\bar{\psi}_2(4)\psi_3(5)\bar{\psi}_3(6)|: \quad \text{a spin configuration}$$

In these expressions we have used ψ_μ for space orbitals only, and have distinguished the spin orbitals $\psi_\mu(1)\alpha(1)$ and $\psi_\mu(1)\beta(1)$ by writing $\psi_\mu(1)$ and $\bar{\psi}_\mu(1)$, that is, by placing a bar over the ψ if the associated spin is β, but no bar if it is α.

A singly excited state of benzene may be described as the configuration

$$\psi_1{}^2\psi_2{}^2\psi_3\psi_4$$

In this case, however, we can now write several spin configurations:

$$|\psi_1(1)\bar{\psi}_1(2)\psi_2(3)\bar{\psi}_2(4)\psi_3(5)\psi_4(6)| \qquad\qquad (a)$$

$$|\psi_1(1)\bar{\psi}_1(2)\psi_2(3)\bar{\psi}_2(4)\psi_3(5)\bar{\psi}_4(6)| \qquad\qquad (b)$$

$$|\psi_1(1)\bar{\psi}_1(2)\psi_2(3)\bar{\psi}_2(4)\bar{\psi}_3(5)\psi_4(6)| \qquad\qquad (c)$$

$$|\psi_1(1)\bar{\psi}_1(2)\psi_2(3)\bar{\psi}_2(4)\bar{\psi}_3(5)\bar{\psi}_4(6)| \qquad\qquad (d)$$

Configurations (a) and (d) obviously have two unpaired (parallel) electrons and must belong to a triplet state. It would appear as if both (b) and (c) had all electrons paired and belonged to a singlet. However, this is not so. Writing either of these functions (and their permutations in the Slater determinant) implies the knowledge that in function (b) the electron in orbital ψ_3 has α spin, that in orbital ψ_4 has β spin, and the reverse in function (c). Actually, there is no way of having this information. Consequently, as usual, we must form two linear combinations, (b') and (c'):

$$\frac{1}{\sqrt{2}}\left\{|\psi_1(1)\bar{\psi}_1(2)\psi_2(3)\bar{\psi}_2(4)\psi_3(5)\bar{\psi}_4(6)| + |\psi_1(1)\bar{\psi}_1(2)\psi_2(3)\bar{\psi}_2(4)\bar{\psi}_3(5)\psi_4(6)|\right\}$$

$$(b')$$

$$\frac{1}{\sqrt{2}}\left\{|\psi_1(1)\bar{\psi}_1(2)\psi_2(3)\bar{\psi}_2(4)\psi_3(5)\bar{\psi}_4(6)| - |\psi_1(1)\bar{\psi}_1(2)\psi_2(3)\bar{\psi}_2(4)\bar{\psi}_3(5)\psi_4(6)|\right\}$$

$$(c')$$

The first four orbitals in each term of these two functions are the same. If we write only the last two, we have

$$\frac{1}{\sqrt{2}}\left\{|\psi_3(5)\bar{\psi}_4(6)| \pm |\bar{\psi}_3(5)\psi_4(6)|\right\}$$

$$= \frac{1}{\sqrt{2}}\left(\psi_3(5)\bar{\psi}_4(6) - \psi_3(6)\bar{\psi}_4(5) \pm [\bar{\psi}_3(5)\psi_4(6) - \bar{\psi}_3(6)\psi_4(5)]\right)$$

Multiplying out the spin functions, this gives

$$\psi_3(5)\alpha(5)\psi_4(6)\beta(6) - \psi_3(6)\alpha(6)\psi_4(5)\beta(5) \pm [\psi_3(5)\beta(5)\psi_4(6)\alpha(6) -$$

$$- \psi_3(6)\beta(6)\psi_4(5)\alpha(5)]$$

$$= [\psi_3(5)\psi_4(6) \pm \psi_3(6)\psi_4(5)]\alpha(5)\beta(6) - [\psi_3(6)\psi_4(5) \pm \psi_3(5)\psi_4(6)]\alpha(6)\beta(5)$$

Using the upper signs, we obtain,

$$[\psi_3(5)\psi_4(6) - \psi_3(6)\psi_4(5)][\alpha(5)\beta(6) + \alpha(6)\beta(5)] \qquad (b'')$$

whereas with the lower signs we obtain

$$[\psi_3(5)\psi_4(6) + \psi_3(6)\psi_4(5)] \, [\alpha(5)\beta(6) - \alpha(6)\beta(5)] \tag{c''}$$

If we examine the behavior of (b'') and (c'') under the operation of exchanging electrons, we note that both change sign, as required by the Pauli principle, but in a different way: in both cases we have factored out the space function (the first term) from the spin function (the second term). In (b'') the change of sign upon exchange of electrons arises from the space term. It is obvious that this is the same behavior shown by (a) and (d), in which the spin function is not changed by the permutations of electrons 5 and 6. This behavior is a general criterion of all triplet wave functions, and thus we have found the three components of the triplet, (a), (d), and (b''), which is $(b) + (c)$. The behavior of (c'') is the opposite; the space function is symmetric, the spin function antisymmetric, and this is the typical behavior of singlets.

Thus we see that a single spin configuration does not adequately describe an excited state. The problem shown here, however, is not serious since by $\psi_\nu(i)\bar{\psi}_\mu(j)$, in speaking of singlets, we always imply

$$|\psi_\nu(i)\bar{\psi}_\mu(j)| - |\bar{\psi}_\nu(i)\psi_\mu(j)|$$

However, in the benzene case, since ψ_2 and ψ_3 are degenerate, as are ψ_4 and ψ_5, we have a further complication because there are four configurations which appear degenerate:

$$\psi_2 \to \psi_4: |\psi_1(1)\bar{\psi}_1(2)\psi_2(3)\bar{\psi}_4(4)\psi_3(5)\bar{\psi}_3(6)|$$
$$\psi_2 \to \psi_5: |\psi_1(1)\bar{\psi}_1(2)\psi_2(3)\bar{\psi}_5(4)\psi_3(5)\bar{\psi}_3(6)|$$
$$\psi_3 \to \psi_4: |\psi_1(1)\bar{\psi}_1(2)\psi_2(3)\bar{\psi}_2(4)\psi_3(5)\bar{\psi}_4(6)|$$
$$\psi_3 \to \psi_5: |\psi_1(1)\bar{\psi}_1(2)\psi_2(3)\bar{\psi}_2(4)\psi_3(5)\bar{\psi}_5(6)|$$

and for each a second term is implied. Unfortunately, each of these spin configurations does not transform separately as single irreducible representations of the point group D_{6h}.

Coming back to the representation of the configurations in symmetry notation given above, we can write all of them as $a_{2u}^2 e_{1g}^3 e_{2u}$. The easiest way to find the proper combinations of these spin configurations is by means of a technique called *configuration interaction*. Before proceeding to a description of this technique, we shall give another rationale for its use.

Consider the energy level diagram of butadiene, Fig. 12.1. The ground state of butadiene is satisfactorily described by the configuration of $\psi_1{}^2\psi_2{}^2$, and the first excited state by $\psi_1{}^2\psi_2\psi_3$. The transition to this state is shown by the arrow labelled 1 in Fig. 12.1. The next higher state, however, is described by either $\psi_1\psi_2{}^2\psi_3$ (arrow 2) or $\psi_1{}^2\psi_2\psi_4$ (arrow 3). In the light of the symmetrical spacing of these orbitals on an energy scale, these two configurations are exactly degenerate. Again, this degeneracy is not real; it is artificially produced by the assumption that we can describe the states of a molecule by configurations, and is removed by configuration interaction.

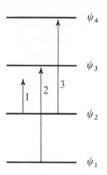

Fig. 12.1   Transitions in butadiene.

Configuration interaction (CI) consists of expressing the final, spectroscopic state of a molecule as a linear combination of spin configurations:

$$\Psi_k = \sum_l A_{lk} \Psi_l^c$$

Where Ψ_l^c stands for a spin configuration, and the A_{lk} are arbitrary coefficients. These coefficients are again evaluated by use of the variational theorem, and lead to a set of secular equations and a secular determinant exactly as outlined in Chapter 10. Here the matrix elements, H_{lm}, are the integrals

$$H_{lm} = \int \Psi_l^c H \Psi_m^c \, d\tau$$

where H is the Hamiltonian operator for the molecule. We shall not carry through the evaluation of these integrals here, but simply indicate that they are readily expressed in terms of the same set of integrals required for the SCF calculation, and that in any particular level of approximation the same approximations are made in the CI as in the SCF treatment.

Thus, in order to obtain spectroscopic information from the PPP method, it is necessary to follow the SCF calculation by a CI calculation. We have shown above, in the case of benzene, that this was necessary even for the lowest excited state; this situation, however, is unusual. More commonly, the splitting pattern is such that, as in butadiene, the lowest excited state is approximated reasonably well without CI, but calculations for higher excited states without CI are all but meaningless. The number of excited spin configurations is very large, and CI can extend only over a limited set. The one most commonly chosen is the set consisting of all the configurations obtained by excitation of a single electron from the ground state (all singly excited configurations).

12.5   The Extended Hückel Method (EHMO)

The great simplicity of the HMO method depends to a large extent on the fact that, in planar molecules, the molecular plane is a plane of symmetry. Since every molecular orbital is built up of the pπ atomic orbitals of the elements involved, all of which are antisymmetric with respect to this plane, all π molecular orbitals have this property, and they are the only ones which do. Consequently the π molecular orbitals transform in different symmetry species than all other MO's, and this simple fact permits the separation of σ and π electrons. If this symmetry plane is absent, no such separation into σ and π electrons (or orbitals) can be made. Then, in nonplanar systems, we must make our calculations for *all* valence electrons of *all* atoms in the molecule.

To perform such a calculation, we require a much larger number of parameters than we did in the HMO. Even in a hydrocarbon, which requires a relatively small number of parameters, we now deal with the 1s orbitals of hydrogen and the 2s and 2p orbitals of carbon, and we thus require three Coulomb integrals (α's) for these. Furthermore, we need resonance integrals (β's) for the following interactions: $1s_H - 1s_H$, $2s_C - 2s_C$, $2s_C - 2p_{\sigma C}$, $2p_{\sigma C} - 2p_{\sigma C}$, $2p_{\pi C} - 2p_{\pi C}$, $1s_H - 2s_C$, and $1s_H - 2p_{\sigma C}$. Whereas, in HMO, we need the two parameters α and β, the former the origin of the energy scale, and the latter the unit of energy, we now need three α's and seven β's, a total of ten parameters. Thus the scheme used so widely and successfully in HMO, of empirically adjusting the parameters for a few compounds, is impractical.

The problem was solved by Hoffmann, who introduced a systematic formulation for all needed integrals:

$$\alpha = VSIP - A$$

$$\beta_{AB} = K\left(\frac{\alpha_A + \alpha_B}{2}\right)S_{AB}$$

where $VSIP$ is the proper valence-state ionization potential for the AO in question, A its electron affinity, S_{AB} the overlap integral, and K the only empirically adjusted constant remaining. For hydrocarbons K is taken to be 1.75. With this parametrization we are then no longer restricted to hydrocarbons, but can make calculations for any arbitrary molecule provided the necessary atomic data (ionization potential and electron affinity) are known or can be "guesstimated".

With this parametrization, the formulation of the EHMO follows identically that introduced in the early part of Chapter 10. Major differences are that overlap integrals are not neglected, and that *all* β's, even for nonbonded atoms, are calculated. In addition, of course, the computational process is much more complicated for the following reasons:

1. Four orbitals per atom (except hydrogen) must be considered, and even nine if d electrons are taken into account.

2. The inclusion of overlap integrals leads to a much more complicated polynomial upon expansion of the secular determinant.

3. The simple definitions of charge density and bond order need modification because S is not neglected.

In practice, the computational problem is such that calculations, even for simple molecules, are impractical without the use of electronic computers. In the computer programs, the calculational methods are entirely different from those used in the HMO method, since the handling of matrices is relatively simple. Nevertheless, the process requires the diagonalization of two determinants. The first is the determinant of the overlap integrals, giving the "solutions" S_D and the coefficients U_{AB}. The results of this calculation are then used to reduce the energy determinant

$$\left| H_{\mu\nu} - S_{\mu\nu}E \right| = 0$$

into the more tractable form

$$\left| H'_{\mu\nu} - \delta_{\mu\nu}E \right| = 0$$

where $\delta_{\mu\nu}$, the Kronecker $\delta = 1$ if $\mu = \nu$ and $\delta = 0$ if $\mu \neq \nu$, and $H'_{\mu\nu}$ is given by

$$\mathbf{H}' = (\mathbf{U}^{\ddagger}\mathbf{S_D}^{-1/2}\mathbf{U})\mathbf{H}(\mathbf{U}^{\ddagger}\mathbf{S_D}^{-1/2}\mathbf{U})$$

where $\mathbf{U}$ is the matrix of the coefficients (eigenvectors) from the diagonalization of the S matrix; $\mathbf{U}^{\ddagger}$ is its transpose, that is, the same matrix with rows and columns reversed; $\mathbf{S_D}^{-1/2}$ is the array (vector) of the reciprocals of the square roots of the solutions (eigenvalues) of the S matrix. This equation involves considerable multiplication of matrices and vectors, but these operations are readily executed by the computer.

The transformation from $\mathbf{H}$ to $\mathbf{H}'$ actually corresponds to a transformation of the set of basis orbitals (which are nonorthogonal since we do not neglect S) to a new, artificial set, the members of which are orthogonal but have no physical meaning. Consequently, in the end we need to reconvert the coefficients obtained from the solution of the transformed determinant, the $\mathbf{C}'$, which are the LCAO expansion coefficients in terms of our orthogonal, artificial basis set, into coefficients of our original, nonorthogonal basis set. This is achieved by the matrix operation

$$\mathbf{C} = (\mathbf{U}^{\ddagger}\mathbf{S_D}^{-1/2}\mathbf{U})\mathbf{C}'$$

The coefficients $C_{\mu\nu}$, the elements of the matrix $\mathbf{C}$, now require interpretation analogous to that given in the HMO method. The difficulty here lies in the fact that the electron population is not as readily distributed to a set of atoms.

Let us take as an example a wave function for H_2:

$$\psi = \frac{1}{\sqrt{2(1 + S)}}(\phi_1 + \phi_2)$$

In HMO, $q_r = \sum_v n_v c_{vr}^2$. If we apply this formula here, we obtain

$$q_1' = q_2' = 2\left[\frac{1}{\sqrt{2(1 + S)}}\right]^2 = 2\frac{1}{2 + 2S}$$

In HMO, if we add the q_r for all atoms we obtain the total number of π electrons. Performing the same addition in the q', we obtain

$$2 \cdot 2 \frac{1}{2 + 2S} = 2\frac{1}{1 + S} =$$

$$2 \cdot \frac{1 + S + S}{1 + S} = 2\left(\frac{1 + S}{1 + S} - \frac{S}{1 + S}\right)$$

$$= 2 - \frac{2S}{1 + S}$$

But the total number of electrons is 2. The fraction $2S/(1 + S)$ is called the *overlap population*, and the value of $q_1' = 1/(1 + S)$ is called the *net orbital population*. But we wish to distribute *all* of our electrons between atomic orbitals. To do this, we arbitrarily divide the overlap population equally between the two orbitals forming the bond. This gives

$$q_1 = \frac{1}{1 + S} + \frac{1}{2}\frac{2S}{1 + S} = 1$$

In the more general case, where the wave function might be

$$\psi = c_1\phi_1 + c_2\phi_2$$

the net orbital populations of atoms 1 and 2 are $q_1' = 2c_1^2$ and $q_2' = 2c_2^2$, the overlap population $q_{12} = 4c_1c_2S_{12}$, and the *gross orbital populations* are

$$q_1'' = 2c_1^2 + \tfrac{4}{2}c_1c_2S_{12}$$
$$= 2c_1(c_1 + c_2S_{12})$$

In the even more general case, where we deal with many centers and many MO's,

$$q_\rho' = \sum_v n_v c_{v\rho}^2$$
$$q_{\rho\sigma} = 2\sum_v n_v c_{v\rho}c_{v\sigma}S_{\rho\sigma}$$
$$q_\rho = \sum_v \left(n_v c_{v\rho}\left(c_{v\rho} + \sum_\sigma c_{v\sigma}S_{\rho\sigma}\right)\right)$$

However, usually we are mainly interested, not in the population of single AO's, but rather in the population on atoms and in the total bond population between atoms. We obtain the total net, Q', and gross atomic population, Q, by summing over all AO's of the same atom:

$$Q_r' = \sum_{\rho \text{ of } r} q_\rho'$$

$$Q_r = \sum_{\rho \text{ of } r} q_\rho$$

and the total overlap population (often called the reduced overlap population)

$$Q_{rs} = \sum_{\rho \text{ of } r} \sum_{\sigma \text{ of } s} q_{\rho\sigma}$$

It is customary in EHMO to use, not the bond orders, as defined in HMO, but the total overlap populations, Q_{rs}, as a measure of the bond character (bond order) between two atoms, r and s.

The extended Hückel method has proved to be quite useful in obtaining much information about molecules in their ground states.

12.6   All-Valence-Electron Self-Consistent Field Methods

Just as the HMO method needed to be refined by taking explicit account, even if not with complete adequacy, of electron-electron repulsion, so a similar refinement was needed for methods including all valence electrons. Such methods have now been introduced under a variety of names, but most prominently under the designation CNDO (Complete Neglect of Differential Overlap), due to Pople and coworkers.

At the heart of CNDO and some of its variants lies the complete SCF treatment, as described in Section 12.2. In its simplest form, CNDO uses the neglect of differential overlap, as introduced in the Pariser-Parr-Pople method (Section 12.3). In contrast to the latter method, however, the core in CNDO is defined as the nucleus and the inner-shell electrons (since most applications to date are restricted to first-row elements, this means the nucleus and the 1s electrons).

In these methods, again, the basic choices of integrals are critical. The $H_{\rho\rho}^{\text{core}}$ are again evaluated from ionization potentials and electron affinities. The $H_{\rho\sigma}^{\text{core}}$ are empirically adjusted. The $\Gamma_{\rho\rho}$ and $\Gamma_{\rho\sigma}$ are evaluated theoretically in some variants of the method, and semiempirically and quite analogously to P^3 in others. For spectroscopic calculations, CI again is essential.

Even for relatively small molecules, CNDO calculations are impossible to carry out without large computers, and programs are available from the Quantum Chemistry Program Exchange. Consequently, no details of the method will be given here.

Many variants exist, some relaxing the differential overlap neglect to include some of the one-center and even the two-center exchange integrals.

12.7   *Ab Initio* Methods

Finally, this account would not be complete without mention of *ab initio* methods. In these methods all electrons are included, and all integrals are evaluated. With Slater-type orbitals (STO) as basis functions, such calculations are extremely time consuming, primarily because of the difficulty of evaluating three- and four-center integrals. Consequently relatively few calculations, and these mostly on small and linear molecules, have been made.

The introduction of Gaussian-type orbitals (GTO) has changed the picture. The GTO has the form

$$\phi_{GTO} = Y_{l,\,m_l} f(r) e^{-\zeta r^2}$$

while the STO has the form

$$\phi_{STO} = Y_{l,\,m_l} f(r) e^{-\zeta r}$$

The $Y_{l,\,m_l}$ are the spherical harmonics, that is, the angular parts of the AO's, and are the same for GTO's and STO's; l and m_l are the azimuthal and magnetic quantum numbers characterizing the AO. The GTO does not represent the behavior of the electron near the nucleus, or at far distances from the nucleus, nearly as well as the STO. Consequently, much larger basis sets are required. Thus an s orbital is often written as anywhere from three to six GTO's. However, the evaluation of all integrals, including many-centered ones, is *very* much easier, and consequently very much larger basis sets can be used. The mathematics is still basically that outlined under the SCF calculations (Section 12.3). Calculations for moderately sized molecules are now becoming possible with the use of GTO's.

12.8   Open-Shell Methods

Most of what we have discussed so far applies to closed shells, that is, to molecules in which all electrons occupy orbitals in pairs. We are, however, often interested in molecules which do not have closed-shell structures, for example, free radicals, biradicals, and triplet states. Although the HMO and EHMO methods, because they neglect electron repulsions, are directly applicable to such molecules, the SCF methods present new problems. Two basically different methods are available for the treatment of open shells.

One method, due to Roothaan, treats the electron population in two parts. The first is a closed-shell part, for which the treatment is quite analogous to that given above under the SCF methods. The other part is concerned only with the open shell. Here, again, the treatment is similar, except that each orbital is occupied by only a single electron. The major complication arises from the fact that the two problems are not independent but are coupled, in that each matrix element of one section must include the interactions with the electrons in the other.

The other method, due to Pople, also treats the electron population in two sections, but the division is made in a different way. All electrons of α spin are included in one part, all those of β spin in the other. Again, two sets of coupled secular equations arise. In this method, two sets of MO's are calculated, one for each spin, and the simplicity of unique MO's is lost.

Neither method is completely satisfactory. The Roothaan method fails to provide proper spin polarization, while the Pople method does not yield states of pure multiplicity.

Both methods can readily be applied at any of the various levels of approximation discussed.

12.9   Comparison of the Results Obtained by the Various Methods Beyond the Hückel Molecular Orbital

The added complications introduced in the methods discussed in this chapter make the interpretation of the calculated quantities more difficult. Throughout, the emphasis has been on energy, but even this quantity has not been treated completely. In the HMO method, the total electronic energy was expressed as

$$E = \sum_{v} n_{v}\epsilon_{v} \tag{12.23}$$

where n_v represents the number of electrons in the vth orbital, ϵ_v is its energy, and the summation extends over all orbitals. In the EHMO method, the energy is expressed in the same form but now E refers to the total electronic energy.

In SCF methods, the problem is more difficult. A general expression is given in equation 12.18. But this equation is valid only for a complete treatment, taking account of all electrons explicitly. In this case, the last term represents the repulsion of all the nuclei, stripped of all electrons. In the semiempirical methods, a discussion of which forms the bulk of this chapter, this term must be replaced by a less precise one; the all-valence-electron method (CNDO, etc.) requires a term which represents the repulsions of the cores consisting of nuclei and inner-shell electrons, but which also must include the repulsions of these electrons. In π-electron methods we can get only π-electron energies, and a constant term must account for the cores, including all σ electrons.

The h_v terms of equation 12.18 represent the binding energy of the electron in the absence of electronic repulsions. A more meaningful term is

$$\epsilon_{v} = h_{v} + \sum_{\mu \neq v} (J_{\mu v} - K_{\mu v})$$

which includes all the repulsions. Unfortunately, if this term is used in equation 12.23 to obtain the total electronic energy, each electron repulsion is counted twice. As a result, the expression for the total energy becomes

$$E = \sum_v \epsilon_v - \sum_{\mu < v} (J_{\mu v} - K_{\mu v}) + \text{term} \qquad (12.24)$$

where the last term is the core repulsion term discussed in the preceding paragraph.

The ϵ_v have another important significance. According to Koopman's theorem, they represent the energy required to remove the electron from the orbital, and consequently provide calculated ionization potentials. Particularly in conjunction with photoelectron spectra, they thus provide a wealth of information.

The energies of spectroscopic transitions are, of course, the energy differences between initial and final (ground and excited) states. These differences can be readily obtained by the use of equation 12.24 for two states. A bit of algebraic manipulation shows that, for the transition in which an electron is excited from ψ_μ to ψ_v,

$$\Delta E = \epsilon_v - \epsilon_\mu + J_{\mu v} \pm K_{\mu v}$$

with the upper or lower sign applicable, depending on whether the spin orbitals ψ_μ and ψ_v represent equal or opposite spins, that is, whether the excited state is a singlet or a triplet.

In all the methods assuming neglect of differential overlap, that is, Pariser-Parr-Pople, CNDO, etc., electron densities have the definition introduced in Chapter 10. In the HMO and P^3 methods, these quantities naturally refer to π electrons only; in the all-valence-electron treatments, however, π-electron and σ-electron quantities can be obtained separately or, more commonly, can be added together to give total electron densities. Bond orders for individual AO pairs are defined in perfect analogy with $p_{\rho\sigma}$ in HMO, and thus permit discussion of π bond orders in all methods. Expression of atom-atom bond orders in all-valence-electron methods is more complicated and will not be pursued here. Quantities calculated for π electrons normally show the same trends in different methods but are not usually identical; this should serve as a reminder that these quantities are not absolute properties of molecules. A comparison of bond orders and electron densities is given in Table 12.2 for butadiene as an example of the types of results obtained.

Whenever overlap (differential or overlap integrals) is not neglected the problem becomes much more difficult. In particular, electron densities calculated by the formulas of Chapter 10 are not additive, that is, they do not add up to the total number of electrons; cf. Section 12.5. As indicated in that section, a Mulliken population analysis takes the place of the calculation of electron densities and bond orders. The numerical values obtained for overlap populations bear no direct numerical relation to the bond orders, but reproduce the same trends. Table 12.2 shows an EHMO population analysis for the π electrons of butadiene.

TABLE 12.2
Comparison of Electron Densities and Bond Orders for the π-Electron System of Butadiene Obtained by Different Semiempirical Methods

Electron Density

Atom	HMO	P^3	CNDO/2	EHMO*	EHMO†
1	1.0	1.0	1.011	0.855	1.039
2	1.0	1.0	0.989	0.728	0.962
3	1.0	1.0	0.989	0.728	0.962
4	1.0	1.0	1.011	0.855	1.039

π Bond Orders

Bond	HMO	P^3	CNDO/2	EHMO‡
1, 2	0.894	0.949	0.951	0.187
2, 3	0.447	0.316	0.311	0.018

* Net orbital population.   † Gross orbital population.   ‡ Overlap population.

PROBLEMS

12.1 Given the Slater determinant for the ethylene π system:

$$^1\Psi_G = \frac{1}{\sqrt{2}} \begin{vmatrix} \psi_1(1) & \psi_1(\bar{1}) \\ \psi_1(2) & \psi_1(\bar{2}) \end{vmatrix}$$

and the Hamiltonian of the system:

$$\bar{H} = T(1) + T(2) + H(1) + H(2) + \frac{e^2}{r_{12}}$$

where $T(i)$ and $H(i)$ represent the kinetic and potential energies of the ith electron, determine the total energy of the π system in the ground state as described by $^1\Psi_G$.

12.2 Describe what kinds of information may be obtained from a molecular orbital calculation and the related molecular quantities.

12.3 Determine the symmetry of the individual molecular orbitals and the symmetry of the six lowest energy molecular states for 1,3,5-hexatriene in C_{2h}.

12.4 Express the following integrals in both integral and Dirac notation: (a) 3-center exchange integral, (b) 4-center exchange integral, (c) 2-center exchange integral, and (d) 2-center hybrid integral.

12.5 Explain the advantage of Gaussian-type orbitals (GTO) over Slater-type orbitals (STO).

GENERAL REFERENCES

1. F. L. Pilar, *Elementary Quantum Chemistry*, McGraw-Hill Book Company, New York, 1968.
2. R. G. Parr, *The Quantum Theory of Molecular Electronic Structure*, W. A. Benjamin, New York, 1964.
3. J. A. Pople and D. L. Beveridge, *Approximate Molecular Orbital Theory*, McGraw-Hill Book Company, New York, 1970.

Appendix 1 Correlation of symmetry species in different point groups

In conjunction with the method of descending symmetries (cf. Chapter 7), it is essential to be able to correlate the irreducible representations (symmetry species) of a higher-order point group with those of its subgroups. In particular, the degenerate irreducible representations of the highly symmetrical point groups (O_h, T_d, I_h) become reducible in the subgroups, and extensive use is made of this behavior in determining the symmetry and multiplicity of various states. Unfortunately, as we shall see, the correlation between the representations of a group and its subgroups cannot always be made uniquely and some confusion exists because of this fact. We shall examine the problems involved in making these correlations in some detail, and then propose a convention to produce a *unique* correlation in the majority of cases.

In order to arrive at the correlations, we shall proceed in three steps. First, we shall examine the subgroups of a given group; next, we shall examine the correlation of symmetry elements (or operations) between the groups and its subgroups; and finally we shall be prepared to make the correlation of representations.

A1.1 Subgroups

A *subgroup* of any group is itself a group, having all the required properties of the group; *all* of its elements are also elements of the larger group (*supergroup*) to which it belongs. A simple example is the group C_{2h}, having elements I, σ, i, and C_2. The group C_{2h} has four (and only four) subgroups, the trivial C_1 (element I only), C_2 (I and C_2), C_s (I and σ), and C_i (I and i). The order of any subgroup of the original (super-) group is always an integral rational fraction of its order; in other words, the order of a supergroup is a rational multiple of the order of each of its subgroups.

Since the method of descending symmetries is particularly useful with octahedral inorganic complexes, we have listed in Fig. A1.1 all the subgroups of O_h and their interrelations. In principle we should construct correlation tables among all of these subgroups; however, once the general methods

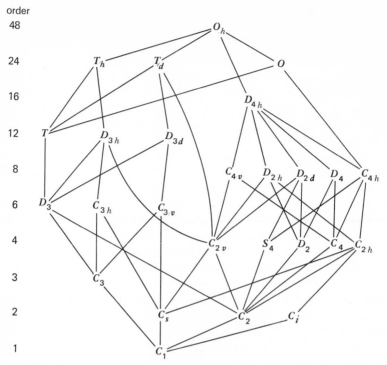

Fig. A1.1 The subgroups of O_h (the numbers in the left column refer to the order of the groups at that level).

are established, new correlations may be worked out, and hence we shall deal only with the most important correlations.

A complete interrelation of all point groups would have to start with K_h, of which all other point groups are subgroups. If we are willing to consider only rotational symmetry (C_p operations), all pure rotational groups are subgroups of K. Besides the hierarchy of groups based on O_h shown in Fig. A1.1, there is a second main one, based on I_h, as well as others generally based on D_{ph}, with different p.

A1.2   Correlation of Symmetry Operations

We shall now examine how a subgroup is formed for various different relations between group and subgroup.

1. The simplest such relation is that between O_h and O, or between D_{4h} and D_4. In either case, destruction of the center of symmetry, i, is automatically accompanied by the disappearance of all planes, σ, and the four-

fold rotation reflection axes, S_4. Alternatively, we may say that we eliminate all planes of symmetry, whereupon the disappearance of i and of the S_4 follows. No real problems arise here. The elements of any one class of symmetry operations either all disappear or all remain.

2. A slightly more difficult case is the relation between T_d and T, or between C_{2v} and C_2. Again, all elements of a given class either disappear or remain; however, the interrelation is not so evident, since it is not readily apparent that destruction of one element of a class necessarily requires the destruction of all elements.

3. The more complicated cases arise when some, but not all, elements of a class are destroyed. The simplest case of this type is the reduction of C_{2v} to C_s. This can be achieved in either of two ways, by destroying $\sigma(xz)$ or $\sigma(yz)$. This leads to two equivalent subgroups C_s, which are distinct in the sense that the σ carries different labels, each relative to a fixed coordinate system. However, both of these subgroups are distinct from the C_s for which we have character tables, since we generally assume the xy-plane as the σ plane. Thus, though all three C_s are equivalent, they differ in orientation.

A more difficult case is the reduction of T_d to C_{2v}. This is accomplished by selecting any one of the three C_2 axes of T_d and treating it as the (unique) C_2 axis of C_{2v}. Each of these axes lies in the intersection of two perpendicular planes (called σ_d in T_d), and these two planes and the axis in which they intersect form the three symmetry elements (other than the trivial I) of a C_{2v} group. There are thus three possible choices for defining the subgroup C_{2v}, depending on which of the three C_2 is chosen. However, the three C_2 as well as the six σ each belong to one class, respectively, and the three possible correlations are all equivalent and may be said to belong to one class; hence there is only one way to reduce T_d to C_{2v}.

The situation is more complicated if we try to reduce O_h to C_{2v}. The group O_h has two classes of C_2, and either class may be used as a basis of a C_{2v} subgroup. These possibilities are more effectively demonstrated by reference to Fig. A1.2. The axes going through opposite corners of the octahedron are four-fold axes, but coincident with each such C_4 is a two-fold axis, usually referred to as C_2''. However, the axes GH and IJ, bisecting opposite edges of the octahedron, are also two-fold axes, called C_2. Each C_2 is the intersection of two planes, a σ_h (which includes four edges) and a σ_d bisecting two opposite edges (of O_h). The C_2'', on the other hand, lie at the intersection of four planes, two σ_h and two σ_d. Thus we can form three classes of C_{2v} from O_h: the first based on C_2; and the other two on C_2'', one in which the C_2'' are taken together with two σ_h, and the other in which a C_2'' is taken with two σ_d. The latter two classes *each* have three equivalent possibilities (there are three C_2'' axes), but the first class has six (there are six C_2 axes) equivalent possibilities.

The situation can become even more difficult. In reducing O_h to D_{4h}, two classes of D_{4h} arise which are similar but distinct. Either class of D_{4h} is based on *any one* of the three C_4 axes of O_h, so that each of the two classes consists of three equivalent members. Let us consider only those members in which we choose the *EF* axis (Fig. A1.2) as the principal axis. Table A1.1 shows the pertinent symmetry elements appropriate to O_h, and to the right and left of this column are columns showing the elements of the two types of D_{4h}; on the far left the elements in each class of symmetry elements are identified according to the labeling scheme of Fig. A1.2.

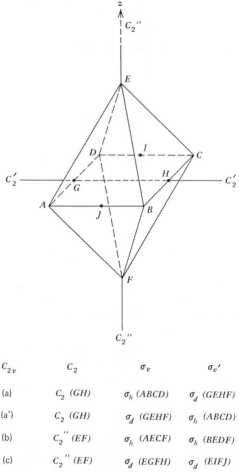

C_{2v}	C_2	σ_v	σ_v'
(a)	C_2 (GH)	σ_h (ABCD)	σ_d (GEHF)
(a')	C_2 (GH)	σ_d (GEHF)	σ_h (ABCD)
(b)	C_2'' (EF)	σ_h (AECF)	σ_h (BEDF)
(c)	C_2'' (EF)	σ_d (EGFH)	σ_d (EIFJ)

Fig. A1.2 The symmetry elements of an octahedron and the choice of these which constitute the various C_{2v}.

TABLE A1.1
Breaking up O_h into Two Different D_{4h} (see Fig. A1.2)

Axes and Planes	$D_{4h}(1)$	O_h	$D_{4h}(2)$
EF*	$2C_4$ ⟵	$6C_4 \begin{cases} EF \\ AC \\ BD \end{cases}$ ⟶	$2C_4$
EF	C_2'' ⟵	$3C_2''$ ⟶	C_2''
AC, BD	$2C_2$ ⟵		$2C_2'$
GH, IJ	$2C_2'$ ⟵	$6C_2'$ GH, IJ, etc. ⟶	$2C_2$
ABCD	σ_h ⟵	$3\sigma_h \begin{cases} ABCD \\ AECF \\ BEDF \end{cases}$ ⟶	σ_h
AECF, BEDF	$2\sigma_v$ ⟵		σ_d
EGFH, EIFJ	$2\sigma_d$ ⟵	$6\sigma_d \begin{matrix} EFJI \\ EFGH \end{matrix}$, etc. ⟶	$2\sigma_v$
0	i ⟵	i ⟶	i
EF	$2S_4$ ⟵	$6S_4 \begin{cases} EF \\ AC \\ BD \end{cases}$ ⟶	$2S_4$

* Axes and planes according to Fig. A1.2. Note that the same axis or plane may be assigned to different symmetry elements in $D_{4h}(1)$ and $D_{4h}(2)$.

Examination of Table A1.1 in detail shows that the difference in the two classes of D_{4h} formed from O_h resides in the arbitrary assignment of two-fold axes to C_2 or C_2', and of planes to σ_v or σ_d. But these assignments depend on an arbitrary assignment of coordinate axes to symmetry axes. By a general consensus, the z-axis is always assigned to the unique, principal axis, the C_4 axis of D_{4h}. By a fairly general consensus also, the x- and y-axes, are identified with the C_2 axes (through corners) rather than the C_2' axes in D_{4h}. The two types of D_{4h} are then converted into one another by a 45° rotation of the coordinate system.

To further complicate matters, reduction of D_{4h} into C_{2v} can be achieved in four different ways, giving four classes. Any of the three classes of two-fold axes of D_{4h} may be used to give a C_{2v}, but the C_2'' (coincident with the C_4) can be used in combination with either a pair of σ_v or a pair of σ_d.

If we examine these four ways of generating C_{2v} from D_{4h}, we note that in C_{2v} a coordinate system is automatically defined since we assume C_2 parallel to the z-axis, and the two σ_v in the xz- and yz-planes. In D_{4h} we assume the z-axis parallel to C_4, and the x- and y-axes parallel to two C_2 (not the C_2')

axes. With this definition, only case (b), Fig. A1.2, leaves the coordinate system unaffected; the other three imply some rotation of coordinates.

We now propose to introduce the convention in the reduction of symmetry which will make the correlations of symmetry elements (operations) and of representations unique in most cases: *In a reduction of symmetry we choose, if possible, the symmetry elements so that no rotation of coordinate axes occurs.* This convention makes $D_{4h}(1)$ the unique choice as the subgroup of O_h. We shall call a correlation made on this basis the *principal* correlation.

The distinction can be made similarly for the other reductions discussed above. Thus, the single class of C_{2v} generated from T_d represents the principal correlation, since S_4 is the principal axis in T_d and the C_2 are coincident with it. However, in going from O_h to C_{2v}, the choice of C_2'' and the two σ_h is unique and the principal correlation, since C_2'' is the z-axis, and the x- and y-axes are chosen to parallel C_2 (not C_2').

4. A special situation is the reduction from T_d to C_{3v}. In this case a rotation of coordinates is mandatory, since the z-axis in T_d is defined along S_4 (C_2), in C_{3v} along C_3. In this particular case no ambiguity arises; nevertheless, no principal correlation exists.

A1.3   Correlation of Representations

We are now prepared to examine the correlations between the irreducible representations of a group and those of its subgroups. First let us examine the simple case of the reduction of C_{2v} to C_s. The character table of C_{2v} is as follows:

	I	$C_2^{\ z}$	$\sigma(xz)$	$\sigma(yz)$
A_1	$+1$	$+1$	$+1$	$+1$
A_2	$+1$	$+1$	-1	-1
B_1	$+1$	-1	$+1$	-1
B_2	$+1$	-1	-1	$+1$

Choosing $\sigma(xz)$ as the σ of C_s, we have A_1 and B_1 of C_{2v} as totally symmetric, that is, A' in C_s, and A_2 and B_2 in C_{2v} become A'' in C_s. On the other hand, choosing $\sigma(yz)$, we obtain A_1 and B_2 going into A', and A_2 and B_1 into A''. *Thus depending on the choice of plane in C_{2v} retained in C_s, different correlations are obtained.*

The most systematic way of obtaining the correlation is to treat the characters of an irreducible representation of the main group as a reducible representation in the subgroup and to perform the reduction. Take, for example, the irreducible representation T_{2u} of O_h, and the question is what symmetry species does T_{2u} decompose into in point group D_{4h}? What we seek is the relationship between the symmetry elements in O_h and the correlation of these with symmetry elements of $D_{4h}(1)$ and $D_{4h}(2)$ of the preceding section (the blocked-out characters focus on the differences between the two D_{4h}):

From O_h

	I	$2C_4$	C_2''	$2C_2$	$2C_2'$	σ_h	$2\sigma_v$	$2\sigma_d$	$2S_4$	i
$D_{4h}(1)\,T_{2u}$	$+3$	-1	-1	-1	$+1$	$+1$	$+1$	-1	$+1$	-3
$D_{4h}(2)\,T_{2u}$	$+3$	-1	-1	$+1$	-1	$+1$	$+1$	-1	$+1$	-3

From $D_{4h}(1)$

E_u	$+2$	0	-2	0	0	$+2$	0	0	0	-2
B_{2u}	$+1$	-1	$+1$	-1	$+1$	-1	$+1$	-1	$+1$	-1
$E_u + B_{2u}$	$+3$	-1	-1	-1	$+1$	$+1$	$+1$	-1	$+1$	-3

From $D_{4h}(2)$

E_u	$+2$	0	-2	0	0	$+2$	0	0	0	-2
B_{1u}	$+1$	-1	$+1$	$+1$	-1	-1	$+1$	-1	$+1$	-1
$E_u + B_{1u}$	$+3$	-1	-1	$+1$	-1	$+1$	$+1$	-1	$+1$	-3

Thus, depending on the definition, T_{2u} of O_h correlates with $E_u + B_{2u}$ or, alternatively, with $E_u + B_{1u}$ of D_{4h}.

The first of these is the *principal* correlation and will be the one used here. However, provided everything is properly defined, either correlation is valid.

There exists a more convenient way of finding the principal correlation. All that is required is to examine the behavior of some functions which form the basis of an irreducible representation in the subgroup. The various atomic orbitals represent such a set of functions. Thus the three p orbitals transform[a] as t_{1u} in O_h, as a_{2u} and e_u in D_{4h}, as a_1, b_1, and b_2 in C_{2v}. Consequently the principal correlation of T_{1u} in O_h is with $A_{2u} + E_u$ in D_{4h} and with $A_1 + B_1 + B_2$ in C_{2v}.

As a final example let us determine the irreducible representations in C_{2v} which correspond to T_{2g} in O_h:

O_h	I	$8C_3$	$3C_2''$	$6C_4$	$6C_2$	i	$8S_6$	$3\sigma_h$	$6S_4$	$6\sigma_d$
T_{2g}	$+3$	0	-1	-1	$+1$	$+3$	0	-1	-1	$+1$

[a]As has been our practice, we indicate by lower-case letters the symmetry species of one-electron functions (orbitals). The symmetry species of the irreducible representations of character tables or of many-electron functions is given in capital letters.

In C_{2v}, we concern ourselves only with one C_2 and two σ_v. We have already seen from Fig. A1.2 that we have three choices in making the reduction from O_h to C_{2v}. The principal correlation based on our convention would involve option (b), consisting of EF as C_2 (C_2'' in O_h) and $AECF$ and $BEDF$ as the two σ_v (both σ_h in O_h). Our reducible representation then becomes

$$
\begin{array}{ccccc}
 & I & C_2^z(C_2'') & \sigma_v(\sigma_h) & \sigma_v(\sigma_h) \\
t_{2g} & 3 & -1 & -1 & -1
\end{array}
$$

This reducible representation corresponds to $A_2 + B_1 + B_2$ in C_{2v}, which is the principal correlation. If we use the "convenient" method (consulting the character tables) for finding the principal correlations, we note that the d orbitals, d_{xy}, d_{xz}, d_{yz}, belong jointly to t_{2g} in O_h. If we now examine our character table, we see that these orbitals in C_{2v} correspond, respectively, to $a_2 + b_1 + b_2$, which is exactly the result arrived at above.

TABLE A1.2
Possible Correlations of O_h with C_{2v}

O_h	$C_2^z, 2\sigma_h$	$C_2^z, 2\sigma_d$	C_2, σ_h, σ_d	C_2, σ_d, σ_h
A_{1g}	A_1	A_1	A_1	A_1
A_{1u}	A_2	A_2	A	A_2
A_{2g}	A_1	A_2	B_1	B_2
A_{2u}	A_2	A_1	B_2	B_1
E_g	$2A_1$	$A_1 + A_2$	$A_1 + B_1$	$A_1 + B_2$
E_u	$2A_2$	$A_1 + A_2$	$A_2 + B_2$	$A_2 + B_1$
T_{1g}	$A_2 + B_1 + B_2$	$A_2 + B_1 + B_2$	$A_2 + B_1 + B_2$	$A_2 + B_1 + B_2$
T_{1u}	$A_1 + B_1 + B_2$	$A_1 + B_1 + B_2$	$A_1 + B_1 + B_2$	$A_1 + B_1 + B_2$
T_{2g}	$A_2 + B_1 + B_2$	$A_1 + B_1 + B_2$	$A_1 + A_2 + B$	$A_1 + A_2 + B_2$
T_{2u}	$A_1 + B_1 + B_2$	$A_2 + B_1 + B_2$	$A_1 + A_2 + B$	$A_1 + A_2 + B_1$

Careful examination of other possible correlations between T_{2g} in O_h and species in C_{2v} reveals that options (a) and (a'), Fig. A1.2, would lead to either $A_1 + A_2 + B_1$ or $A_1 + A_2 + B_2$, depending on which plane of the σ_h, σ_d combination in O_h we consider the xz- or yz-plane in C_{2v}. Option (c) leads to the result $A_1 + B_1 + B_2$. Thus the correlation of T_{2g} in O_h can theoretically lead to four possibilities in C_{2v}:

$$
\begin{array}{ll}
A_1 + A_2 + B_1 & \text{(a)} \\
A_1 + A_2 + B_2 & \text{(a')} \\
A_2 + B_1 + B_2 & \text{(b), the principal correlation} \\
A_1 + B_1 + B_2 & \text{(c)}
\end{array}
$$

The principal correlations for the irreducible representations of a variety of point groups with those of C_{2v} and C_{2h} are given in Tables 7.16 and 7.17, respectively. As an example of the serious complications that a lack of consistency can generate, we list in Table A1.2 all the possible correlations of the species with the O_h species.

Appendix 2 Character tables of various point groups

Point Group	Table No.	Point Group	Table No.	Point Group	Table No.
C_2	1	D_2	2	I_h	23
C_s	1	D_{2h}	3	O	20
C_i	1	D_{2d}	10	O_h	21
C_{2v}	2	D_3	6	S_4	9
C_{2h}	2	D_{3h}	7	T	22
C_3	4	D_{3d}	8	T_d	20
C_{3h}	5	D_4	10	V	2
C_{3v}	6	D_{4d}	14	V_h	3
C_4	9	D_{4h}	11	V_d	10
C_{4v}	10	D_5	12		
C_{4h}	13	D_{5h}	16		
C_5	12	D_{5d}	16		
C_{5v}	12	D_6	18		
C_{5h}	15	D_{6h}	19		
C_6	17	$D_{\infty h}$	25		
C_{6h}	17				
C_{6v}	18				
$C_{\infty v}$	24				

TABLE A2.1
Symmetry Species and Characters for the Point Groups C_2, C_s, and $C_i = S_2$

C_2	I	$C_2(z)$			C_s	I	$\sigma(xy)$		
A	$+1$	$+1$	z, R_z	$\alpha_{xx}, \alpha_{yy}, \alpha_{zz}, \alpha_{xy}$	A'	$+1$	$+1$	x, y, R_z	$\alpha_{xx}, \alpha_{yy}, \alpha_{zz}, \alpha_{xy}$
B	$+1$	-1	x, y, R_x, R_y α_{xz}, α_{yz}		A''	$+1$	-1	z, R_x, R_y α_{xz}, α_{yz}	

$C_i \equiv S_2$	I	i	
A_g	$+1$	$+1$	R_x, R_y, R_z; all α
A_u	$+1$	-1	x, y, z

Symmetry Species and Characters for the Point Groups C_{2v}, C_{2h}, and $D_2 \equiv V$

C_{2v}	I	$C_2(z)$	$\sigma_v(xz)$	$\sigma_v(yz)$		
A_1	+1	+1	+1	+1	z	$\alpha_{xx}, \alpha_{yy}, \alpha_{zz}$
A_2	+1	+1	−1	−1	R_z	α_{xy}
B_1	+1	−1	+1	−1	x, R_y	α_{xz}
B_2	+1	−1	−1	+1	y, R_x	α_{yz}

C_{2h}	I	$C_2(z)$	$\sigma_h(xy)$	i		
A_g	+1	+1	+1	+1	R_z	$\alpha_{xx}, \alpha_{yy}, \alpha_{zz}, \alpha_{xy}$
A_u	+1	+1	−1	−1	z	—
B_g	+1	−1	−1	+1	R_x, R_y	α_{xz}, α_{yz}
B_u	+1	−1	+1	−1	x, y	—

$D_2 \equiv V$	I	$C_2(z)$	$C_2(y)$	$C_2(x)$		
A	+1	+1	+1	+1	—	$\alpha_{xx}, \alpha_{yy}, \alpha_{zz}$
B_1	+1	+1	−1	−1	z, R_z	α_{xy}
B_2	+1	−1	+1	−1	y, R_y	α_{xz}
B_3	+1	−1	−1	+1	x, R_x	α_{yz}

Symmetry Species and Characters for the Point Group $D_{2h} \equiv V_h$

$D_{2h} \equiv V_h$	I	$\sigma(xy)$	$\sigma(xz)$	$\sigma(yz)$	i	$C_2(z)$	$C_2(y)$	$C_2(x)$		
A_g	+1	+1	+1	+1	+1	+1	+1	+1	—	$\alpha_{xx}, \alpha_{yy}, \alpha_{zz}$
A_u	+1	−1	−1	−1	−1	+1	+1	+1	—	—
B_{1g}	+1	+1	−1	−1	+1	+1	−1	−1	R_z	α_{xy}
B_{1u}	+1	−1	+1	+1	−1	+1	−1	−1	z	—
B_{2g}	+1	−1	+1	−1	+1	−1	+1	−1	R_y	α_{xz}
B_{2u}	+1	+1	−1	+1	−1	−1	+1	−1	y	—
B_{3g}	+1	−1	−1	+1	+1	−1	−1	+1	R_x	α_{yz}
B_{3u}	+1	+1	+1	−1	−1	−1	−1	+1	x	—

Symmetry Species and Characters for the Point Group C_3

C_3	I	$2C_3(z)$		
A	+1	+1	z, R_z	$\alpha_{xx} + \alpha_{yy}, \alpha_{zz}$
E	+2	−1	x, y, R_x, R_y	$\alpha_{xy}, \alpha_{xz}, \alpha_{yz}, \alpha_{xx} - \alpha_{yy}$

TABLE A2.5
Symmetry Species and Characters for the Point Group C_{3h}

C_{3h}	I	$2C_3(z)$	$\sigma_h(xy)$	$2S_3(z)$		
A'	$+1$	$+1$	$+1$	$+1$	R_z	$\alpha_{xx}+\alpha_{yy},\alpha_{zz}$
A''	$+1$	$+1$	-1	-1	z	—
E'	$+2$	-1	$+2$	-1	x,y	$\alpha_{xx}-\alpha_{yy},\alpha_{xy}$
E''	$+2$	-1	-2	$+1$	R_x,R_y	α_{xz},α_{yz}

TABLE A2.6
Symmetry Species and Characters for the Point Groups C_{3v} and D_3

C_{3v}	I	$2C_3(z)$	$3\sigma_v$		
A_1	$+1$	$+1$	$+1$	z	$\alpha_{xx}+\alpha_{yy},\alpha_{zz}$
A_2	$+1$	$+1$	-1	R_z	—
E	$+2$	-1	0	x,y,R_x,R_y	$\alpha_{xx}-\alpha_{yy},\alpha_{xy},\alpha_{xz},\alpha_{yz}$

D_3	I	$2C_3(z)$	$3C_2$		
A_1	$+1$	$+1$	$+1$	—	$\alpha_{xx}+\alpha_{yy},\alpha_{zz}$
A_2	$+1$	$+1$	-1	z,R_z	—
E	$+2$	-1	0	x,y,R_x,R_y	$\alpha_{xx}-\alpha_{yy},\alpha_{xy},\alpha_{yz},\alpha_{xz}$

TABLE A2.7
Symmetry Species and Characters for the Point Group D_{3h}

D_{3h}	I	$2C_3(z)$	$3C_2$	$\sigma_h(xy)$	$2S_3(z)$	$3\sigma_v$		
A_1'	$+1$	$+1$	$+1$	$+1$	$+1$	$+1$	—	$\alpha_{xx}+\alpha_{yy},\alpha_{zz}$
A_1''	$+1$	$+1$	$+1$	-1	-1	-1	—	—
A_2'	$+1$	$+1$	-1	$+1$	$+1$	-1	R_z	—
A_2''	$+1$	$+1$	-1	-1	-1	$+1$	z	—
E'	$+2$	-1	0	$+2$	-1	0	x,y	$\alpha_{xx}-\alpha_{yy},\alpha_{xy}$
E''	$+2$	-1	0	-2	$+1$	0	R_x,R_y	α_{xz},α_{yz}

Symmetry Species and Characters for the Point Group D_{3d}

D_{3d}	I	$2S_6(z)$	$2S_6^2 \equiv 2C_3$	$S_6^3 \equiv S_2 \equiv i$	$3C_2$	$3\sigma_d$		
A_{1g}	$+1$	$+1$	$+1$	$+1$	$+1$	$+1$	—	$\alpha_{xx} + \alpha_{yy}, \alpha_{zz}$
A_{1u}	$+1$	-1	$+1$	-1	$+1$	-1	—	—
A_{2g}	$+1$	$+1$	$+1$	$+1$	-1	-1	R_z	—
A_{2u}	$+1$	-1	$+1$	-1	-1	$+1$	z	—
E_g	$+2$	-1	-1	$+2$	0	0	R_x, R_y	$\alpha_{xx} - \alpha_{yy}, \alpha_{xy}, \alpha_{xz}, \alpha_{yz}$
E_u	$+2$	$+1$	-1	-2	0	0	x, y	—

Symmetry Species and Characters of the Point Groups C_4 and S_4

C_4	I	$2C_4(z)$	$C_4^2 \equiv C_2''*$		
S_4	I	$2S_4(z)$	$S_4^2 \equiv C_2''*$		
A	$+1$	$+1$	$+1$	z for C_4, R_z	$\alpha_{xx} + \alpha_{yy}, \alpha_{zz}$
B	$+1$	-1	$+1$	z for S_4	$\alpha_{xx} - \alpha_{yy}, \alpha_{xy}$
E	$+2$	0	-2	x, y, R_x, R_y	α_{xz}, α_{yz}

* The C_2 are identified as x- and y-axes.

Symmetry Species and Characters for the Point Groups C_{4v}, D_4, and $D_{2d} \equiv V_d$

C_{4v}	I	$2C_4(z)$	$C_4^2 \equiv C_2''$	$2\sigma_v*$	$2\sigma_d$		
D_4	I	$2C_4(z)$	$C_4^2 \equiv C_2''$	$2C_2\dagger$	$2C_2'$		
$D_{2d} \equiv V_d$	I	$2S_4(z)$	$S_4^2 \equiv C_2''$	$2C_2\dagger$	$2\sigma_d$		
A_1	$+1$	$+1$	$+1$	$+1$	$+1$	z for C_{4v}	$\alpha_{xx} + \alpha_{yy}, \alpha_{zz}$
A_2	$+1$	$+1$	$+1$	-1	-1	z for D_4, R_z	—
B_1	$+1$	-1	$+1$	$+1$	-1	—	$\alpha_{xx} - \alpha_{yy}$
B_2	$+1$	-1	$+1$	-1	$+1$	z for D_{2d}	α_{xy}
E	$+2$	0	-2	0	0	x, y, R_x, R_y	α_{xz}, α_{yz}

* The xz- and yz-planes.
† The x- and y-axes.

TABLE A2.11
Symmetry Species and Characters for the Point Group D_{4h}

D_{4h}	I	$2C_4(z)$	$C_4^2 \equiv C_2''$	$2C_2^*$	$2C_2'$	$\sigma_h(xy)$	$2\sigma_v$†	$2\sigma_d$	$2S_4$	$S_2 \equiv i$		
A_{1g}	+1	+1	+1	+1	+1	+1	+1	+1	+1	+1	—	$\alpha_{xx}+\alpha_{yy},\ \alpha_{zz}$
A_{1u}	+1	+1	+1	+1	+1	−1	−1	−1	−1	−1	—	—
A_{2g}	+1	+1	+1	−1	−1	+1	−1	−1	+1	+1	R_z	—
A_{2u}	+1	+1	+1	−1	−1	−1	+1	+1	−1	−1	z	—
B_{1g}	+1	−1	+1	+1	−1	+1	+1	−1	−1	+1	—	$\alpha_{xx}-\alpha_{yy}$
B_{1u}	+1	−1	+1	+1	−1	−1	−1	+1	+1	−1	—	—
B_{2g}	+1	−1	+1	−1	+1	+1	−1	+1	−1	+1	—	α_{xy}
B_{2u}	+1	−1	+1	−1	+1	−1	+1	−1	+1	−1	—	—
E_g	+2	0	−2	0	0	−2	0	0	0	+2	R_x, R_y	α_{xz}, α_{yz}
E_u	+2	0	−2	0	0	+2	0	0	0	−2	x, y	—

* The x- and y- axes.

† The xz- and yz-planes.

TABLE A2.12
Symmetry Species and Characters for the Point Groups C_5, C_{5v}, D_5 *

C_{5v}	I	$2C_5$	$2C_5^2$	$5\sigma_v$†		
A_1	+1	+1	+1	+1	z for C_{5v}	$\alpha_{xx}+\alpha_{yy},\ \alpha_{zz}$
A_2	+1	+1	+1	−1	z for D_5, R_z	—
E_1	+2	$2\cos 72°$	$2\cos 144°$	0	x, y, R_x, R_y	α_{xz}, α_{yz}
E_2	+2	$2\cos 144°$	$2\cos 72°$	0	—	$\alpha_{xx}-\alpha_{yy},\ \alpha_{xy}$

* In C_5, A_1 and A_2 coalesce to form A, since there is no σ_v.

† In D_5, replace $5\sigma_v$ by $5C_2$.

384

TABLE A2.13
Symmetry Species and Characters for the Point Group C_{4h}

C_{4h}	I	$2C_4(z)$	$C_4{}^2 \equiv C_2''(z)$	$\sigma_h(xy)$	$2S_4(z)$	$S_2 \equiv i$		
A_g	$+1$	$+1$	$+1$	$+1$	$+1$	$+1$	R_z	$\alpha_{xx}+\alpha_{yy}, \alpha_{zz}$
A_u	$+1$	$+1$	$+1$	-1	-1	-1	z	—
B_g	$+1$	-1	$+1$	$+1$	-1	$+1$	—	$\alpha_{xx}-\alpha_{yy}, \alpha_{xy}$
B_u	$+1$	-1	$+1$	-1	$+1$	-1	—	—
E_g	$+2$	0	-2	-2	0	$+2$	R_x, R_y	α_{xz}, α_{yz}
E_u	$+2$	0	-2	$+2$	0	-2	x, y	—

TABLE A2.14
Symmetry Species and Characters for the Point Group D_{4d}

D_{4d}	I	$2S_8(z)$	$2S_8{}^2 \equiv 2C_4(z)$	$2S_8{}^3$	$S_8{}^4 \equiv C_2''(z)$	$4C_2$	$4\sigma_d$		
A_1	$+1$	$+1$	$+1$	$+1$	$+1$	$+1$	$+1$	—	$\alpha_{xx}+\alpha_{yy}, \alpha_{zz}$
A_2	$+1$	$+1$	$+1$	$+1$	$+1$	-1	-1	R_z	—
B_1	$+1$	-1	$+1$	-1	$+1$	$+1$	-1	—	—
B_2	$+1$	-1	$+1$	-1	$+1$	-1	$+1$	z	—
E_1	$+2$	$+\sqrt{2}$	0	$-\sqrt{2}$	-2	0	0	x, y	—
E_2	$+2$	0	-2	0	$+2$	0	0	—	$\alpha_{xx}-\alpha_{yy}, \alpha_{xy}$
E_3	$+2$	$-\sqrt{2}$	0	$+\sqrt{2}$	-2	0	0	R_x, R_y	α_{xz}, α_{yz}

TABLE A2.15
Symmetry Species and Characters for the Point Group C_{5h}

C_{5h}	I	$2C_5(z)$	$2C_5{}^2$	$\sigma_h(xy)$	$2S_5{}^3$	$2S_5{}^2$		
A'	$+1$	$+1$	$+1$	$+1$	$+1$	$+1$	R_z	$\alpha_{xx}+\alpha_{yy}, \alpha_{zz}$
A''	$+1$	$+1$	$+1$	-1	-1	-1	z	—
E_1'	$+2$	$2\cos 72°$	$2\cos 144°$	$+2$	$2\cos 72°$	$2\cos 144°$	x, y	—
E_1''	$+2$	$2\cos 72°$	$2\cos 144°$	-2	$-2\cos 72°$	$-2\cos 144°$	R_x, R_y	α_{xz}, α_{yz}
E_2'	$+2$	$2\cos 144°$	$2\cos 72°$	$+2$	$2\cos 144°$	$2\cos 72°$	—	$\alpha_{xx}-\alpha_{yy}, \alpha_{xy}$
E_2''	$+2$	$2\cos 144°$	$2\cos 72°$	-2	$-2\cos 144°$	$-2\cos 72°$	—	—

TABLE A2.16
Symmetry Species and Characters for the Point Groups D_{5h} and D_{5d}

D_{5d} / D_{5h}	I	$2C_5(z)$	$2C_5{}^2$	$5C_2$	i / $\sigma_h(xy)$	$2S_{10}{}^3$ / $2S_5{}^3$	$2S_{10}$ / $2S_5$	$5\sigma_d$ / $5\sigma_v$		
A_{1g} / A_1'	$+1$	$+1$	$+1$	$+1$	$+1$	$+1$	$+1$	$+1$		$\alpha_{xx}+\alpha_{yy},\ \alpha_{zz}$
A_{1u} / A_1''	$+1$	$+1$	$+1$	$+1$	-1	-1	-1	-1		—
A_{2g} / A_2'	$+1$	$+1$	$+1$	-1	$+1$	$+1$	$+1$	-1	R_z	—
A_{2u} / A_2''	$+1$	$+1$	$+1$	-1	-1	-1	-1	$+1$	z	—
E_{1g} / E_1'	$+2$	$2\cos 72°$	$2\cos 144°$	0	$+2$	$+2\cos 72°$	$+2\cos 144°$	0	R_x, R_y; (x, y) for D_{5h}	$(\alpha_{xz}, \alpha_{yz})$ for D_{5d}
E_{1u} / E_1''	$+2$	$2\cos 72°$	$2\cos 144°$	0	-2	$-2\cos 72°$	$-2\cos 144°$	0	(x, y) for D_{5d}	$(\alpha_{xz}, \alpha_{yz})$ for D_{5h}
E_{2g} / E_2'	$+2$	$2\cos 144°$	$2\cos 72°$	0	$+2$	$+2\cos 144°$	$+2\cos 72°$	0		$\alpha_{xy},\ \alpha_{xx}-\alpha_{yy}$
E_{2u} / E_2''	$+2$	$2\cos 144°$	$2\cos 72°$	0	-2	$-2\cos 144°$	$-2\cos 72°$	0		—

TABLE A2.17
Symmetry Species and Characters for the Point Groups C_6 and C_{6h}*

C_{6h}	I	$2C_6(z)$	$2C_6{}^2 \equiv C_3(z)$	$C_6{}^3 \equiv C_2''(z)$	$\sigma_h(xy)$	$2S_6(z)$	$2S_3(z)$	$S_2 \equiv i$		
A_g	$+1$	$+1$	$+1$	$+1$	$+1$	$+1$	$+1$	$+1$	R_z	$\alpha_{xx}+\alpha_{yy},\ \alpha_{zz}$
A_u	$+1$	$+1$	$+1$	$+1$	-1	-1	-1	-1	z	—
B_g	$+1$	-1	$+1$	-1	-1	$+1$	-1	$+1$		—
B_u	$+1$	-1	$+1$	-1	$+1$	-1	$+1$	-1		—
E_{1g}	$+2$	$+1$	-1	-2	-2	-1	$+1$	$+2$	R_x, R_y	α_{xz}, α_{yz}
E_{1u}	$+2$	$+1$	-1	-2	$+2$	$+1$	-1	-2	x, y	—
E_{2g}	$+2$	-1	-1	$+2$	$+2$	-1	-1	$+2$		$\alpha_{xx}-\alpha_{yy},\ \alpha_{xy}$
E_{2u}	$+2$	-1	-1	$+2$	-2	$+1$	$+1$	-2		—

* For C_6, absence of σ_h, S_6, S_3 and $S_2 \equiv i$ eliminates the g, u classification and reduces the species to A, B, E_1, and E_2.

TABLE A2.18
Symmetry Species and Characters for the Point Groups C_{6v} and D_6

C_{6v} / D_6	I	$2C_6(z)$	$2C_6^2\equiv2C_3$	$C_6^3\equiv C_2''$ / $C_6^3\equiv C_2'$	$3\sigma_v$ / $3C_2$	$3\sigma_d$ / $3C_2'$		
A_1	+1	+1	+1	+1	+1	+1	z for C_{6v}	$\alpha_{xx}+\alpha_{yy},\ \alpha_{zz}$
A_2	+1	+1	+1	+1	−1	−1	z for D_6, R_z	—
B_1	+1	−1	+1	−1	+1	−1	—	—
B_2	+1	−1	+1	−1	−1	+1	—	—
E_1	+2	+1	−1	−2	0	0	x, y, R_x, R_y	$\alpha_{xz},\ \alpha_{yz}$
E_2	+2	−1	−1	+2	0	0	—	$\alpha_{xx}-\alpha_{yy},\ \alpha_{xy}$

TABLE A2.19
Symmetry Species and Characters for the Point Group D_{6h}

D_{6h}	I	$2C_6(z)$	$2C_6^2\equiv2C_3(z)$	$C_6^3\equiv C_2''(z)$	$3C_2$	$3C_2'$	$\sigma_h(xy)$	$3\sigma_v$	$3\sigma_d$	$2S_6$	$2S_3$	$S_6^3\equiv S_2\equiv i$		
A_{1g}	+1	+1	+1	+1	+1	+1	+1	+1	+1	+1	+1	+1		$\alpha_{xx}+\alpha_{yy},\ \alpha_{zz}$
A_{1u}	+1	+1	+1	+1	+1	+1	−1	−1	−1	−1	−1	−1		—
A_{2g}	+1	+1	+1	+1	−1	−1	+1	−1	−1	+1	+1	+1	R_z	—
A_{2u}	+1	+1	+1	+1	−1	−1	−1	+1	+1	−1	−1	−1	z	—
B_{1g}	+1	−1	+1	−1	+1	−1	−1	−1	+1	+1	−1	+1		—
B_{1u}	+1	−1	+1	−1	+1	−1	+1	+1	−1	−1	+1	−1		—
B_{2g}	+1	−1	+1	−1	−1	+1	−1	+1	−1	+1	−1	+1		—
B_{2u}	+1	−1	+1	−1	−1	+1	+1	−1	+1	−1	+1	−1		—
E_{1g}	+2	+1	−1	−2	0	0	−2	0	0	−1	+1	+2	R_x, R_y	$\alpha_{xz},\ \alpha_{yz}$
E_{1u}	+2	+1	−1	−2	0	0	+2	0	0	+1	−1	−2	x, y	—
E_{2g}	+2	−1	−1	+2	0	0	+2	0	0	−1	−1	+2		$\alpha_{xx}-\alpha_{yy},\ \alpha_{xy}$
E_{2u}	+2	−1	−1	+2	0	0	−2	0	0	+1	+1	−2		—

TABLE A2.20
Symmetry Species and Characters for the Point Groups T_d and O

T_d O	I I	$8C_3$ $8C_3$	$6\sigma_d$ $6C_2$	$6S_4{}^*$ $6C_4{}^*$	$3S_4{}^2 \equiv 3C_2''$ $3C_4{}^2 \equiv 3C_2''$		
A_1	+1	+1	+1	+1	+1		$\alpha_{xx} + \alpha_{yy} + \alpha_{zz}$
A_2	+1	+1	−1	−1	+1		—
E	+2	−1	0	0	+2		$\alpha_{xx} + \alpha_{yy} - 2\alpha_{zz}, \alpha_{xx} - \alpha_{yy}$
T_1	+3	0	−1	+1	−1	x, y, z for O, R_x, R_y, R_z	—
T_2	+3	0	+1	−1	−1	x, y, z for T_d	$\alpha_{xy}, \alpha_{xz}, \alpha_{yz}$

*The x-, y-, and z-axes.

TABLE A2.21
Symmetry Species and Characters for the Point Group O_h

O_h	I	$8C_3$	$6C_2'$	$6C_4{}^*$	$3C_4{}^2 \equiv 3C_2''$	$S_2 \equiv i$	$6S_4$	$8S_6$	$3\sigma_h$	$6\sigma_d$		
A_{1g}	+1	+1	+1	+1	+1	+1	+1	+1	+1	+1	—	$\alpha_{xx} + \alpha_{yy} + \alpha_{zz}$
A_{1u}	+1	+1	+1	+1	+1	−1	−1	−1	−1	−1	—	
A_{2g}	+1	+1	−1	−1	+1	+1	−1	+1	+1	−1	—	
A_{2u}	+1	+1	−1	−1	+1	−1	+1	−1	−1	+1	—	
E_g	+2	−1	0	0	+2	+2	0	−1	+2	0	—	$\alpha_{xx} + \alpha_{yy} - 2\alpha_{zz}, \alpha_{xx} - \alpha_{yy}$
E_u	+2	−1	0	0	+2	−2	0	+1	−2	0	—	
T_{1g}	+3	0	−1	+1	−1	+3	+1	0	−1	−1	R_x, R_y, R_z	
T_{1u}	+3	0	−1	+1	−1	−3	−1	0	+1	+1	x, y, z	
T_{2g}	+3	0	+1	−1	−1	+3	−1	0	−1	+1	—	$\alpha_{xy}, \alpha_{xz}, \alpha_{yz}$
T_{2u}	+3	0	+1	−1	−1	−3	+1	0	+1	−1	—	

*The x-, y-, and z-axes.

TABLE A2.22
Symmetry Species and Characters for the Point Group T

T	I	$8C_3$	$3C_2$*		
A	$+1$	$+1$	$+1$	—	$\alpha_{xx}+\alpha_{yy}+\alpha_{zz}$
E	$+2$	-1	$+2$	—	$\alpha_{xx}+\alpha_{yy}-2\alpha_{zz},\ \alpha_{xx}-\alpha_{yy}$
T	$+3$	0	-1	x,y,z,R_x,R_y,R_z	$\alpha_{xy},\alpha_{xz},\alpha_{yz}$

* The x-, y-, and z-axes.

TABLE A2.23
Symmetry Species and Character Tables for the Point Groups I and I_h*

I_h	I	$12C_5$	$12C_5{}^2$	$20C_3$	$15C_2$	i	$12S_{10}$	$12S_{10}{}^3$	$20S_6$	15σ		
A_g	1	1	1	1	1	1	1	1	1	1	$\alpha_{xx}+\alpha_{yy}+\alpha_{zz}$	
A_u	1	1	1	1	1	-1	-1	-1	-1	-1		
T_{1g}	3	$(1+\sqrt5)/2$	$(1-\sqrt5)/2$	0	-1	3	$(1-\sqrt5)/2$	$(1+\sqrt5)/2$	0	-1	R_x, R_y, R_z	
T_{1u}	3	$(1+\sqrt5)/2$	$(1-\sqrt5)/2$	0	-1	-3	$-(1-\sqrt5)/2$	$-(1+\sqrt5)/2$	0	1	x, y, z	
T_{2g}	3	$(1-\sqrt5)/2$	$(1+\sqrt5)/2$	0	-1	3	$(1+\sqrt5)/2$	$(1-\sqrt5)/2$	0	-1		
T_{2u}	3	$(1-\sqrt5)/2$	$(1+\sqrt5)/2$	0	-1	-3	$-(1+\sqrt5)/2$	$-(1-\sqrt5)/2$	0	1		
G_g†	4	-1	-1	1	0	4	-1	-1	1	0		
G_u†	4	-1	-1	1	0	-4	1	1	-1	0		
H_g‡	5	0	0	-1	1	5	0	0	-1	1	$(\alpha_{xy},\alpha_{yz},\alpha_{xz},$ $\alpha_{xx}+\alpha_{yy}-2\alpha_{zz})\alpha_{xx}-\alpha_{yy}$	
H_u‡	5	0	0	-1	1	-5	0	0	1	-1		

* For I, absence of i, the S_{10}, S_6, and σ eliminates the g, u classification and leaves species A, T_1, T_2, G, and H.
† A four-fold degenerate species.
‡ A five-fold degenerate species.

TABLE A2.24
Symmetry Species and Characters for the Point Group $C_{\infty v}$

$C_{\infty v}$	I	$2C_\infty^{\varphi}(z)$	$2C_\infty^{2\varphi}$	$2C_\infty^{3\varphi}$	$\cdots$	$\infty\sigma_v$		
Σ^+	$+1$	$+1$	$+1$	$+1$	$\cdots$	$+1$	z	$\alpha_{xx}+\alpha_{yy},\ \alpha_{zz}$
Σ^-	$+1$	$+1$	$+1$	$+1$	$\cdots$	-1	R_z	—
Π	$+2$	$2\cos\varphi$	$2\cos 2\varphi$	$2\cos 3\varphi$	$\cdots$	0	x, y, R_x, R_y	$\alpha_{xz},\ \alpha_{yz}$
Δ	$+2$	$2\cos 2\varphi$	$2\cos 2\cdot 2\varphi$	$2\cos 3\cdot 2\varphi$	$\cdots$	0	—	$\alpha_{xx}-\alpha_{yy},\ \alpha_{xy}$
Φ	$+2$	$2\cos 3\varphi$	$2\cos 2\cdot 3\varphi$	$2\cos 3\cdot 3\varphi$	$\cdots$	0	—	—
$\vdots$						$\vdots$		$\cdots$

TABLE A2.25
Symmetry Species and Characters for the Point Group $D_{\infty h}$

$D_{\infty h}$	I	$2C_\infty^{\varphi}(z)$	$2C_\infty^{2\varphi}$	$2C_\infty^{3\varphi}$	$\cdots$	σ_h	∞C_2	$\infty\sigma_v$	$2S_\infty^{\varphi}$	$2S_\infty^{2\varphi}$	$\cdots$	$S_2 \equiv i$		
Σ_g^+	$+1$	$+1$	$+1$	$+1$	$\cdots$	$+1$	$+1$	$+1$	$+1$	$+1$	$\cdots$	$+1$		$\alpha_{xx}+\alpha_{yy},\ \alpha_{zz}$
Σ_u^+	$+1$	$+1$	$+1$	$+1$	$\cdots$	-1	-1	$+1$	-1	-1	$\cdots$	-1	z	—
Σ_g^-	$+1$	$+1$	$+1$	$+1$	$\cdots$	$+1$	-1	-1	$+1$	$+1$	$\cdots$	$+1$	R_z	—
Σ_u^-	$+1$	$+1$	$+1$	$+1$	$\cdots$	-1	$+1$	-1	-1	-1	$\cdots$	-1		—
Π_g	$+2$	$2\cos\varphi$	$2\cos 2\varphi$	$2\cos 3\varphi$	$\cdots$	-2	0	0	$-2\cos\varphi$	$-2\cos 2\varphi$	$\cdots$	$+2$	R_x, R_y	$\alpha_{xz},\ \alpha_{yz}$
Π_u	$+2$	$2\cos\varphi$	$2\cos 2\varphi$	$2\cos 3\varphi$	$\cdots$	$+2$	0	0	$+2\cos\varphi$	$+2\cos 2\varphi$	$\cdots$	-2	x, y	—
Δ_g	$+2$	$2\cos 2\varphi$	$2\cos 4\varphi$	$2\cos 6\varphi$	$\cdots$	$+2$	0	0	$+2\cos 2\varphi$	$+2\cos 4\varphi$	$\cdots$	$+2$	—	$\alpha_{xx}-\alpha_{yy},\ \alpha_{xy}$
Δ_u	$+2$	$2\cos 2\varphi$	$2\cos 4\varphi$	$2\cos 6\varphi$	$\cdots$	-2	0	0	$-2\cos 2\varphi$	$-2\cos 4\varphi$	$\cdots$	-2	—	—
Φ_g	$+2$	$2\cos 3\varphi$	$2\cos 6\varphi$	$2\cos 9\varphi$	$\cdots$	-2	0	0	$-2\cos 3\varphi$	$-2\cos 4\varphi$	$\cdots$	$+2$	—	—
Φ_u	$+2$	$2\cos 3\varphi$	$2\cos 6\varphi$	$2\cos 9\varphi$	$\cdots$	$+2$	0	0	$+2\cos 3\varphi$	$+2\cos 4\varphi$	$\cdots$	-2	—	—
$\vdots$														$\cdots$

Index